# Mathematics Study Resources

Volume 17

**Series Editors**

Kolja Knauer, Departament de Matemàtiques Informàtic, Universitat de Barcelona, Barcelona, Barcelona, Spain

Elijah Liflyand, Department of Mathematics, Bar-Ilan University, Ramat-Gan, Israel

This series comprises direct translations of successful foreign language titles, especially from the German language.

Powered by advances in automated translation, these books draw on global teaching excellence to provide students and lecturers with diverse materials for teaching and study.

Norbert Henze

# Very First Steps in Random Walks

## The Power of Combinatorial Methods and Generating Functions

Norbert Henze
KIT
Karlsruhe, Germany

ISSN 2731-3824  ISSN 2731-3832 (electronic)
Mathematics Study Resources
ISBN 978-3-658-46312-0  ISBN 978-3-658-46313-7 (eBook)
https://doi.org/10.1007/978-3-658-46313-7

The translation was done with the help of an artificial intelligence machine translation tool. A subsequent human revision was done primarily in terms of content.
Translation from the German language edition: "Irrfahrten – Faszination der Random Walks" by Norbert Henze, © The Editor(s) (if applicable) and The Author(s), under exclusive license to Springer Fachmedien Wiesbaden GmbH, part of Springer Nature 2024. Published by Springer Spektrum Wiesbaden. All Rights Reserved.

© The Editor(s) (if applicable) and The Author(s), under exclusive license to Springer Fachmedien Wiesbaden GmbH, part of Springer Nature 2025

This work is subject to copyright. All rights are solely and exclusively licensed by the Publisher, whether the whole or part of the material is concerned, specifically the rights of translation, reprinting, reuse of illustrations, recitation, broadcasting, reproduction on microfilms or in any other physical way, and transmission or information storage and retrieval, electronic adaptation, computer software, or by similar or dissimilar methodology now known or hereafter developed.
The use of general descriptive names, registered names, trademarks, service marks, etc. in this publication does not imply, even in the absence of a specific statement, that such names are exempt from the relevant protective laws and regulations and therefore free for general use.
The publisher, the authors and the editors are safe to assume that the advice and information in this book are believed to be true and accurate at the date of publication. Neither the publisher nor the authors or the editors give a warranty, expressed or implied, with respect to the material contained herein or for any errors or omissions that may have been made. The publisher remains neutral with regard to jurisdictional claims in published maps and institutional affiliations.

This Springer imprint is published by the registered company Springer Fachmedien Wiesbaden GmbH, part of Springer Nature.
The registered company address is: Abraham-Lincoln-Str. 46, 65189 Wiesbaden, Germany

If disposing of this product, please recycle the paper.

# Preface

This book is based on the third edition of a book first written in German about random walks. Unlike other books on the topic, this one is positioned at the most basic technical level, as indicated by the title. The target readership is mathematics students in their second or third year of study who have attended introductory courses on calculus and stochastics. As much of the book deals with various aspects of the symmetric Bernoulli random walk on the integers, which can be generated by repeatedly tossing a fair coin, the book is particularly aimed at students aspiring to become high school math teachers.

Considering the target readership, I have deliberately chosen classic topics that do not require advanced technical tools like Fourier analysis and measure theory. Using only combinatorial considerations, primarily employing the reflection principle, as well as generating functions and one-dimensional calculus, it is possible to introduce this fascinating subfield of stochastics with its numerous counterintuitive phenomena and manifold applications. The book is well-suited for an introductory seminar, often called a *sophomore seminar*, or for a seminar specific to a major or department. The German-language edition was successfully tested with students at several universities for this purpose. The text includes 53 self-assessment questions, abbreviated as SAQ, with answers provided at the end of each chapter. Additionally, 74 exercises with solutions assist in understanding the material deeply.

I have placed special emphasis on an approach focused on problem-solving, a comprehensive discussion of the results, and the inclusion of limit theorems to clarify the long-term behavior of random walks. Numerous graphics illustrate results and employed methods. In some places, mathematical rigor (e.g., introduction of stopping times, strong Markov property) has been intentionally omitted in favor of conceptual understanding.

After a brief introductory chapter, which outlines the historical origins of random walks and introduces some standard notation, the extended second chapter focuses exclusively on the simple symmetric random walk on the integers, i.e., to the Bernoulli random walk with a success probability of $\frac{1}{2}$. Here, numerous unexpected and counterintuitive phenomena are revealed, first comprehensively published in [FE1]. This chapter concludes with an outlook on the Brown–Wiener process and Donsker's invariance principle as well as the law of the iterated logarithm. Chapter 3 addresses topics connected with the tied-down simple symmetric random walk, also called a bridge. An alternative concept involves an urn containing an equal number of red and black balls, and all these balls are drawn randomly without replacement, noting the color of the drawn balls each time. This chapter concludes with an outlook on the Brownian bridge.

Chapter 4 is devoted to questions related to the *asymmetric* or *biased* Bernoulli random walk on $\mathbb{Z}$, the gambler's ruin problem, the distribution of the length of the longest run in Bernoulli sequences, and the simple Galton–Watson branching process. Chapter 5 addresses problems related to the simple random walk on the integer lattice $\mathbb{Z}^d$ in higher dimensions. Here, a main result is Pólya's theorem [PO], which states that a simple symmetric random walk on $\mathbb{Z}^d$ returns to its starting point with probability one if $d \leq 2$, but this property is lost if $d \geq 3$. This chapter also deals with the number of distinct visited sites and the discrete Dirichlet problem.

Chapter 6 briefly covers several additional topics such as self-avoiding walks, walks with an arbitrary distribution for the single steps, random walks on graphs, and the connection between random walks and electrical networks. The final Chap. 7 consolidates various terms and results from probability theory, combinatorics, and analysis referenced in earlier chapters.

One compelling aspect of random walks is that even at the simplest technical level, many limit theorems with nonstandard limit distributions can be derived regarding their long-term behavior. Thus, in a fruitful interplay between calculus and combinatorics using elementary methods, various continuous distributions such as the arcsine distribution, the half-normal distribution, the Lévy distribution, the Rényi distribution, the Weibull distribution, and the Kolmogorov distribution emerge.

I am deeply grateful to the following individuals for their assistance: Ludwig Baringhaus, Bruno Ebner, Daniel Hug, and Steffen Winter for numerous suggestions for improvement. Special thanks are due to Peter Eichelsbacher, who pointed out certain didactic deficiencies, as well as to Nina Gantert and Felizitas Weidner, who brought to my attention an error in the proof of Theorem 2.16 in the first German edition. I thank Lothar Heinrich for a biographical correction. Special thanks also go to Andreas Rüdinger from Springer Spektrum, who encouraged me to consider an English edition of this book and who noticed a not insignificant number of errors even in a seemingly perfect manuscript.

I am always grateful for comments and criticism.

Pfinztal, Germany  Norbert Henze
August 2024

# Contents

| | | |
|---|---|---|
| **1** | **Introduction** | 1 |
| **2** | **The Simple Symmetric Random Walk on $\mathbb{Z}$** | 5 |
| | 2.1 Basic Concepts and Reflection Principle | 12 |
| | 2.2 The Main Lemma | 16 |
| | 2.3 Last Visits to Zero | 19 |
| | 2.4 The Number of Visits to Zero | 29 |
| | 2.5 First Return to Zero and Recurrence | 39 |
| | 2.6 Sojourn Times | 45 |
| | 2.7 Maximum and Minimum | 50 |
| | 2.8 Number and Positions of Maximizers | 57 |
| | 2.9 First-Passage Times | 68 |
| | 2.10 Collisions of Random Walks | 77 |
| | 2.11 Changes of Sign | 85 |
| | 2.12 The Maximum Modulus | 93 |
| | 2.13 A Test of Symmetry | 99 |
| | 2.14 Duality: New Insights | 103 |
| | 2.15 Outlook: Brown–Wiener Process and Law of the Iterated Logarithm | 106 |
| **3** | **Bridges: The Tied-down Random Walk** | 121 |
| | 3.1 The Number of Interior Zeros | 124 |
| | 3.2 Sojourn Times | 138 |
| | 3.3 Last Visit to Zero and First Return Time | 141 |
| | 3.4 Maximum and Minimum | 145 |
| | 3.5 Changes of Sign | 151 |
| | 3.6 Maximum Modulus, Kolmogorov Distribution | 158 |
| | 3.7 The Kolmogorov–Smirnov Test | 164 |
| | 3.8 Outlook: The Brownian Bridge | 169 |
| **4** | **Asymmetric Random Walks on $\mathbb{Z}$ and Related Topics** | 177 |
| | 4.1 First-passage Times | 178 |
| | 4.2 Visits to Zero | 185 |
| | 4.3 Random Walks with Absorbing Boundaries: The Gambler's Ruin Problem | 190 |

|  |  |  |  |
|---|---|---|---|
| | 4.4 | Longest Upward and Downward Runs | 197 |
| | 4.5 | The Galton–Watson Process | 201 |
| **5** | **Random Walks on the Integer Lattice $\mathbb{Z}^d$** | | **217** |
| | 5.1 | Recurrence and Transience | 218 |
| | 5.2 | The Number of Visited Sites | 223 |
| | 5.3 | The Discrete Dirichlet Problem | 228 |
| **6** | **Outlook** | | **243** |
| | 6.1 | Self-Avoiding Walks | 243 |
| | 6.2 | Walks with Arbitrary Distributions for $X_1$ | 244 |
| | 6.3 | Random Walks on Graphs | 246 |
| | 6.4 | Random Walks and Electrical Networks | 248 |
| **7** | **Tools from Stochastics, Combinatorics, and Analysis** | | **251** |
| | 7.1 | Probability Spaces | 251 |
| | 7.2 | A Canonical Probability Space | 253 |
| | 7.3 | Convergence in Distribution | 254 |
| | 7.4 | Central Limit Theorems | 256 |
| | 7.5 | Inequalities for the Natural Logarithm | 256 |
| | 7.6 | Expectation and Variance of Non-negative Integer-valued Random Variables | 257 |
| | 7.7 | The Wallis Product Representation for $\pi$ | 258 |
| | 7.8 | Stirling's Formula | 260 |
| | 7.9 | Generating Functions | 262 |
| | 7.10 | Some Identities involving Binomial Coefficients | 264 |
| | 7.11 | The Binomial Series | 267 |
| | 7.12 | Legendre Polynomials | 268 |
| | 7.13 | The Borel–Cantelli Lemma | 270 |

**Solutions to the Exercises** ..... 273

**Bibliography** ..... 299

**Index** ..... 303

# Introduction 1

It was apparently Karl Pearson [PE] who introduced the term *random walk* (German: *Irrfahrt*, French: *promenade au hazard*), when he asked the readers of the journal *Nature* in 1905 the following question: "Can any of your readers refer me to a work wherein I should find a solution of the following problem, or failing the knowledge of any existing solution provide me with an original one? I should be extremely grateful for aid in the matter.

A man starts from a point $O$ and walks $\ell$ yards in a straight line; he then turns through any angle whatever and walks another $\ell$ yards in a second straight line. He repeats this process $n$ times. I require the probability that after $n$ of these stretches he is at distance between $r$ and $r + \delta r$ from his starting point, $O$."

Pearson's interest in this problem stemmed from the desire to find a model for random migrations of biological species, and it was probably surprising when just a week later Lord Rayleigh provided an answer to Pearson's problem by conceptually equating the random walk with the superposition of $n$ randomly phase-shifted oscillations of the same amplitude and frequency. In his thanks to Rayleigh (again just a week later!) Pearson summarized: "The lesson of Lord Rayleigh's solution is that in open country the most probable place of finding a drunken man who is at all capable of keeping on his feet is somewhere near his starting point."

Figure 1.1 shows the result of a drunkard's random walk in a "grid city," where the streets are arranged like lines on a chessboard. Regardless of the path he has taken up to that point, the drunkard randomly chooses one of the four possible directions at each intersection, and he continues in that direction to the next intersection.

**SAQ 1**  Can you reconstruct the exact sequence of this random walk?

**Fig. 1.1** The first 21 steps of a random walk on $\mathbb{Z}^2$

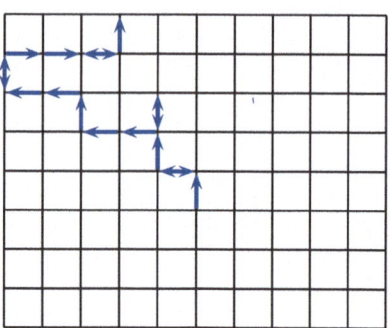

**SAQ 2** At which positions can the above random walk be after 3 steps?

In the year 1921, George Pólya (see [PO]) proved that a purely random walk on the integer lattice in the plane eventually returns to the starting point with probability one, but not in higher dimensions (see Chap. 5). His investigations were groundbreaking for further research. If you search today (August 26, 2024) in the *Mathscinet* database under the keyword "random walk," you will find 14,539 results, exclusively in the form of journal articles. This impressive number illustrates the international research activity in this area. While there are numerous journal articles and books on random walks, none are specifically aimed at second-year mathematics bachelor students. This book aims to fill that gap. Tailored to this target group, the next two chapters extensively examine the one-dimensional case, with geometric and combinatorial arguments playing prominent roles.

As for the notation used in this book, there should be few surprises. We write the equal sign with a colon preceding it, so we write ":=", when the expression on the left-hand side of the equal sign is defined by the right-hand side. In this sense, $\mathbb{N} := \{1, 2, \ldots\}$ and $\mathbb{N}_0 := \mathbb{N} \cup \{0\}$ represent set of positive and the set of non-negative integers, respectively. Furthermore, $\mathbb{R}$ denotes the set of real numbers, and $\mathbb{Z} := \{\ldots, -2, -1, 0, 1, 2, \ldots\}$ stands for the set of integers. For any positive integer $k$ and any non-empty set $M$, $M^k := \{(m_1, \ldots, m_k) : m_j \in M \text{ for } j = 1, \ldots, k\}$ is the $k$-fold Cartesian product of $M$ with itself, and $M^{\mathbb{N}} := \{(m_j)_{j \geq 1} : m_j \in M \text{ for every } j \in \mathbb{N}\}$ is the set of all sequences with elements from $M$. If $M$ is a finite set, $|M|$ denotes the number of elements in $M$. For disjoint sets $M$ and $N$, we write $M \uplus N$ for the union of $M$ and $N$. Similarly, we use this notation for the union of more than two pairwise disjoint sets.

For a real number $x$, let $\lfloor x \rfloor := \max\{k \in \mathbb{Z} : k \leq x\}$ denote the greatest integer less than or equal to $x$ (called the *floor* of $x$). Similarly, $\lceil x \rceil := \min\{k \in \mathbb{Z} : k \geq x\}$ stands for the smallest integer greater than or equal $x$ (called *ceiling* of $x$). The function log : $(0, \infty) \to \mathbb{R}$ represents the *natural logarithm*. If $(a_n)$ and $(b_n)$ are sequences of real numbers with $b_n > 0$ for every $n \geq 1$, we define

$$a_n \sim b_n :\Longleftrightarrow \lim_{n \to \infty} \frac{a_n}{b_n} = 1 \tag{1.1}$$

and say that the sequences $(a_n)$ and $(b_n)$ are *asymptotically equal*.

If $X$ is a random variable with $\mathbb{E}(X^2) < \infty$, we write $\mathbb{E}(X)$ for the expectation of $X$ and $\mathbb{V}(X)$ for the variance of $X$. The indicator function of an event $A$ is denoted by $\mathbf{1}\{A\}$ or sometimes by $\mathbf{1}_A$. If $X$ and $Y$ are random variables, then $X \sim Y$ means that $X$ and $Y$ have the same distribution. Similarly, $X \sim N(0, 1)$ means that the random variable $X$ follows a standard normal distribution. The binomial distribution with parameters $n$ and $p$ is abbreviated as Bin$(n, p)$. Here, $n$ is an integer and $p$ is a number in the unit interval. It holds that $X \sim$ Bin$(n, p)$, if $\mathbb{P}(X = k) = \binom{n}{k} p^k (1-p)^{n-k}$, where $k \in \{0, 1, \ldots, n\}$.

**Answers to the Self-Assessment Questions**

**Answer 1** Since the random walk has completed 21 steps, after the first upward step, a left step followed by a right step followed by another left step must have occurred. The next upward step is followed by an upward step, then a downward step, followed by two left steps and an upward step, followed again by two left steps. Afterwards, there is an upward and a downward step, followed by an upward step and three right steps. Finally, three more steps are taken: left, right, and upward.

**Answer 2** From the point marked by a larger filled circle, the random walk can arrive at any of the points marked by a cross after three steps.

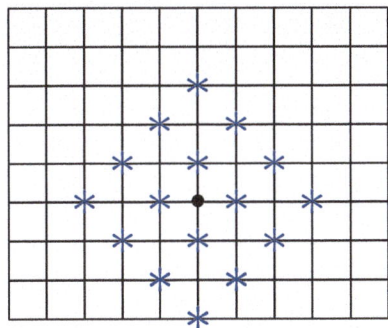

# The Simple Symmetric Random Walk on $\mathbb{Z}$     2

This extensive chapter deals with the simple symmetric random walk on the set $\mathbb{Z} = \{\ldots, -2, -1, 0, 1, 2, \ldots\}$ of integers, or equivalently, the *symmetric Bernoulli random walk*. A "particle" performs such a random walk when it starts at 0 and, to put it vividly, moves without memory in equally probable unit steps back and forth on $\mathbb{Z}$, where, for each positive integer $j$, the $j$-th step is completed at time $j$ (Fig. 2.1).

For each integer $j$, we model the $j$-th step of the random walk by a random variable $X_j$ with the property that $\mathbb{P}(X_j = 1) = \mathbb{P}(X_j = -1) = \frac{1}{2}$. Here, the event $\{X_j = 1\}$ or $\{X_j = -1\}$ means that the particle moves to the right or to the left, respectively. The random walk is thus *simple* in the sense that it cannot skip over any integer. The equal probabilities of the two directions are expressed by the attribute *symmetric*. Later, we will also consider *asymmetric* or, what is synonymous, *biased* random walks, for which $\mathbb{P}(X_j = 1) = p = 1 - \mathbb{P}(X_j = -1)$, where $p \neq \frac{1}{2}$. Such random walks exhibit a trend in a particular direction. The fact that the random walk has no memory is taken into account by assuming that $X_1, X_2, \ldots$ are *independent* random variables. This means that for every $k \geq 2$ and any choice of step directions $(a_1, \ldots, a_k) \in \{-1, 1\}^k$,

$$\mathbb{P}(X_1 = a_1, \ldots, X_k = a_k) = \left(\frac{1}{2}\right)^k \left(= \prod_{j=1}^{k} \mathbb{P}(X_j = a_j)\right). \tag{2.1}$$

A suitable sample space, on which $X_1, X_2, \ldots$ are defined as (measurable) mappings, is guaranteed by general theorems of measure theory; a possible choice can be found in Sect. 7.2.

If we set $S_0 := 0$ and

$$S_n := X_1 + X_2 + \ldots + X_n \tag{2.2}$$

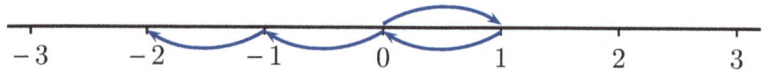

**Fig. 2.1** The beginning of a random walk on $\mathbb{Z}$

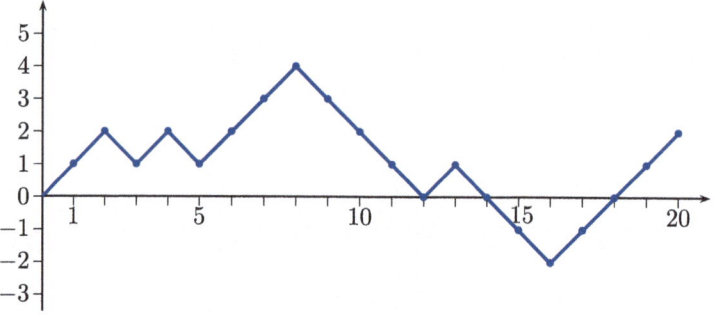

**Fig. 2.2** The first 20 steps of a random walk as a polygonal line

for each $n \geq 1$, the random variable $S_n$ indicates the position of the particle at time $n$, at which the $n$-th step is completed.

**SAQ 1** Can you derive $\mathbb{P}(S_7 = 3)$?

To illustrate the trajectory of a random walk, we depart from the inadequate representation shown in Fig. 2.1. The random walk depicted there may have already completed 1000 steps, comprising 499 sequences of a right step followed by a left step, followed by two left steps. Instead, we represent random walks in a Cartesian coordinate system, with the $x$-axis indicating the number of steps taken since the beginning as the abscissa. The $y$-axis serves to indicate the particle's position on the set of integers. In this coordinate system, the points $(0, S_0), (1, S_1), (2, S_2), \ldots$ describe the trajectory over time of the random walk. It is customary to connect successive points $(j, S_j)$ and $(j + 1, S_{j+1})$, forming a polygonal chain known as the *path*. The *length* of a path is the number of steps it has completed.

Figure 2.2 illustrates a random walk of length 20 that initially moves from 0 to 1, then from 1 to 2, then from 2 to 1, and so on. After 20 steps—at time 20—it is located at position 2. Synonyms for *position* include *height*, *site*, or *state*.

Figure 2.3 shows the progression of a random walk of length 200. As in Fig. 2.2, the same scale for labeling both axes was chosen here. However, it is evident that the path depicted in Fig. 2.3 remains flat, and this effect becomes even more pronounced as more steps are shown. Is this phenomenon explained by chance, or does the stochastic behavior of random walks over longer periods require a different choice of scale for both axes in graphical representation? To explore this, we note

## 2 The Simple Symmetric Random Walk on $\mathbb{Z}$

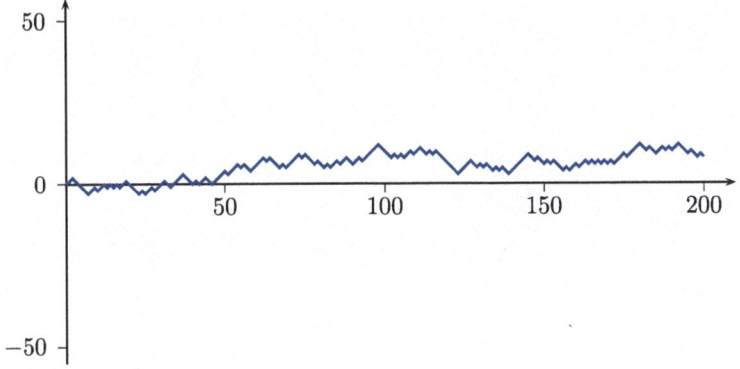

**Fig. 2.3** Random walk of length 200, shown on the same scale

that according to (2.2), the height $S_n$ of the random walk after $n$ steps is a sum of independent and identically distributed (i.i.d.) random variables. If we define

$$Y_j := \frac{X_j + 1}{2}, \qquad j = 1, 2, \ldots,$$

we record a right step $X_j = 1$ as a *success* ($Y_j = 1$) and a left step $X_j = -1$ as a *failure* ($Y_j = 0$). Consequently, $Y_1, Y_2, \ldots$ are independent random variables with $\mathbb{P}(Y_j = 1) = \mathbb{P}(Y_j = 0) = \frac{1}{2}$. Therefore, the sum

$$B_n := Y_1 + Y_2 + \ldots + Y_n$$

follows the binomial distribution $\text{Bin}(n, \frac{1}{2})$. According to the de Moivre[1]–Laplace[2] central limit theorem (see Sect. 7.4), for real numbers $a$ and $b$ with $a < b$,

$$\lim_{n \to \infty} \mathbb{P}\left(a \leq \frac{B_n - \frac{n}{2}}{\sqrt{n \frac{1}{2} \cdot \frac{1}{2}}} \leq b\right) = \Phi(b) - \Phi(a). \tag{2.3}$$

---

[1] Abraham de Moivre (1667–1754), considered the most significant probabilist before P.S. Laplace, was admitted to the Royal Society in 1697 and to the Berlin Academy in 1735.

[2] Pierre Simon Laplace (1749–1827), physicist and mathematician, was a professor at the École Polytechnique in Paris. His book, *Théorie analytique des probabilités*, published in 1812, summarized the probabilistic knowledge of his time.

Here,

$$\Phi(x) = \int_{-\infty}^{x} \frac{1}{\sqrt{2\pi}} \exp\left(-\frac{t^2}{2}\right) dt \qquad (2.4)$$

denotes the cumulative distribution function of the standard normal distribution. By the definition of $S_n$, $Y_j$, and $B_n$,

$$S_n = 2B_n - n. \qquad (2.5)$$

If we insert this relation into (2.3), it follows that

$$\lim_{n \to \infty} \mathbb{P}\left(a \leq \frac{S_n}{\sqrt{n}} \leq b\right) = \Phi(b) - \Phi(a). \qquad (2.6)$$

This result may also be obtained with the more general Lindeberg[3]–Lévy[4] central limit theorem (see Sect. 7.4).

In particular, it follows that $\mathbb{P}(-1.96\sqrt{n} \leq S_n \leq 1.96\sqrt{n}) \approx 0.95$ for large $n$, and thus approximately $\mathbb{P}(-98 \leq S_{2,500} \leq 98) \approx 0.95$. In Sect. 2.12, we will see that, for large $n$, $\mathbb{P}(-2.242\sqrt{n} \leq \max_{k \leq n} |S_k| \leq 2.242\sqrt{n})$ is approximately equal to 0.95. The range of the stochastic fluctuations of a simple symmetric random walk up to time $n$ is therefore *of the order* $\sqrt{n}$, and the labeling of the ordinate should reflect this $n/\sqrt{n}$ scale ratio. For example, a random walk of length $n = 2500$ can take any of the values $-2500, -2498, \ldots, -2, 0, 2, \ldots, 2498$, and $2500$ at the end. However, the corresponding polygonal line will remain within the bounds $\pm 2.242\sqrt{2500} \approx \pm 113$ with a probability of 0.95 throughout its entire temporal course. Figure 2.4 and all further plots showing random walks that take many steps account for this fact.

According to (2.5), the event $\{S_n = k\}$ is equivalent to $\{B_n = \frac{n+k}{2}\}$. With the convention that

$$\binom{n}{\ell} := 0, \qquad \text{if } \ell \notin \mathbb{Z},$$

we obtain

$$\mathbb{P}(S_n = k) = \frac{\binom{n}{\frac{n+k}{2}}}{2^n}, \qquad n \in \mathbb{N}, \ |k| \leq n. \qquad (2.7)$$

---

[3] Jarl Waldemar Lindeberg (1876–1932), farmer and mathematician.

[4] Paul Lévy (1886–1971), one of the main founders of modern probability theory, was a professor at the École Polytechnique in Paris from 1919 to 1959.

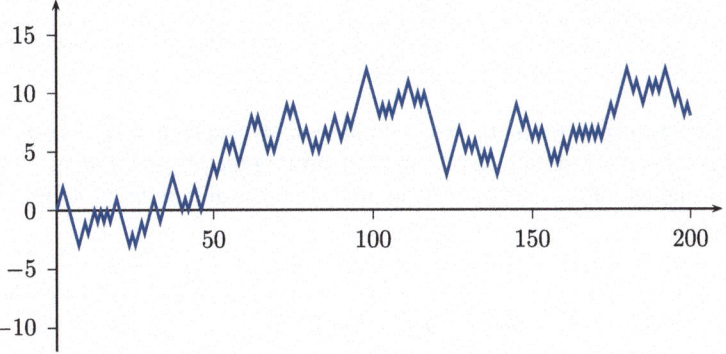

**Fig. 2.4** The random walk from Fig. 2.3 in adjusted scale

In particular, for any positive integer $n$,

$$\mathbb{P}(S_{2n} = 2k) = \frac{\binom{2n}{n+k}}{2^{2n}}, \quad k \in \mathbb{Z}, \ |k| \leq n. \tag{2.8}$$

From this, it follows that

$$\mathbb{E}|S_{2n}| = 2n \frac{\binom{2n}{n}}{2^{2n}}.$$

With the help of (7.16) and the notation introduced in (1.1), we then get

$$\mathbb{E}|S_{2n}| \sim \sqrt{\frac{2}{\pi}} \sqrt{2n} \tag{2.9}$$

as $n \to \infty$ (Exercise 2.1). Since $\mathbb{E}(S_{2n})$ and $\mathbb{E}(S_{2n+1})$ differ in absolute value by at most 1, it follows that $\mathbb{E}|S_n| \sim \sqrt{2n/\pi}$ as $n \to \infty$.

**SAQ 2** Why does $\left|\mathbb{E}|S_{2n}| - \mathbb{E}|S_{2n+1}|\right| \leq 1$ hold?

After $n$ steps, the expected value of the distance of a particle starting at 0 and performing a simple symmetric random walk is approximately equal to $\sqrt{2n/\pi}$ and is thus of the order of $\sqrt{n}$.

In this book, random walks often (especially in proofs) start at a general height $S_0 := a$, where $a \in \mathbb{Z}$. The position of such a random walk after $n$ steps is given by

$$S_n = a + \sum_{j=1}^{n} X_j, \quad n \geq 0. \tag{2.10}$$

The random walk defined in this way is both spatially and temporally *homogeneous*. *Spatial homogeneity*, intuitively, means that shifting the initial height $S_0 = a$ by $b$ causes the height of the random walk at time $n$ to also shift by $b$; formally,

$$\mathbb{P}(S_n = \ell | S_0 = a) = \mathbb{P}(S_n = \ell + b | S_0 = a + b), \quad n \in \mathbb{N}, \, a, b, \ell \in \mathbb{Z}.$$

Note that on both sides of the equality sign, $\mathbb{P}\left(\sum_{j=1}^{n} X_j = \ell - a\right)$ appears.

*Temporal homogeneity*, loosely speaking, means that the stochastic behavior of a random walk that starts at the height $a$ at time $m \geq 0$ and is observed after $n$ additional steps is the same as if a random walk starting at the height $a$ at time 0 were observed after $n$ steps. Formally, the property of temporal homogeneity means that

$$\mathbb{P}(S_n = \ell | S_0 = a) = \mathbb{P}(S_{m+n} = \ell | S_m = a), \quad a, \ell \in \mathbb{Z}, \, m \geq 0, n \geq 1.$$

**SAQ 3** Why does this equation hold?

Both homogeneity properties also apply to the more general situation of a potentially asymmetric random walk, as considered in Chap. 4, where $\mathbb{P}(X_j = 1) = p = 1 - \mathbb{P}(X_j = -1)$ for a $p$ with $0 < p < 1$.

Another fundamental characteristic of simple random walks is their *Markov property*. Intuitively, this means that the position of the random walk at time $n + 1$ depends only on its position at time $n$ and not on its "entire history." Formally, the Markov property, named after the Russian mathematician A.A. Markov,[5] states that for any $n \geq 1$ and any choice of integers $a_0, \ldots, a_n$ with $\mathbb{P}(S_0 = a_0, \ldots, S_n = a_n) > 0$, the equation

$$\mathbb{P}(S_{n+1} = j | S_0 = a_0, S_1 = a_1, \ldots, S_n = a_n) = \mathbb{P}(S_{n+1} = j | S_n = a_n), \quad j \in \mathbb{Z},$$

holds. This is obvious because, according to (2.10), $S_{n+1} = S_n + X_{n+1}$, where $S_n$ and $X_{n+1}$ are independent random variables.

---

[5] Andrei Andreyevich Markov (1856–1922) was a professor at the University of St. Petersburg from 1893. Markov was politically progressive; among other things, he resigned all orders and decorations in protest when the election of Maxim Gorky as a member of the Academy was rejected on tsarist orders. His main field of work was probability theory.

## 2 The Simple Symmetric Random Walk on $\mathbb{Z}$

The simple symmetric random walk exhibits numerous unexpected phenomena that defy common intuition, which were first rigorously formulated by W. Feller[6] [FE1]. In the following sections, we address and answer several key questions. Section 2.1 provides necessary notation and explains the fundamental reflection principle. This principle is then applied in Sect. 2.2 to demonstrate that different sets of paths have the same cardinality. Examining the random walk depicted in Fig. 2.4, it intersects the $x$-axis after 46 steps but does not return to zero (at least not until time 200).

In Sect. 2.3, we demonstrate that the random time $L_{2n}$, representing the last visit to zero of a random walk taking $2n$ steps, exhibits a counterintuitive U-shaped distribution. Moreover, as $n \to \infty$, the distribution of $L_{2n}/(2n)$ converges to the famous arcsine distribution. Section 2.4 investigates $N_{2n}$, the number of visits to zero in a random walk of length $2n$. Surprisingly, $N_{2n}$ does not increase linearly with the number $2n$ of steps as $n$ grows, but rather increases approximately in proportion to $\sqrt{2n}$. This finding is connected to the result presented in Sect. 2.5, which asserts that, although a particle will eventually return to the starting point 0 with probability one, it takes, on average, an infinite number of steps to do so.

In Sect. 2.6, we discover that a random walk tends to remain either very short or very long above the $x$-axis (see Fig. 2.4), and that the time $O_{2n}$ spent above this axis by a random walk of length $2n$ follows the same distribution as $L_{2n}$. In Sect. 2.7, we analyze the distribution of the maximum height $M_n$ of a simple symmetric random walk of length $n$ and establish an interesting connection with the number of visits to zero. Section 2.8 reveals, in particular, that for every fourth random walk, this maximum height is attained exactly at two distinct points in time.

Section 2.9 addresses the question of how long it takes to reach a specified integer $k$ for the first time. Among other findings, it reveals that, on average, every second random walk requires approximately 22,000 steps to first hit the height of 100. Similarly surprising phenomena arise in the problem discussed in Sect. 2.10, which examines when two independent random walks starting at different heights first intersect.

In Sect. 2.11, we investigate the number of changes of sign, which represent passages through zero in the random walk. Section 2.12 addresses the maximum absolute value, specifically examining the distribution of the largest absolute number visited by a simple symmetric random walk of length $n$. These findings are applied to a statistical problem discussed in Sect. 2.13. A duality principle presented in Sect. 2.14 unveils intriguing connections to prior investigations and offers new insights. The chapter closes with a discussion of broader perspectives.

---

[6] William Feller (1906–1970) graduated in 1926 under R. Courant in Göttingen and became a lecturer at the University of Kiel in 1928. Fleeing from the Nazis in Germany in 1933, he became a US citizen in 1944 and, from 1950, was a professor at Princeton University. His two-volume work, *An Introduction to Probability Theory and Its Applications*, is considered one of the best mathematical textbooks of the twentieth century.

To conclude, random walks can be interpreted in various ways. One interpretation liken them to a game between two individuals, A and B, who flip a fair coin repeatedly. A bets on heads and B on tails. If heads appear, B pays one euro to A, and vice versa. In scenarios where both players have unlimited capital, a simple symmetric random walk on $\mathbb{Z}$ is generated. However, when financial resources are limited, the random walk is constrained by so-called *absorbing boundaries* (see Sect. 4.3).

## 2.1 Basic Concepts and Reflection Principle

Based on the preceding explanations, we illustrate the realization of a random walk taking $k$ steps as a path in a Cartesian coordinate system. The path starts at the origin $(0, 0)$ and is defined by the vector $(a_1, \ldots, a_k) \in \{-1, 1\}^k$ of step directions. It connects the points $(0, s_0), (1, s_1), \ldots (k, s_k)$ to form a polygonal chain. Here, $s_0 := 0$ and for $m \geq 1$,

$$s_m := a_1 + a_2 + \ldots + a_m.$$

We use lowercase instead of uppercase letters here, because $s_m$ is considered a *realization* of the random variable $S_m$.

Figure 2.5 shows such a path for the case $k = 20$. The total number $k$ of steps is called the *length* of the path. A path of length $k$ is briefly termed a $k$-*path*.

We can identify a path both with $(a_1, \ldots, a_k)$ and with $(s_1, \ldots, s_k)$; due to $a_1 = s_1$ and $a_j = s_j - s_{j-1}$ for $j = 2, \ldots, k$, there is a one-to-one correspondence between the tuple $(a_1, \ldots, a_k)$ of upward and downward steps of the path and the tuple $(s_1, \ldots, s_k)$, which describes the sequence of *heights* of the path measured on the vertical axis after each step. Thus, the set of all $k$-paths can be formally defined by

$$W_k := \{(s_0, s_1, \ldots, s_k) : s_0 = 0, s_j - s_{j-1} \in \{1, -1\} \text{ for } j = 1, \ldots, k\}.$$

Since each step of a $k$-path can occur in two ways, there are a total of $2^k$ different $k$-paths. A $k$-path is called

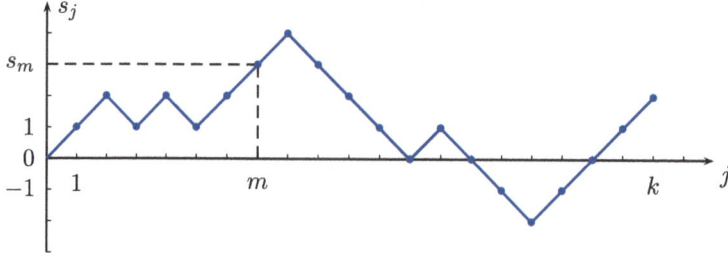

**Fig. 2.5** Path of length 20

## 2.1 Basic Concepts and Reflection Principle

- (*strictly*) *positive*, if $s_1 > 0$, $s_2 > 0, \ldots, s_k > 0$,
- (*strictly*) *negative*, if $s_1 < 0$, $s_2 < 0, \ldots, s_k < 0$,
- *non-negative*, if $s_1 \geq 0$, $s_2 \geq 0, \ldots, s_k \geq 0$,
- *non-positive*, if $s_1 \leq 0$, $s_2 \leq 0, \ldots, s_k \leq 0$,
- *zero-avoiding*, if $s_1 \neq 0$, $s_2 \neq 0, \ldots, s_k \neq 0$,
- *a bridge*, if $s_k = 0$ (this case can only occur for even $k$).

In the following, let

- $W_{k,>0}$ denote the set of (strictly) positive $k$-paths,
- $W_{k,<0}$ denote the set of (strictly) negative $k$-paths,
- $W_{k,\geq 0}$ denote the set of non-negative $k$-paths,
- $W_{k,\leq 0}$ denote the set of non-positive $k$-paths,
- $W_{k,\neq 0}$ denote the set of zero-avoiding $k$-paths,
- $W_k^\circ$ denote the set of bridges of length $k$.

The superscript symbol " $\circ$ " is meant to resemble a zero, signifying that a bridge returns to zero at the end. In what follows, a bridge of length $k$ is referred to as a $k$-*bridge*.

Due to the independence condition (2.1), all $2^k$ paths of length $k$ are equally probable. As such, probabilities of subsets of $W_k$ are determined by counting the favorable cases. In this section, we introduce various counting methods for paths and begin by investigating the number of $k$-paths that reach a height $b \geq 0$ at the end, i.e., paths leading from the point $(0, 0)$ to the point $(k, b)$. Evidently, for such a path to exist, $b$ must satisfy $b \leq k$. Additionally, $k$ and $b$ must have the same *parity*, meaning both must be either even or odd. Since a path is determined by the tuple $(a_1, \ldots, a_k)$ of upward and downward steps, the question arises: how many components of this tuple are equal to 1? Let $c$ be the number of 1's, and hence $k - c$ denotes the number of $-1$'s. Then, $b = c - (k - c) = 2c - k$, so $c = \frac{k+b}{2}$. The height at time $k$ is precisely the difference between upward and downward steps. Since there are $\binom{k}{c}$ ways to choose $c$ components from a $k$-tuple and assign each a value of 1 (with the remaining components assigned $-1$), we arrive at the following result:

**Lemma 2.1 (Number of Paths from $(0, 0)$ to $(k, b)$)** *Let $k$ be a positive integer and $b$ an integer such that $0 \leq b \leq k$. If $k$ and $b$ have the same parity, then there are*

$$\binom{k}{\frac{k+b}{2}}$$

*paths from the point $(0, 0)$ to the point $(k, b)$, that is, $k$-paths with the property $s_k = b$.*

To determine the number of paths leading from the point $(u, v) \in \mathbb{Z} \times \mathbb{Z}$ to the point $(r, s) \in \mathbb{Z} \times \mathbb{Z}$, where $r > u$ and $s \geq v$, and each step of the path can only move right or upward, we translate the problem by shifting $(u, v)$ to the origin. This involves counting paths that start at the origin and end at $(r-u, s-v)$. If $r-u$ and $s-v$ have the same parity, then by Lemma 2.1, this count is given by the binomial coefficient

$$\binom{r-u}{\frac{r-u+s-v}{2}}. \tag{2.11}$$

**The Reflection Principle**

Many combinatorial counting problems concerning paths can be simplified using the so-called *reflection principle*. The reflection principle[7] refers synonymously to a bijective mapping between two sets of paths denoted by $M_1$ and $M_2$, where the mapping rule involves a (somehow defined) reflection across a problem-specific chosen axis of reflection. If, for example, the number of elements in $M_2$ can be easily determined, then due to the bijectivity of the mapping, one also knows the cardinality of the set $M_1$. A simple example of a reflection principle is the mapping that reflects a path across the $x$-axis, assigning the tuple $(a_1, \ldots, a_k)$ to the tuple $(-a_1, \ldots, -a_k)$. Using this reflection, every $k$-path with $s_k = b > 0$ is mapped to a $k$-path with $s_k = -b < 0$ and vice versa. Therefore, the number of $k$-paths given in Lemma 2.1 is also the number of paths from $(0, 0)$ to $(k, -b)$.

**SAQ 4** Does the equation $\binom{n}{k} = \binom{n}{n-k}$ follow any reflection principle?

Figure 2.6 illustrates the application of a commonly used reflection principle. No coordinate system is shown intentionally, as often only segments of paths undergo subsequent reflections, and the axis of reflection does not necessarily have to be the $x$-axis; it can also be vertical. If there is a path from point $P$ to point $Q$ that intersects the marked axis $A$, reflecting the segment of the path up to its first encounter with $A$ (denoted as point $S$ in the figure) creates a new path from point $P^*$ to $Q$. Conversely, every path from $P^*$ to $Q$ intersects axis $A$ for the first time at some point. Reflecting this initial segment of the path from $P^*$ to $S$ across $A$, while leaving the second segment unchanged, reconstructs the original path from $P$ to $Q$. This mapping of paths from $P$ to $Q$ that intersect axis $A$ at least once, to paths from $P^*$ to $Q$, is clearly bijective, thus justifying the caption.

**The Ballot Problem**

The reflection principle leads to a quick solution[8] to the following classical problem (see [WI] or [BE]): An election has taken place between two individuals, A and B. As each ballot is counted one by one, it is always known who is currently leading. In

---

[7] The reflection principle is commonly attributed to the French mathematician Désiré André (1840–1918). However, the cited work [AND] contains no geometric arguments.

[8] The presented solution can be found in a 1923 article by J. Aebly (see [AEB]).

## 2.1 Basic Concepts and Reflection Principle

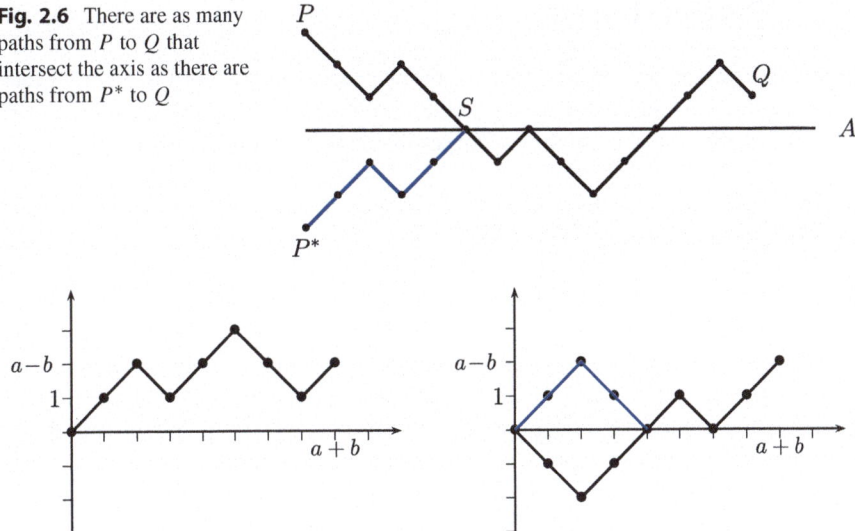

**Fig. 2.6** There are as many paths from $P$ to $Q$ that intersect the axis as there are paths from $P^*$ to $Q$

**Fig. 2.7** Illustrating a vote-counting process using a path and reflection principle

the end, A has won with $a$ votes, compared to $b$ votes for B (where $a > b$). What is the probability of event $C$, the event that A led throughout the entire vote-counting process?

We assign paths to the counting sequences by recording each vote for A or B as an upward or downward step, respectively. Each possible vote-counting process then corresponds to a path leading from $(0, 0)$ to $(a + b, a - b)$ (Fig. 2.7 left). According to (2.11), there are $\binom{a+b}{a}$ such paths, which we assume are equally probable.

The paths *unfavorable* to the occurrence of event $C$ are those that initially move downward on the first step, i.e., all paths leading from $(1, -1)$ to $(a + b, a - b)$. There are $\binom{a+b-1}{a}$ such paths, according to (2.11). Additionally, there are paths that start with an upward step but subsequently intersect the $x$-axis. However, by the reflection principle (see Fig. 2.7, right, and Fig. 2.6), the number of such paths is equal to the number of paths leading from $(1, -1)$ to $(a + b, a - b)$. Thus, A leads throughout the entire vote-counting process with the probability

$$\mathbb{P}(C) = 1 - 2 \frac{\binom{a+b-1}{a}}{\binom{a+b}{a}} = \frac{a-b}{a+b}.$$

A variant of the question is found in Exercise 2.3.

## 2.2 The Main Lemma

In this section, we will count $k$-paths with the properties described in Sect. 2.1 and encounter various surprising results. Since bridges play a significant role, $k$ will often be an even integer. The following result can be found in [FE1], p. 76.

**Main Lemma 2.2** *For the numbers of bridges, non-negative paths, and zero-avoiding paths of the same length $2n$,*

$$|W_{2n}^\circ| \;=\; |W_{2n,\geq 0}| \;=\; |W_{2n,\neq 0}|, \qquad n \geq 1.$$

Therefore, there are as many paths of length $2n$ that hit zero at the end as there are non-negative $2n$-paths. This count is equivalent to the number of $2n$-paths that avoid hitting zero. Viewing the random walk as a game between two people, A and B, we find an equal number of sequences after $2n$ coin tosses resulting in a tie as there are sequences where B *never takes the lead*. Similarly, there are as many sequences where *one player consistently leads*, ensuring no tie until the $2n$-th toss.

***Proof*** We establish the main lemma by constructing bijective mappings between the sets of $2n$-paths involved. To prove the first equality, we begin by considering an arbitrary $2n$-bridge. If this bridge is non-negative, we leave it unchanged. Otherwise, we locate the point $M$ where the minimum value first occurs, reflect this segment of the path across the $y$-axis, and shift the reflected part to reach the point $(2n, 0)$ (see Fig. 2.8). In a new coordinate system shown in blue in Fig. 2.8, with its origin at $M$, a non-negative path[9] is then generated.

This verbal description of mapping bridges to non-negative paths corresponds to the following function $f : W_{2n}^\circ \to W_{2n,\geq 0}$: If $w = (a_1, \ldots, a_{2n}) \in \{-1, 1\}^{2n}$ with

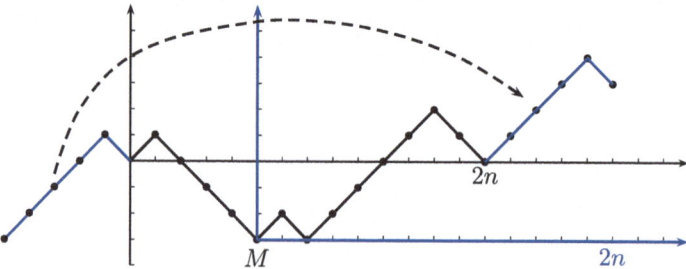

**Fig. 2.8** A $2n$-bridge (black) is cut at the first point at which the minimum is attained. This path segment is reflected across the $y$-axis, and the reflected path is slid to $(2n, 0)$. In the blue coordinate system, a non-negative path, starting at $M$, is created

---

[9] This idea goes back to Edward Nelson (1932–2014), who was a professor at Princeton University from 1964 to 2013.

## 2.2 The Main Lemma

$a_1 + \ldots + a_{2n} = 0$ represents a bridge, then

$$\min(w) := \min\{a_1 + \ldots + a_k : k \in \{1, 2, \ldots, 2n\}\}$$

denotes the minimum value of $w$, and

$$k_0 = k_0(w) := \min\{k \in \{1, \ldots, 2n\} : a_1 + \ldots + a_k = \min(w)\}$$

is the time at which this minimum is first attained. For the path $w$ shown in Fig. 2.8, $\min(w) = -3$ and $k_0 = 5$.

We define $f(w) := w$, if $\min(w) = 0$, and

$$f(w) := (a_{k_0+1}, \ldots, a_{2n}, -a_{k_0}, -a_{k_0-1}, \ldots, -a_1), \tag{2.12}$$

if $\min(w) < 0$. Note that if $\min(w) = -r$ with $r \in \mathbb{N}_0$, the image path $f(w)$ leads from $(0, 0)$ to $(2n, 2r)$, due to $a_1 + \ldots + a_{k_0} = -r$ and $a_1 + \ldots + a_{2n} = 0$. From this, we directly infer the injectivity of $f$: If two different bridges $w = (a_1, \ldots, a_{2n})$ and $v = (b_1, \ldots, b_{2n})$ have different minima, they are mapped to different paths. In the case $\min(w) = \min(v)$, either $k_0(w) = k_0(v)$ or $k_0(w) \neq k_0(v)$. Each of these two subcases also ensures $f(w) \neq f(v)$ according to (2.12): In the second subcase, without loss of generality, if $k_0(w) < k_0(v)$, the tuples $f(w)$ and $f(v)$ differ at the $k_0(w)$-th position.

However, the function $f$ is also surjective. To demonstrate this, we reverse the construction just defined: Begin with an arbitrary non-negative path. If this is a bridge, where $s_{2n} = 0$, leave it unchanged. Otherwise, let $s_{2n} = 2r$ for some $r \in \{1, \ldots, n\}$. We then divide the path at the last occurrence of height $r$, denoted as point $P$, reflect the cut segment across the vertical axis passing through the point $(2n, 0)$, and reattach it to the origin on the left (see Fig. 2.9). This procedure results in a $2n$-bridge.

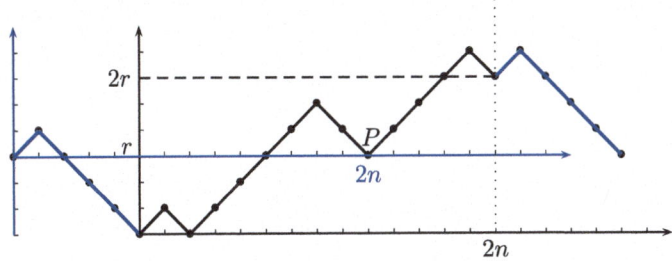

**Fig. 2.9** A non-negative $2n$-path with $s_{2n} = 2r$ and $r > 0$ is reflected across the vertical line passing through $(2n, 0)$ from the last time it hits the height $r$. The resulting segment, shown in blue, is attached to the left at the origin. In the blue coordinate system, a $2n$-bridge is created

To establish the second equality in the main lemma, we partition the set of all non-negative paths based on whether they are strictly positive or visit zero at least once. Thus, we express it as

$$W_{2n,\geq 0} = W_{2n,>0} \uplus (W_{2n,\geq 0} \setminus W_{2n,>0}). \tag{2.13}$$

The plus sign inside the union symbol signifies that we are combining *disjoint* sets. Next, we demonstrate that the set $W_{2n,\geq 0} \setminus W_{2n,>0}$ can be bijectively mapped onto $W_{2n,>0}$. With Eq. (2.13), it follows that

$$|W_{2n,\geq 0}| = 2|W_{2n,>0}|.$$

Since, due to symmetry, the right-hand side equals the count of the number of all zero-avoiding paths, the main lemma would thereby be fully established. To construct a bijective mapping between $W_{2n,\geq 0} \setminus W_{2n,>0}$ and $W_{2n,>0}$, we consider an arbitrary path from $W_{2n,\geq 0} \setminus W_{2n,>0}$. This path is mapped to a strictly positive path by converting the last downward step before the first visit to zero (denoted by $(a, 0)$) into an upward step. Subsequently, we shift the sub-path starting at $(a, 0)$ two units upward, thereby attaching it at the point $(a, 2)$ (see Fig. 2.10, left). This transformation results in a strictly positive path. Given that different paths either feature distinct first visits to zero or differ in their trajectories prior to or following this first visit, this mapping assigns different paths to different paths and is thus injective.

Conversely, each positive path is mapped to a non-negative path as follows: The positive path reaches the height 1 for the last time (no later than time $2n - 1$; in Fig. 2.10, this time is marked as $b$). We convert the subsequent upward step into a downward step and shift the following segment of the path two units downward. This process creates a non-negative path (Fig. 2.10, right), demonstrating that the mapping is surjective. Hence, the main lemma is fully proved. ∎

**Corollary 2.3** *For each $n \geq 1$,*

$$|W^{\circ}_{2n}| = |W_{2n,\geq 0}| = |W_{2n,\leq 0}| = |W_{2n,\neq 0}| = \binom{2n}{n}.$$

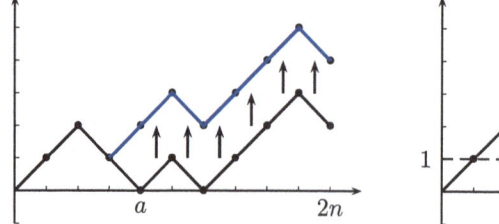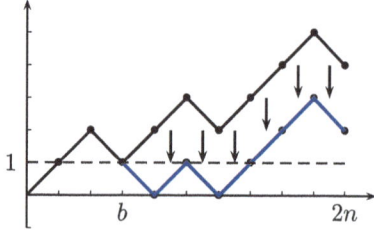

**Fig. 2.10** Transition from a non-negative to a positive path (left) and its inverse (right)

## 2.3 Last Visits to Zero

**Proof** According to Lemma 2.1, there are $\binom{2n}{n}$ paths from $(0,0)$ to $(2n,0)$, as exactly $n$ upward steps must be chosen out of a total of $2n$ steps. ∎

### 2.3 Last Visits to Zero

In the following, we consider all $2^{2n}$ paths of length $2n$ as equally likely, so we apply the uniform distribution $\mathbb{P}$ on the sample space

$$W_{2n} = \{w := (s_0, s_1, \ldots, s_{2n}) : s_0 = 0, \, s_j - s_{j-1} \in \{1, -1\} \text{ for } j = 1, \ldots, 2n\} \quad (2.14)$$

We are initially interested in the distribution of the random variable $L_{2n} : W_{2n} \to \mathbb{N}_0$, defined by

$$L_{2n}(w) := \max\{j : j \in \{0, 1, \ldots, n\} \text{ and } s_{2j} = 0\},$$

where $w = (s_0, s_1, \ldots, s_{2n})$. Apparently, $L_{2n}$ models the (even) *time of the last visit to zero* of a simple symmetric random walk of length $2n$, see Fig. 2.11. Such a random walk always has a (trivial) visit to zero at time 0 according to the above definition. The event $\{L_{2n} = 0\}$ (i.e., $\{w \in W_{2n} : L_{2n}(w) = 0\}$) therefore occurs if and only if the random walk is zero-avoiding.

Setting

$$u_{2m} := \frac{\binom{2m}{m}}{2^{2m}}, \quad m = 0, 1, 2, \ldots, \quad (2.15)$$

the main result of this section reads as follows (see [FE1], p. 79):

**Theorem 2.4 (Distribution of $L_{2n}$)** *For the time $L_{2n}$ of the last visit to zero of a simple symmetric random walk on $\mathbb{Z}$ of length $2n$, starting at 0, the following hold:*

(a) $\mathbb{P}(L_{2n} = 2k) = u_{2k} u_{2(n-k)} = \dfrac{\binom{2k}{k} \binom{2(n-k)}{n-k}}{2^{2n}}, \quad k = 0, 1, \ldots, n,$

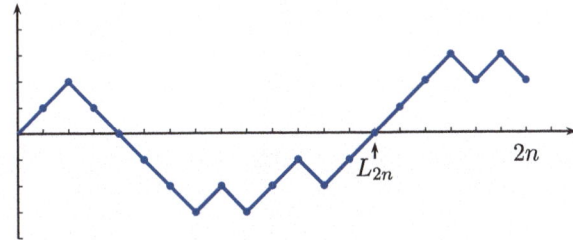

**Fig. 2.11** Illustrating the time of the last visit to zero of a $2n$-path

(b) $\mathbb{E}(L_{2n}) = n$,
(c) $\mathbb{V}(L_{2n}) = \binom{n+1}{2}$.

*Proof*

(a) We first address the special cases where $k = 0$ and $k = n$. The event $\{L_{2n} = 0\}$ occurs if and only if the path belongs to the set $W_{2n, \neq 0}$, i.e., it is zero-avoiding. Using the main lemma and Corollary 2.3, we obtain

$$\mathbb{P}(L_{2n} = 0) = \frac{|W_{2n, \neq 0}|}{2^{2n}} = \frac{\binom{2n}{n}}{2^{2n}} = u_{2n} = u_0 u_{2n}.$$

If $L_{2n} = 2n$, a bridge is present, so $\{L_{2n} = 2n\} = W_{2n}^\circ$. The main lemma and Corollary 2.3 yield

$$\mathbb{P}(L_{2n} = 2n) = \frac{|W_{2n}^\circ|}{2^{2n}} = \frac{\binom{2n}{n}}{2^{2n}} = u_{2n} = u_{2n} u_0. \tag{2.16}$$

If $k \in \{1, \ldots, n-1\}$, the $2n$-path visits zero at time $2k$ and is zero-avoiding afterwards. The number of all paths with these properties can be counted using the multiplication rule of combinatorics and the main lemma. There are $|W_{2k}^\circ|$ paths of length $2k$ that visit zero at time $2k$. For each of these paths, there are $|W_{2(n-k), \neq 0}|$ zero-avoiding continuations from $(2k, 0)$ to time $2n$, i.e., from $(2k, 0)$ to every point $(2n, \ell)$ with $\ell \neq 0$ that can possibly be reached. Consequently,

$$\mathbb{P}(L_{2n} = 2k) = \frac{|W_{2k}^\circ| \cdot |W_{2(n-k), \neq 0}|}{2^{2n}} = \frac{|W_{2k}^\circ|}{2^{2k}} \cdot \frac{|W_{2(n-k), \neq 0}|}{2^{2(n-k)}} = u_{2k} u_{2(n-k)}.$$

(b) Because

$$\mathbb{P}(L_{2n} = 2k) = \mathbb{P}(L_{2n} = 2(n-k)) = \mathbb{P}(2n - L_{2n} = 2k), \quad k = 0, 1, \ldots, n,$$

the distribution of $L_{2n}$ is symmetric around $n$. Therefore, the random variables $L_{2n}$ and $2n - L_{2n}$ follow the same distribution and thus have the same expectation. This implies that $\mathbb{E}(L_{2n}) = 2n - \mathbb{E}(L_{2n})$, leading to $\mathbb{E}(L_{2n}) = n$.

(c) We provide two proofs: one using the Legendre polynomials from Sect. 7.12 (see [REN], p. 510) and a purely combinatorial one. Both proofs are unsatisfactory, as the simple formula $\mathbb{V}(L_{2n}) = \binom{n+1}{2}$ practically demands a short proof. Given that $\mathbb{E}(L_{2n}) = n$ and $\mathbb{V}(L_{2n}) = \mathbb{E}(L_{2n} - \mathbb{E}(L_{2n}))^2$, we must show that

$$\frac{1}{2^{2n}} \sum_{k=0}^{n} (2k-n)^2 \binom{2k}{k} \binom{2(n-k)}{n-k} = \binom{n+1}{2}. \tag{2.17}$$

## 2.3 Last Visits to Zero

By using Eq. (7.35) for $z = -xe^{iy}$ and for $z = -xe^{-iy}$, where i denotes the imaginary unit and $x$ and $y$ are real numbers with $|x| < 1$, it follows that

$$\left(1 - xe^{iy}\right)^{-1/2} = \sum_{k=0}^{\infty} \binom{2k}{k} \frac{1}{2^{2k}} e^{iky} x^k,$$

$$\left(1 - xe^{-iy}\right)^{-1/2} = \sum_{k=0}^{\infty} \binom{2k}{k} \frac{1}{2^{2k}} e^{-iky} x^k.$$

If we form the Cauchy product of these two series and use the equation

$$1 - 2x \cos y + x^2 = (1 - xe^{iy})(1 - xe^{-iy}), \qquad (2.18)$$

we obtain

$$\frac{1}{\sqrt{1 - 2x \cos y + x^2}} = \sum_{n=0}^{\infty} \left( \sum_{k=0}^{n} \frac{\binom{2k}{k}\binom{2(n-k)}{n-k}}{2^{2n}} e^{i(2k-n)y} \right) x^n.$$

**SAQ 5** Can you derive Eq. (2.18)?

A comparison with Eq. (7.36) shows that

$$P_n(\cos y) = \sum_{k=0}^{n} \frac{\binom{2k}{k}\binom{2(n-k)}{n-k}}{2^{2n}} e^{i(2k-n)y}.$$

Here, $P_n$ denotes the $n$-th Legendre polynomial, see (7.37). If we differentiate both sides of this equation twice with respect to $y$, taking into account the chain rule, the expression $\sin^2 y \cdot P_n''(\cos y) - \cos y \cdot P_n'(\cos y)$ appears on the left-hand side. On the right-hand side, the derivative of the exponential function introduces the additional factor $-(2k-n)^2$. If we evaluate the derivative at $y = 0$, we obtain (2.17) with (7.39).

For a combinatorial proof (credit to Daniel Hug), we define

$$F(n,k) := k^2 \binom{2k}{k}\binom{2n-2k}{n-k}, \quad n \in \mathbb{N}, \; k \in \{0, 1, \ldots, n\}, \qquad (2.19)$$

and $F(n, k) := 0$ for $k > n$. We further define a function $G : \mathbb{N} \times \mathbb{N}_0 \to \mathbb{R}$ by

$$G(n, k) := 2\big[k(3n + 2) - (n + 1)\big] (k - 1)k \binom{2k}{k}\binom{2n - 2k + 1}{n + 1 - k}$$

for $k \in \{0, 1, \ldots, n\}$,

$$G(n, n + 1) := n(3n + 1)F(n + 1, n + 1) = n(3n + 1)(n + 1)^2 \binom{2n + 2}{n + 1},$$

and $G(n, k) := 0$ for $k \geq n + 2$. In particular, $G(n, 0) = G(n, 1) = 0$.
We claim the validity of

$$G(n, k+1) - G(n, k) = -n(3n+1)F(n+1, k) + 4(3n+4)(n+1)F(n, k) \qquad (2.20)$$

for $n \in \mathbb{N}$ and $k \in \{0, \ldots, n + 1\}$. With this, the function $f : \mathbb{N} \to \mathbb{R}$, defined by

$$f(n) := \sum_{k=0}^{n} F(n, k) = 2^{2n} \frac{\mathbb{E}(L_{2n}^2)}{4}, \qquad (2.21)$$

would have the recursion

$$-n(3n + 1)f(n + 1) + 4(3n + 4)(n + 1)f(n) = G(n, n + 2) - G(n, 0) = 0,$$

and therefore,

$$f(n + 1) = 4 \cdot \frac{n + 1}{n} \cdot \frac{3n + 4}{3n + 1} \cdot f(n), \quad n \in \mathbb{N}. \qquad (2.22)$$

Using mathematical induction, one would then obtain

$$f(n) = 2^{2n-3} n (3n + 1), \quad n \in \mathbb{N}, \qquad (2.23)$$

and due to $f(n)2^{-2n} = \mathbb{E}(L_{2n}^2)/4$ and $\mathbb{E}(L_{2n}) = n$, the assertion follows.

**SAQ 6** Can you provide the proof of (2.23)?

## 2.3 Last Visits to Zero

To prove Eq. (2.20), we observe that it is valid for $k = 0$, as both sides are equal to zero in this case. Furthermore, (2.20) is true for $k = n + 1$ by the definition of $G(n, n + 1)$ and because $F(n, n + 1) = 0$. For the case $k = n \geq 1$, the assertion is

$$-n(3n + 1)n^2 \binom{2n}{n} 2 + 4(3n + 4)(n + 1)n^2 \binom{2n}{n}$$
$$= n(3n + 1)(n + 1)^2 \binom{2n + 2}{n + 1} - 2(3n^2 + n - 1)(n - 1)n \binom{2n}{n},$$

which, after dividing by $2n\binom{2n}{n}$, is equivalent to

$$-(3n+1)n^2 + 2(n+1)(3n+4)n = (3n+1)(2n+1)(n+1) - (3n^2+n-1)(n-1)$$

and can be confirmed by direct calculation. Now, let $1 \leq k \leq n - 1$, and thus $n \geq 2$ and $F(n, k) \neq 0$. In this case,

$$F(n + 1, k) = 2F(n, k) \cdot \frac{2n - 2k + 1}{n - k + 1},$$

$$G(n, k) = 2F(n, k) \cdot \frac{2n - 2k + 1}{n - k + 1} \cdot \frac{k - 1}{k} \cdot \left[(3n + 2)k - (n + 1)\right],$$

$$G(n, k + 1) = 2F(n, k) \left[(3n + 2)(k + 1) - (n + 1)\right] \frac{2k + 1}{k}.$$

Hence, Eq. (2.20), after cancelling $2F(n, k)$, is equivalent to

$$-n(3n + 1)\frac{2n - 2k + 1}{n - k + 1} + 2(3n + 4)(n + 1)$$
$$= \left[(3n+2)(k+1) - (n+1)\right]\frac{2k+1}{k} - \frac{2n-2k+1}{n-k+1}\frac{k-1}{k}\left[(3n+2)k - (n+1)\right].$$

This assertion is also confirmed by simple calculation. With this, (2.20) is generally proved. ∎

The consequences of Theorem 2.4 are striking. For the quotients

$$q_{n,k} := \frac{\mathbb{P}(L_{2n} = 2(k + 1))}{\mathbb{P}(L_{2n} = 2k)} = \frac{(2k + 1)(n - k)}{(k + 1)(2n - 2k - 1)},$$

a direct calculation yields:

$$q_{n,k} \begin{Bmatrix} < \\ = \\ > \end{Bmatrix} 1 \iff 2k \begin{Bmatrix} < \\ = \\ > \end{Bmatrix} n - 1.$$

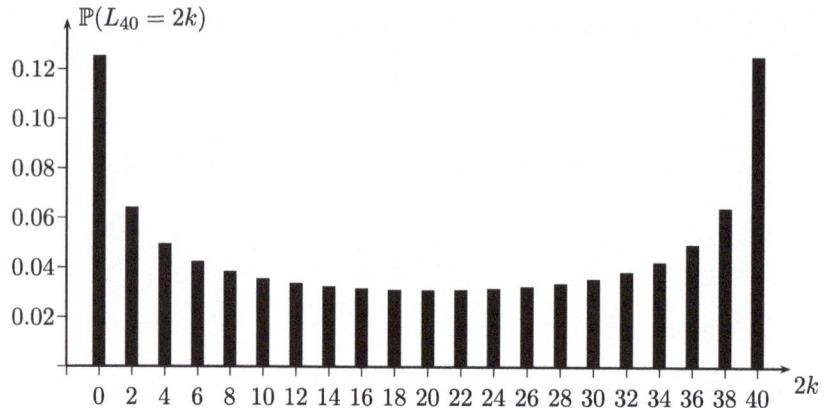

**Fig. 2.12** Bar chart of the distribution of $L_{40}$

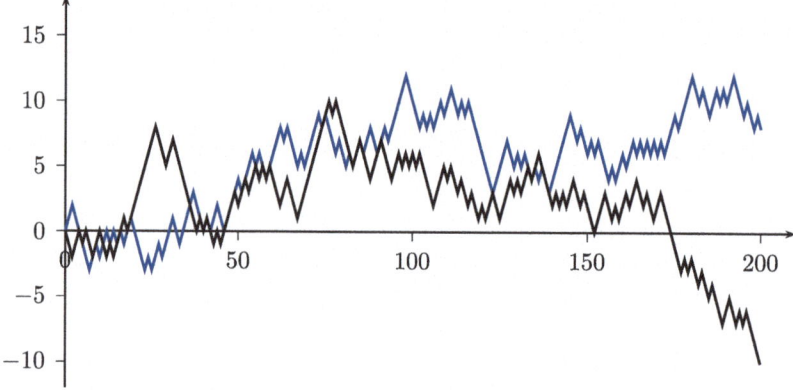

**Fig. 2.13** Random walks of length 200 with last visits to 0 at times 46 and 174

Thus, the probabilities $\mathbb{P}(L_{2n} = 2k)$, for $k = 0, 1, 2, \ldots$, initially decrease strictly monotonically, reach a unique minimum when $n$ is even, and for odd $n$, have two minima (attained at times $2k = n$ or $2k \in \{n-1, n+1\}$). They then increase strictly monotonically. Consequently, the bar chart of the distribution of $L_{2n}$ is U-shaped (see Fig. 2.12 for the case $n = 20$).

Note that it is most likely that the random walk avoids zero until time $2n$, or, with the same probability, last visits zero at time $2n$. The least likely scenario is that the last visit to zero occurs at time $n$ (for even $n$) or at time $n-1$ or $n+1$ (for odd $n$). Thus, the last visit to zero is expected to occur either very early or very late, see also Fig. 2.13.

In the interpretation of the random walk as a game between two people, it is most likely that the last tie occurs either at the beginning or at the very end. For example, if $2n = 40$, then $\mathbb{P}(L_{40} = 0) + \mathbb{P}(L_{40} = 40) \approx 0.2508$. This means that at least one

## 2.3 Last Visits to Zero

of these extreme cases (one person leads all the time or a tie occurs for the first time after 40 games) occurs, on average, in every fourth game sequence of length 40.

The distribution of $L_{2n}$ is known as the *discrete arcsine distribution*. The following result (cf. [FE1], p. 81) clarifies the reason for this naming. It pertains to the limit behavior of $L_{2n}/(2n)$ as $n \to \infty$, which represents the *proportion of time* that elapses until the last visit to zero occurs.

**Theorem 2.5 (Arcsine Law for the Time of the Last Visit to Zero)** *Let $L_{2n}$ be the time of the last visit to zero of a symmetric Bernoulli random walk. Then, for every $x$ with $0 \leq x \leq 1$:*

$$\lim_{n \to \infty} \mathbb{P}\left(\frac{L_{2n}}{2n} \leq x\right) = \int_0^x \frac{1}{\pi \sqrt{t(1-t)}} \, dt = \frac{2}{\pi} \arcsin \sqrt{x}. \qquad (2.24)$$

***Proof*** We briefly set $L_{2n}^* := \frac{L_{2n}}{2n}$ and

$$g(t) := \frac{1}{\pi \sqrt{t(1-t)}}, \qquad 0 < t < 1. \qquad (2.25)$$

The idea of the proof is to estimate the probability $\mathbb{P}(L_{2n}^* \leq x)$ sufficiently well using Stirling's formula (7.17) and a Riemann sum to approximate the integral $\int_0^x g(t) \, dt$. In this regard, we first show the convergence

$$\lim_{n \to \infty} \mathbb{P}\left(a \leq L_{2n}^* \leq b\right) = \int_a^b g(t) \, dt \qquad (2.26)$$

for any choice of $a, b$ with $0 < a < b < 1$. Setting $I_n := \{k \in \mathbb{N} : a \leq \frac{k}{n} \leq b\}$, it follows that

$$\mathbb{P}\left(a \leq L_{2n}^* \leq b\right) = \sum_{k \in I_n} u_{2k} u_{2(n-k)} \qquad (2.27)$$

with $u_{2m}$ as in (2.15). We define

$$a_m := \frac{m! \, e^m}{m^m \sqrt{2\pi m}}, \qquad m \in \mathbb{N}, \qquad (2.28)$$

and note the equality $a_{2m}/a_m^2 = u_{2m} \sqrt{\pi m}$, which can be derived by direct calculation. With this, Eq. (7.16) gives

$$\lim_{m \to \infty} u_{2m} \sqrt{\pi m} = 1. \qquad (2.29)$$

For a given $\varepsilon$ with $0 < \varepsilon \leq 1$ there is thus an integer $m_0$ such that $|u_{2m}\sqrt{\pi m} - 1| \leq \varepsilon$ for every $m \geq m_0$. For $k \in I_n$ we have $k \geq na$ and $n - k \geq n(1-b)$. Therefore, if the inequality $n \geq C := \max\left(\frac{m_0}{a}, \frac{m_0}{1-b}\right)$ is satisfied, then $k \geq m_0$ as well as $n - k \geq m_0$, and thus both $|u_{2k}\sqrt{\pi k} - 1| \leq \varepsilon$ and $|u_{2(n-k)}\sqrt{\pi(n-k)} - 1| \leq \varepsilon$. Since the inequalities $|y - 1| \leq \varepsilon$ and $|z - 1| \leq \varepsilon$ (with $y, z \in \mathbb{R}$ and $\varepsilon \leq 1$) imply the upper bound

$$|yz - 1| = |(y-1)(z-1) + y - 1 + z - 1| \leq \varepsilon^2 + 2\varepsilon \leq 3\varepsilon,$$

we obtain for every $n \geq C$ and every $k \in I_n$

$$\left| u_{2k} u_{2(n-k)} \pi \sqrt{k(n-k)} - 1 \right| \leq 3\varepsilon,$$

and thus, after division by $\pi\sqrt{k(n-k)}$ and using (2.25),

$$\left| u_{2k} u_{2(n-k)} - g\left(\frac{k}{n}\right)\frac{1}{n} \right| \leq 3\varepsilon\, g\left(\frac{k}{n}\right)\frac{1}{n}, \qquad n \geq C,\ k \in I_n.$$

With (2.27) and the triangle inequality, we obtain for every $n \geq C$,

$$\left| \mathbb{P}(a \leq L^*_{2n} \leq b) - \sum_{k \in I_n} g\left(\frac{k}{n}\right)\frac{1}{n} \right| \leq 3\varepsilon \sum_{k \in I_n} g\left(\frac{k}{n}\right)\frac{1}{n}. \tag{2.30}$$

Except for at most two terms that become negligible as $n \to \infty$, the sum appearing here is a Riemann sum $R_n$ (say) for the integral on the right-hand side of (2.26), corresponding to the partition $\{a, b\} \cup \{k/n : k \in I_n\}$ of $[a, b]$ (note that neither $a$ nor $b$ need to belong to $I_n$). Since $g$ is continuous on $[a, b]$ and the partition's fineness is $\frac{1}{n}$, we have $\lim_{n \to \infty} R_n = \int_a^b g(t)\,dt$. Therefore, (2.30) implies

$$\limsup_{n \to \infty} \mathbb{P}(a \leq L^*_{2n} \leq b) \leq (1 + 3\varepsilon) \int_a^b g(t)\,dt,$$

$$\liminf_{n \to \infty} \mathbb{P}(a \leq L^*_{2n} \leq b) \geq (1 - 3\varepsilon) \int_a^b g(t)\,dt.$$

Since $\varepsilon > 0$ was arbitrary, (2.26) follows.

To prove (2.24) we can assume without loss of generality $x < 1$.

> **SAQ 7** Why can we assume this without loss of generality?

In addition, we can assume $x > 0$ because $\arcsin 0 = 0$ and $\mathbb{P}(L_{2n}/(2n) \leq 0) = \mathbb{P}(L_{2n} = 0) = u_{2n} \to 0$. For a given $\varepsilon \in (0, 1)$, we choose an $a \in (0, \frac{1}{2})$ with $a \leq x$ and $\int_a^{1-a} g(t)\,dt \geq 1 - \varepsilon$. According to what has already been proved,

## 2.3 Last Visits to Zero

$\liminf_{n\to\infty} \mathbb{P}(a \le L_{2n}^* \le 1-a) \ge 1-\varepsilon$, and thus $\mathbb{P}(L_{2n}^* < a) \le 2\varepsilon$ for every sufficiently large $n$. Because

$$\mathbb{P}(L_{2n}^* \le x) = \mathbb{P}(L_{2n}^* < a) + \mathbb{P}(a \le L_{2n}^* \le x),$$

it follows that

$$\limsup_{n\to\infty} \mathbb{P}(L_{2n}^* \le x) \le 2\varepsilon + \int_a^x g(t)\,dt \le 2\varepsilon + \int_0^x g(t)\,dt$$

and thus, as $\varepsilon \downarrow 0$, the inequality $\limsup_{n\to\infty} \mathbb{P}(L_{2n}^* \le x) \le \int_0^x g(t)\,dt$. On the other hand, $\mathbb{P}(L_{2n}^* \le x) \ge \mathbb{P}(u \le L_{2n}^* \le x)$ for every $u \in (0,x)$, and thus $\liminf_{n\to\infty} \mathbb{P}(L_{2n}^* \le x) \ge \int_u^x g(t)\,dt$, so also $\liminf_{n\to\infty} \mathbb{P}(L_{2n}^* \le x) \ge \int_0^x g(t)\,dt$, which was to be shown. ∎

Theorem 2.5 provides a statement about the proportion of time $L_{2n}^* = L_{2n}/(2n)$ (as a value between zero and one) until the occurrence of the last visit to zero. Since the left-hand side of Eq. (2.24), as a function of $x$, is the distribution function of $L_{2n}^*$, Theorem 2.5 asserts that, as $n \to \infty$, the sequence of distribution functions of $L_{2n}^*$ converges to the function

$$A(x) := \frac{2}{\pi} \arcsin\sqrt{x}, \quad 0 \le x \le 1.$$

This convergence clearly holds for every $x \in \mathbb{R}$, provided the definition of $A$ is extended by setting $A(x) := 0$ for $x < 0$ and $A(x) := 1$ for $x > 1$. The function $A$, thus defined on all of $\mathbb{R}$, is a distribution function—monotonically non-decreasing and right-continuous—with the properties $\lim_{x\to\infty} A(x) = 1$ and $\lim_{x\to-\infty} A(x) = 0$. If $Y$ is a random variable with distribution function $A$, then Theorem 2.5 can also be succinctly expressed in the form of Sect. 7.3 as

$$\frac{L_{2n}}{2n} \xrightarrow{\mathcal{D}} Y \quad \text{as } n \to \infty. \tag{2.31}$$

The distribution of a random variable $Y$ with the distribution function $A$, illustrated on the right in Fig. 2.14, is known as the *(continuous) arcsine distribution*. The density $a$ of this distribution, given by

$$a(x) := \frac{1}{\pi\sqrt{x(1-x)}}, \quad 0 < x < 1,$$

is shown on the left in Fig. 2.14. The $p$-quantile $A^{-1}(p)$ of the arcsine distribution is obtained by solving the equation $A(x) = p$ for $x$; the result is $A^{-1}(p) = \sin^2\left(\frac{p\pi}{2}\right)$. In particular, $A^{-1}(0.1) \approx 0.0245$ and $A^{-1}(0.9) \approx 0.9755$. This means that in simple symmetric random walks on $\mathbb{Z}$ that take many steps, the time of the last

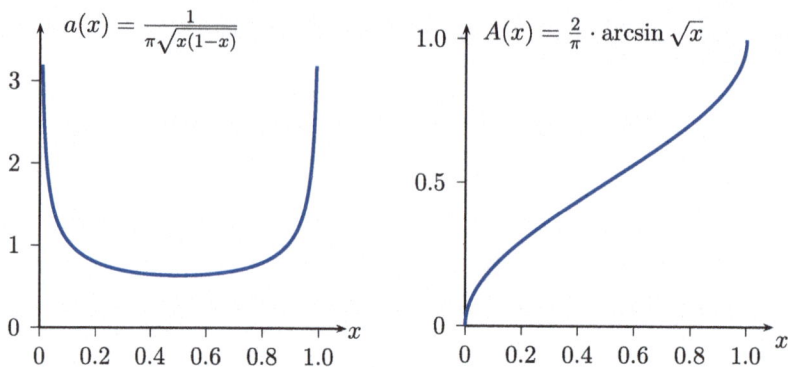

**Fig. 2.14** Density (left) and distribution function (right) of the arcsine distribution

**Table 2.1** Values of the distribution function $A(x)$ of the arcsine distribution

| $x$ | $A(x)$ | $x$ | $A(x)$ | $x$ | $A(x)$ | $x$ | $A(x)$ | $x$ | $A(x)$ |
|---|---|---|---|---|---|---|---|---|---|
| 0.00 | 0.000 | 0.10 | 0.205 | 0.20 | 0.295 | 0.30 | 0.369 | 0.40 | 0.436 |
| 0.01 | 0.064 | 0.11 | 0.215 | 0.21 | 0.303 | 0.31 | 0.376 | 0.41 | 0.442 |
| 0.02 | 0.090 | 0.12 | 0.225 | 0.22 | 0.311 | 0.32 | 0.383 | 0.42 | 0.449 |
| 0.03 | 0.111 | 0.13 | 0.235 | 0.23 | 0.318 | 0.33 | 0.390 | 0.43 | 0.455 |
| 0.04 | 0.128 | 0.14 | 0.244 | 0.24 | 0.326 | 0.34 | 0.396 | 0.44 | 0.462 |
| 0.05 | 0.144 | 0.15 | 0.253 | 0.25 | 0.333 | 0.35 | 0.403 | 0.45 | 0.468 |
| 0.06 | 0.158 | 0.16 | 0.262 | 0.26 | 0.341 | 0.36 | 0.410 | 0.46 | 0.474 |
| 0.07 | 0.171 | 0.17 | 0.271 | 0.27 | 0.348 | 0.37 | 0.416 | 0.47 | 0.481 |
| 0.08 | 0.183 | 0.18 | 0.279 | 0.28 | 0.355 | 0.38 | 0.423 | 0.48 | 0.487 |
| 0.09 | 0.194 | 0.19 | 0.287 | 0.29 | 0.362 | 0.39 | 0.429 | 0.49 | 0.494 |
| | | | | | | | | 0.50 | 0.500 |

For $x > 1/2$ use $A(1-x) = 1 - A(x)$

visit to zero occurs with about a 20% probability within the first or last 2.5% of the total duration, i.e., very early or very late.

Table 2.1 presents some numerical values of the function $A$. Since $A(0.15) = 0.253$, Theorem 2.5 implies that in a long simple symmetric random walk on $\mathbb{Z}$, there is more than a 50% probability that the last visit to zero occurs either within the first 15% or the last 15% of its total duration. See also the remark made prior to Fig. 2.12.

**SAQ 8** Why does $A(1-x) = 1 - A(x), 0 < x < 1$, hold?

## 2.4 The Number of Visits to Zero

In this section, we are interested in the random variable

$$N_k := \sum_{j=1}^{\lfloor k/2 \rfloor} \mathbf{1}\{S_{2j} = 0\},$$

which represents the *number of visits to zero* (or, for short, *the number of zeros*) of a simple symmetric random walk of length $k$ on $\mathbb{Z}$. Here, $\lfloor x \rfloor = \max\{k \in \mathbb{Z} : k \leq x\}$ denotes the greatest integer less than or equal to a real number $x$. Formally, $N_k$ is a mapping on the sample space $W_k$ of all $k$-paths, as defined in (2.14), and is given by

$$N_k(w) := \left|\{j \in \{1, 2, \ldots, \lfloor k/2 \rfloor\} : s_{2j} = 0\}\right|,$$

where $w = (s_0, s_1, \ldots, s_k) \in W_k$. In this sense, the path sketched in Fig. 2.5 visits zero three times. As in the last section, we assume a uniform distribution $\mathbb{P}$ on $W_k$, so we consider all paths from $W_k$ to be equally likely. Since visits to zero can only occur at even times, we set $k = 2n$ with $n \in \mathbb{N}$ in what follows.

**Theorem 2.6 (Distribution of the Number of Zeros)** *For the number $N_{2n}$ of zeros of a simple symmetric random walk of length $2n$, the following holds (for part (a) see, e.g., [FE1], p. 96; for part (b), (c) see [FE2]):*

(a)

$$\mathbb{P}(N_{2n} = j) = \binom{2n-j}{n} \frac{1}{2^{2n-j}}, \quad j = 0, 1, \ldots, n, \qquad (2.32)$$

(b) $\mathbb{E}(N_{2n}) = (2n+1) \dfrac{\binom{2n}{n}}{2^{2n}} - 1,$

(c) $\lim\limits_{n \to \infty} \dfrac{\mathbb{E}(N_{2n})}{\sqrt{2n}} = \sqrt{\dfrac{2}{\pi}} \approx 0.798,$

(d) $\mathbb{V}(N_{2n}) = 2(n+1) - (2n+1) \dfrac{\binom{2n}{n}}{2^{2n}} \left(1 + (2n+1) \dfrac{\binom{2n}{n}}{2^{2n}}\right).$

*Proof*

(a) The case $j = 0$ of no zero follows directly from the main lemma and Corollary 2.3, since the event $\{N_{2n} = 0\}$ is equal to the set $W_{2n, \neq 0}$. In the case

$j = n$, the $2n$-path visits zero after every second step, which means that at each of the $n$ times $0, 2, 4, \ldots, 2n - 2$, a choice must be made between an upward or a downward step (the next step then inevitably is in the opposite direction). Since there are thus $2^n$ favorable paths, the assertion follows for $j = n$. In what follows, we thus assume $1 \leq j \leq n - 1$.

The idea for proving (2.32) is to rewrite that equation in the form

$$\mathbb{P}(N_{2n} = j) = 2^j \binom{2n - j}{n} \frac{1}{2^{2n}}. \qquad (2.33)$$

Since $2^{2n}$ is the number of all $2n$-paths, we need to show that there are $2^j \binom{2n-j}{n}$ paths of length $2n$ with exactly $j$ zeros. Each such path can either take an upward or a downward step both at the beginning (at the origin) and directly after the first, second, ..., $(j - 1)$-th return to zero. *One of these $2^j$ possibilities is that the path leaves the x-axis upward both at the beginning and directly after each visit to zero and thus becomes non-negative until the $j$-th visit to zero.*

After the combinatorial meaning of the factor $2^j$ in (2.33) has become clear, we only need to show that there are $\binom{2n-j}{n}$ paths of length $2n$ that have exactly $j$ zeros and remain non-negative until the last such visit to zero (Fig. 2.15 shows such a path for the case $j = 4$ and $n = 10$). To this end, we construct a bijective mapping from this set of $2n$-paths, denoted by

$$M_1 := \{w = (s_0, s_1, \ldots, s_{2n}) \in W_{2n} : s_i \geq 0 \, \forall i \leq L_{2n}(w), N_{2n}(w) = j\},$$

to the set

$$M_2 := \{w = (s_0, s_1, \ldots, s_{2n-j}) \in W_{2n-j} : s_{2n-j} = -j\}$$

of all paths of length $2n - j$, which lead from the point $(0, 0)$ to $(2n - j, -j)$. Since every path that leads from $(0, 0)$ to $(2n - j, -j)$ comprises $n - j$ upward steps and $n$ downward steps (in total there are $2n - j$ steps, and the difference between the numbers of upward and downward steps must be $-j$, since the path

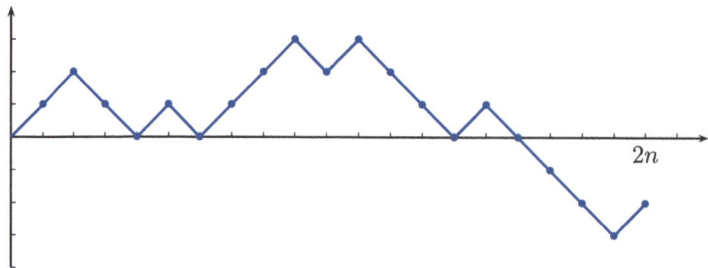

**Fig. 2.15** Path with 4 visits to zero, which is non-negative up to the fourth such visit

### 2.4 The Number of Visits to Zero

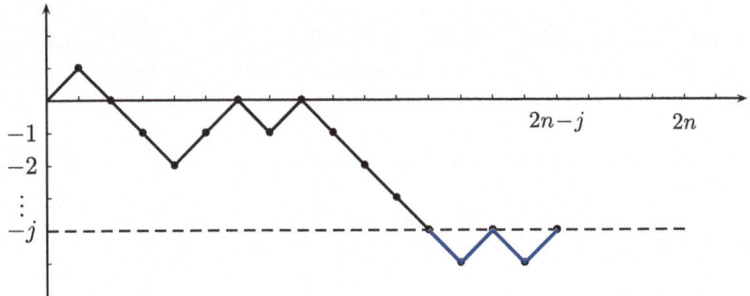

**Fig. 2.16** Path from $(0, 0)$ to $(2n - j, -j)$

ends at height $-j$), and since there are $\binom{2n-j}{n}$ ways to choose the downward steps from the total of $2n - j$ steps, it follows that

$$|M_2| = \binom{2n - j}{n}.$$

A path from the set $M_1$ is assigned a path from $M_2$ as follows: We omit the first upward step and each of the steps directly following the first $j - 1$ visits to zero and reassemble the rest (up to the $j$-th such visit). The zero-avoiding sub-path starting at the $j$-th return to zero is replaced by a bridge of the same length and appended (see Fig. 2.16, where the bridge is marked in blue).

**SAQ 9** Why is the replacement by a bridge possible?

It is evident that this mapping from paths from $M_1$ to paths in $M_2$ is injective. On the other hand, consider an arbitrary path from $M_2$, i.e., a path that leads from $(0, 0)$ to the point $(2n - j, -j)$. This path will, at some point, *first* hit the height $-1$, then subsequently the height $-2$, and so on, until it finally reaches the height $-j$ for the *first* time.

The path thus breaks down into sub-paths. The first sub-path leads from $(0, 0)$ to the first hitting of the height $-1$, the $i$-th sub-path from the first reaching of the height $-(i - 1)$ to the first hitting of the height $-i$ ($i = 2, \ldots, j$). The last sub-path is a bridge that starts from the first reaching of the height $-j$ and leads to the point $(2n - j, -j)$. If we precede each of the first $j$ sub-paths with an upward step and attach these sub-paths one after the other, starting at the origin, we create a non-negative path that visits zero $j$ times. According to the main lemma, the bridge path (last sub-path) corresponds uniquely to a zero-avoiding path of the same length (we can use this correspondence as the inverse mapping in the transition from paths in $M_1$ to paths in $M_2$). In this way, we obtain a path from $M_1$, which shows that a bijection between $M_1$ and $M_2$ exists.

(b) The result (cf. [FE2], Theorem 4) can be obtained in various ways. One possibility is to exploit a recursive relationship between the probabilities $p_j := \mathbb{P}(N_{2n} = j)$, for $j = 0, \ldots, n$. If we briefly set $N := N_{2n}$, then

$$\mathbb{E}(N) = \sum_{j=1}^{n} j\, p_j. \tag{2.34}$$

An elementary calculation yields $p_{j+1}/p_j = \frac{2n-2j}{2n-j}$, for $j = 0, \ldots, n-1$, which is equivalent to

$$j p_j + (2n - j)(p_{j+1} - p_j) = 0, \quad j = 0, \ldots, n-1,$$

Summing these equations over $j$ and taking into account (2.34) as well as the normalization condition $\sum_{j=0}^{n} p_j = 1$, we obtain

$$0 = \mathbb{E}(N) - n p_n + 2n(p_n - p_0) - \sum_{j=0}^{n-1} j p_{j+1} + \mathbb{E}(N) - n p_n$$

$$= 2\mathbb{E}(N) - 2n p_0 - \sum_{j=0}^{n-1} (j+1) p_{j+1} + \sum_{j=0}^{n-1} p_{j+1}$$

$$= 2\mathbb{E}(N) - 2n p_0 - \mathbb{E}(N) + 1 - p_0.$$

From this, the assertion follows. An alternative proof uses the representation $N = \sum_{j=1}^{n} \mathbf{1}\{S_{2j} = 0\}$ and the linearity of expectation as well as the relationship $\mathbb{E}\mathbf{1}\{A\} = \mathbb{P}(A)$ for an event $A$. With $\mathbb{P}(S_{2j} = 0) = \binom{2j}{j}/2^{2j}$, it follows that

$$\mathbb{E}(N_{2n}) = \sum_{j=1}^{n} \frac{\binom{2j}{j}}{2^{2j}}, \tag{2.35}$$

so that one can now conduct a simple proof by induction (Exercise 2.4).

(c) Because $p_0 = u_{2n}$ with $u_{2n}$ as in (2.15), according to (b),

$$\frac{\mathbb{E}(N_{2n})}{\sqrt{2n}} = \frac{2n+1}{\sqrt{\pi n} \cdot \sqrt{2n}} \cdot \sqrt{\pi n} \cdot u_{2n} - \frac{1}{\sqrt{2n}}.$$

With (2.29), the assertion follows.

(d) The proof of the variance formula proceeds analogously to the derivation of the expectation, now using the recursive relationship

$$(2n-j)(2n-j-1) p_{j+2} = 4(n-j)(n-j-1) p_j, \quad j = 0, \ldots, n-2. \tag{2.36}$$

## 2.4 The Number of Visits to Zero

We exploit these equations by summing them over $j$. With a little patience, the result is

$$\mathbb{E}(N_{2n}^2) = \sum_{j=1}^{n} j^2 p_j = 2n + 3 - (6n+3)\frac{\binom{2n}{n}}{2^{2n}}, \qquad (2.37)$$

yielding the assertion. The details are left to the interested reader (Exercise 2.5). The latter also holds for the limit relation

$$\lim_{n \to \infty} \frac{\mathbb{V}(N_{2n})}{2n} = 1 - \frac{2}{\pi} \qquad (2.38)$$

(Exercise 2.6).

∎

Statement (c) of Theorem 2.6 is startling. It suggests that the expected time span until the first return to zero of the random walk does not exist (is infinite). Otherwise, one would expect that the number of visits to zero of a $2n$-path to increase proportionally with $n$ and the path length. However, this is not the case; the average number of visits to zero only grows approximately proportional to the square root of the path length. Also, note Eq. (2.9), which indicates that the expected distance of the height of the random walk increases approximately proportional to the square root of the path length. Figure 2.17 illustrates the phenomenon of the "comparatively rare visits to zero" using two random walks of length 2500.

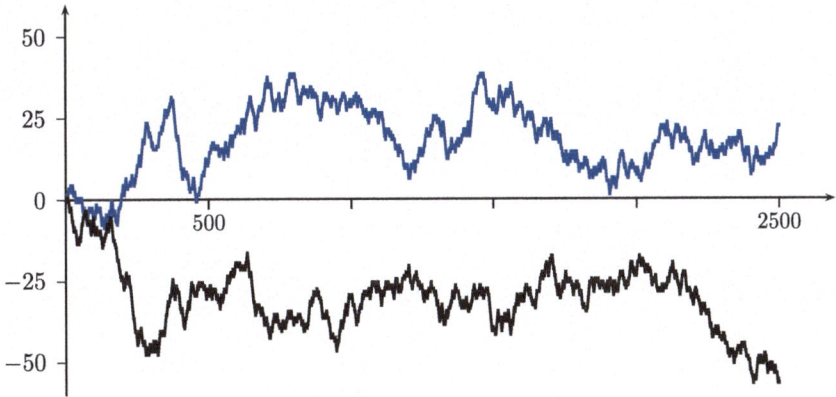

**Fig. 2.17** Random walks have surprisingly few zeros

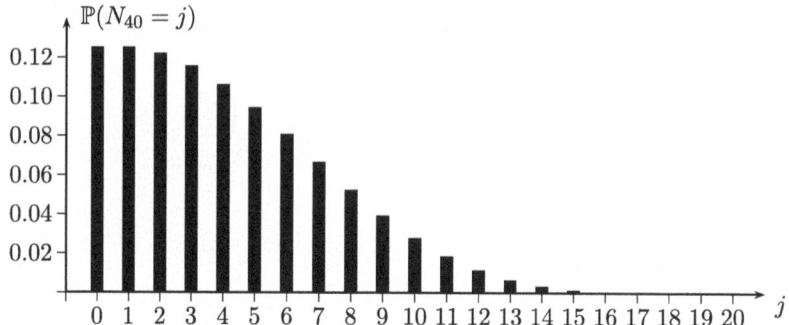

**Fig. 2.18** Bar chart of the distribution of $N_{40}$

Figure 2.18 shows the bar chart of the distribution of $N_{2n}$ for the case $n = 20$. If one considers the quotients already used in the proof of part (b) of Theorem 2.6,

$$\frac{\mathbb{P}(N_{2n} = j+1)}{\mathbb{P}(N_{2n} = j)} = \frac{2n - 2j}{2n - j}, \quad j = 0, 1, \ldots, n-1,$$

it is easy to see that $\mathbb{P}(N_{2n} = 0) = \mathbb{P}(N_{2n} = 1)$, and that, for $j \geq 2$, the probabilities $\mathbb{P}(N_{2n} = j)$ strictly decrease monotonically. The bar chart evokes associations with the "right half" of the Gaussian bell curve (density of the standard normal distribution). The following result shows that there is indeed a connection here.

**Theorem 2.7 (Limit Distribution for the Number of Zeros)** *For every $x > 0$,*

$$\lim_{n \to \infty} \mathbb{P}\left(\frac{N_{2n}}{\sqrt{2n}} \leq x\right) = 2\int_0^x \varphi(t)\, dt = 2\Phi(x) - 1, \quad (2.39)$$

*where*

$$\varphi(t) := \frac{1}{\sqrt{2\pi}} \exp\left(-\frac{t^2}{2}\right), \quad t \in \mathbb{R}, \quad (2.40)$$

*denotes the density of the standard normal distribution* N(0, 1).

**Proof** Fix any $x > 0$. Setting $I_n := \{k \in \mathbb{N}_0 : k \leq x\sqrt{2n}\}$ gives

$$\mathbb{P}\left(\frac{N_{2n}}{\sqrt{2n}} \leq x\right) = \sum_{k \in I_n} p_{n,k},$$

## 2.4 The Number of Visits to Zero

where

$$p_{n,k} := \binom{2n-k}{n} \frac{1}{2^{2n-k}}.$$

The idea of the proof is to approximate the factorials $(2n-k)!$, $n!$, and $(n-k)!$ that make up the above binomial coefficients using Stirling's formula (7.17), and then, except for an asymptotically negligible term, to obtain a Riemann approximation sum for the integral in (2.39). In the following considerations, it is important that $k \in I_n$ is bounded from above by $x\sqrt{2n}$ and thus, for sufficiently large $n$ (depending on $x$), is at most equal to $\frac{n}{2}$. With $a_m$ as in (2.28), a direct calculation yields

$$\frac{a_{2n-k}}{a_n a_{n-k}} = p_{n,k} \sqrt{2\pi} \frac{n^n (n-k)^{n-k} \sqrt{n(n-k)}}{(n-\frac{k}{2})^{2n-k} \sqrt{2n-k}}$$

$$= p_{n,k} \sqrt{2\pi n} \left(\frac{n-k}{n-k/2} \cdot \frac{n}{n-k/2}\right)^n \left(\frac{n-k}{n-k/2}\right)^{-k} \sqrt{\frac{n-k}{2n-k}}.$$

The first bracket expression on the right-hand side equals

$$b_{n,k} := \left(1 - \frac{k^2}{(2n-k)^2}\right)^n = \exp\left[n \log\left(1 - \frac{k^2}{(2n-k)^2}\right)\right],$$

and the second one equals

$$c_{n,k} := \left(1 - \frac{k/2}{n-k/2}\right)^{-k} = \exp\left[-k \log\left(1 - \frac{k}{2n-k}\right)\right].$$

Using the inequalities $1 - \frac{1}{t} \leq \log t \leq t - 1$, $t > 0$ (see Sect. 7.5), and setting

$$x_{n,k} := \frac{k}{\sqrt{2n}},$$

it follows that

$$e^{-x_{n,k}^2/2} \exp\left(-\frac{x_{n,k}^2}{2} \cdot \frac{4nk}{(2n-k)^2 - k^2}\right) \leq b_{n,k} \leq e^{-x_{n,k}^2/2} \exp\left(-\frac{x_{n,k}^2}{2} \cdot \frac{4nk - k^2}{(2n-k)^2}\right)$$

$$e^{x_{n,k}^2} \exp\left(x_{n,k}^2 \frac{k}{2n-k}\right) \leq c_{n,k} \leq e^{x_{n,k}^2} \exp\left(x_{n,k}^2 \frac{k}{n-k}\right).$$

Because $0 \leq x_{n,k} \leq x$, the four highlighted exponential expressions converge uniformly for $k \in I_n$ to one as $n \to \infty$. Since $\sqrt{(n-k)/(2n-k)}$ converges uniformly for $k \in I_n$ to $1/\sqrt{2}$ as $n \to \infty$ and, according to Stirling's formula (7.17), $\lim_{n \to \infty} a_{2n-k}/(a_n a_{n-k}) = 1$, where this convergence is also uniform in $k \in I_n$, we obtain

$$\lim_{n \to \infty} \sup_{k \in I_n} \left| p_{n,k} \sqrt{2\pi n} \exp\left(\frac{x_{n,k}^2}{2}\right) \frac{1}{\sqrt{2}} - 1 \right| = 0.$$

For a given $\varepsilon > 0$, there is therefore a positive integer $n_0$ with

$$\left| p_{n,k} \sqrt{2\pi n} \exp\left(\frac{x_{n,k}^2}{2}\right) \frac{1}{\sqrt{2}} - 1 \right| \leq \varepsilon, \qquad k \in I_n,$$

for every $n \geq n_0$. If we divide both sides of this inequality by $\sqrt{2\pi n} \exp\left(x_{n,k}^2/2\right)/\sqrt{2}$, it follows with (2.40) that

$$\left| p_{n,k} - 2\varphi(x_{n,k}) \frac{1}{\sqrt{2n}} \right| \leq \varepsilon \, 2\varphi(x_{n,k}) \frac{1}{\sqrt{2n}}, \qquad k \in I_n, \; n \geq n_0.$$

Summing over $k \in I_n$ yields

$$\left| \mathbb{P}\left(\frac{N_{2n}}{\sqrt{2n}} \leq x\right) - 2 \sum_{k \in I_n} \varphi(x_{n,k}) \frac{1}{\sqrt{2n}} \right| \leq \varepsilon \, 2 \sum_{k \in I_n} \varphi(x_{n,k}) \frac{1}{\sqrt{2n}}, \qquad n \geq n_0.$$

Because $1/\sqrt{2n} = x_{n,k+1} - x_{n,k}$, the sum appearing here twice, except for at most one term that is negligible as $n \to \infty$, equals a Riemann approximation sum for the integral $\int_0^x \varphi(t) dt$ for the partition $\{k/\sqrt{2n} : k \in I_n\} \cup \{x\}$ of the interval $[0, x]$. Note that $x \in \{k/\sqrt{2n} : k \in I_n\}$ can apply, in which case an additional term appears compared to a Riemann approximation sum. Since the partition has the fineness $1/\sqrt{2n}$, we obtain

$$\liminf_{n \to \infty} \mathbb{P}\left(\frac{N_{2n}}{\sqrt{2n}} \leq x\right) \geq (1 - \varepsilon) \, 2 \int_0^x \varphi(t) \, dt,$$

$$\limsup_{n \to \infty} \mathbb{P}\left(\frac{N_{2n}}{\sqrt{2n}} \leq x\right) \leq (1 + \varepsilon) \, 2 \int_0^x \varphi(t) \, dt.$$

Letting $\varepsilon$ tend to 0, the assertion follows. ∎

**Remark 2.8** If $Z$ is a standard normally distributed random variable, then for every $x \geq 0$,

$$\mathbb{P}(|Z| \leq x) = \mathbb{P}(-x \leq Z \leq x) = \Phi(x) - \Phi(-x) = \Phi(x) - (1 - \Phi(x))$$
$$= 2\Phi(x) - 1.$$

## 2.4 The Number of Visits to Zero

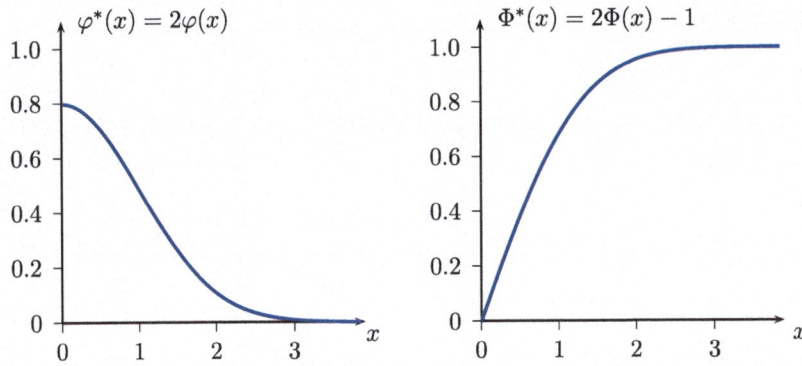

**Fig. 2.19** Density (left) and distribution function (right) of the half-normal distribution

Since $\mathbb{P}(N_{2n}/\sqrt{2n} \leq x)$, as a function of $x$, is the distribution function of $N_{2n}/\sqrt{2n}$, Theorem 2.7 states that the sequence of distribution functions of $N_{2n}/\sqrt{2n}$ converges pointwise to the distribution function of $|Z|$ sketched on the right in Fig. 2.19. Note that the statement $\lim_{n \to \infty} \mathbb{P}(N_{2n}/\sqrt{2n} \leq x) = \mathbb{P}(|Z| \leq x) \,(= 0)$ is valid also for $x \leq 0$. Therefore, according to Sect. 7.3,

$$\frac{N_{2n}}{\sqrt{2n}} \xrightarrow{\mathcal{D}} |Z| \quad \text{as } n \to \infty. \tag{2.41}$$

The distribution of $|Z|$ is referred to as the *half-normal distribution*. Since $|Z|$ has the density $\varphi^*(x) := 2\varphi(x)$ for $x \geq 0$, and $\varphi^*(x) := 0$ for $x < 0$, as shown on the left in Fig. 2.19,

$$\mathbb{E}|Z| = \int_0^\infty x\,\varphi^*(x)\,\mathrm{d}x = \frac{2}{\sqrt{2\pi}} \left[ -\exp\left(-\frac{x^2}{2}\right) \right]_0^\infty = \sqrt{\frac{2}{\pi}}. \tag{2.42}$$

Thus, statement (c) of Theorem 2.6 is perfectly consistent in light of the convergence (2.41). However, it does not follow directly from Theorem 2.7 without additional considerations. The same holds for the asymptotic behavior of the variance of $N_{2n}$ given in (2.38): The variance of $N_{2n}/\sqrt{2n}$ also converges to the variance of the limit distribution. This is because (cf. (2.38))

$$\lim_{n \to \infty} \mathbb{V}\left(\frac{N_{2n}}{\sqrt{2n}}\right) = 1 - \frac{2}{\pi} = \mathbb{V}(|Z|) = \mathbb{E}(Z^2) - (\mathbb{E}|Z|)^2.$$

**SAQ 10** Why does the second equality hold?

**Table 2.2** Values of the distribution function $\Phi^*(x) = 2\Phi(x) - 1$ of the half-normal distribution

| $x$ | $\Phi^*(x)$ | $x$ | $\Phi^*(x)$ | $x$ | $\Phi^*(x)$ |
|---|---|---|---|---|---|
| 0.00 | 0.000 | 0.80 | 0.576 | 1.645 | 0.900 |
| 0.10 | 0.080 | 0.90 | 0.632 | 1.70 | 0.911 |
| 0.20 | 0.159 | 1.00 | 0.683 | 1.80 | 0.928 |
| 0.30 | 0.236 | 1.10 | 0.729 | 1.90 | 0.943 |
| 0.40 | 0.311 | 1.20 | 0.770 | 1.96 | 0.950 |
| 0.50 | 0.383 | 1.30 | 0.806 | 2.00 | 0.954 |
| 0.60 | 0.451 | 1.40 | 0.838 | 2.20 | 0.972 |
| 0.675 | 0.500 | 1.50 | 0.866 | 2.40 | 0.984 |
| 0.70 | 0.516 | 1.60 | 0.890 | 2.60 | 0.991 |

Table 2.2 provides some values of the distribution function of the half-normal distribution, denoted by $\Phi^*(x) = 2\Phi(x) - 1$. Because $\Phi^*(1) = 0.683$, Theorem 2.39 states, for example, that $\mathbb{P}(N_{10,000} \leq 100) \approx 0.683$. This implies there is approximately a 68% probability of observing at most 100 visits to zero in a random walk of length 10,000. Furthermore, the probability that more than 200 visits to zero occur in such a random walk, is less than 5% given that $\Phi^*(2) = 0.954$.

**Banach's Matchbox Problem**

Banach's matchbox problem is a classic problem in probability theory, named after the Polish mathematician Stefan Banach.[10] Imagine you have two identical matchboxes, each containing $n$ matches. You carry one matchbox in your left pocket and the other in your right pocket. Each time you need a match, you randomly choose one of the two pockets with equal probability and take a match from the matchbox in that pocket (if it is not empty). The problem is to find the probability that when you first discover that one of the matchboxes is empty, there are exactly $j$ matches remaining in the other matchbox.

We can represent this stochastic process as a path that starts at the origin of a coordinate system. Here, an upward or downward step means reaching into the matchbox in the right or left pocket, respectively, which results in the numbers of remaining matches in the two boxes decreasing accordingly, as plotted on the two diagonals in Fig. 2.20. A reach into an empty box occurs when a path, as drawn in Fig. 2.20, hits the line $y = 2n - x$ and the next step goes up, or the line $y = x - 2n$ is hit and then the left pocket is reached into. If $U_{2n}$ denotes the number of matches in the other box when a box is found empty, then $U_{2n} = j$ holds if, as in Fig. 2.20, the path leads from $(0, 0)$ to the point $(2n - j, j)$ and then an upward step occurs,

---

[10] Stefan Banach (1892–1945) was a professor in Lwów (formerly Lemberg) from 1922 and the founder of the so-called *Polish School*. He is known for his contributions to mathematics, including Banach spaces and Banach's fixed point theorem.

## 2.5 First Return to Zero and Recurrence

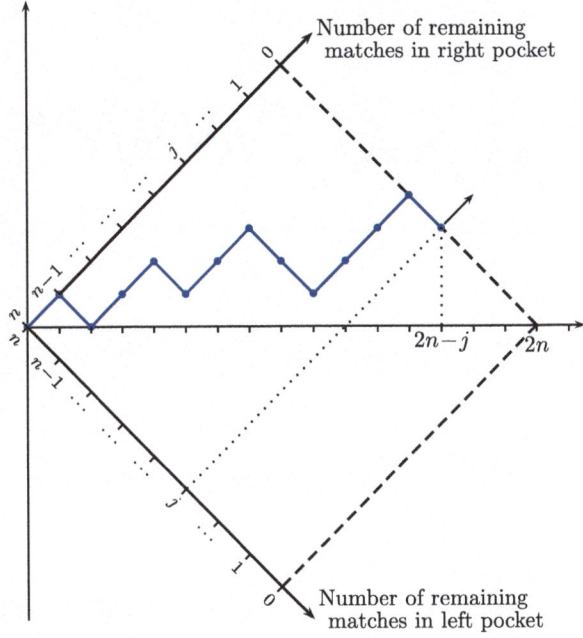

**Fig. 2.20** Illustrating Banach's matchbox problem: A path from $(0,0)$ to $(2n-j, j)$ with subsequent upward step yields $U_{2n} = j$

or—which happens with equal probability—the reflected image of this across the $x$-axis occurs. Using Lemma 2.1, it follows that

$$\mathbb{P}(U_{2n} = j) = 2 \frac{\binom{2n-j}{n}}{2^{2n-j}} \cdot \frac{1}{2} = \mathbb{P}(N_{2n} = j), \quad j = 0, 1, \ldots, n.$$

The random variable $U_{2n}$ thus has the same distribution as the number $N_{2n}$ of visits to zero in a simple random walk of length $2n$.

### 2.5 First Return to Zero and Recurrence

When does a symmetric random walk starting at zero return to zero for the first time? Interpreting the random walk as a game between two players, we are asking for the number of individual games until the first tie occurs. Obviously, we have to allow for arbitrarily long walks because, according to the main lemma and Corollary 2.3, a symmetric random walk of length $2n$ has a probability of $u_{2n} = \binom{2n}{n} 2^{-2n}$ of being zero-avoiding. This means that the time of the return to zero is greater than $2n$.

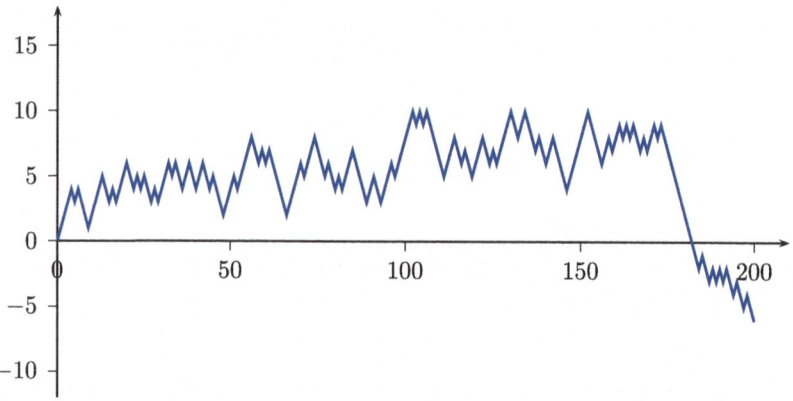

**Fig. 2.21** Random walk with first visit to zero after 182 steps

Figure 2.21 shows the beginning of a random walk of length 200, where the first return to zero occurs after 182 steps.

In this section, we use the probability space $(\Omega, \mathcal{A}, \mathbb{P})$ introduced in Sect. 7.2 as the basis for the random variables $X_1, X_2, \ldots$, introduced at the beginning of this chapter. Here, $\Omega$ is the set of all sequences $\omega = (a_j)_{j \geq 1}$ with $a_j \in \{1, -1\}$ for each $j \geq 1$. The sequence $(a_j)_{j \geq 1} \in \Omega$ represents an (conceptually) infinitely long path, where the $j$-th sequence member $a_j$ indicates the direction of the $j$-th step of the path. The random variable $X_k$ assigns $\omega = (a_j)_{j \geq 1}$ the value $a_k$; so $X_k$ models the $k$-th step of the path. With $S_0 = 0$ and $S_k = X_1 + \ldots + X_k$ (cf. (2.2)), we can define the *first return time*, that is, the random time of the first visit to zero of an infinitely long random walk, in the form

$$W(\omega) := \inf\{2k : k \in \mathbb{N} \text{ and } S_{2k}(\omega) = 0\}, \quad \omega = (a_j)_{j \geq 1} \in \Omega,$$

or more succinctly as

$$W := \inf\{2k : k \in \mathbb{N} \text{ and } S_{2k} = 0\}. \tag{2.43}$$

Here, as is common, the infimum over the empty set is defined as $\infty$. This value is attained, for example, for the constant sequence $(a_j)$ with $a_j = 1$ for each $j \geq 1$, which represents a zero-avoiding path of infinite length. Part (b) of the following theorem states that this case only occurs with probability zero (see, e.g., [FE1], p. 78).

**Theorem 2.9 (Distribution of the First Return Time)** *Let $W$ be the first return time to zero of a simple symmetric random walk on $\mathbb{Z}$ starting at 0. Then:*

## 2.5 First Return to Zero and Recurrence

(a) $\mathbb{P}(W = 2n) = \dfrac{\binom{2(n-1)}{n-1}}{2^{2(n-1)}} \cdot \dfrac{1}{2n} = \dfrac{u_{2(n-1)}}{2n}, \quad n \geq 1,$

(b) $\mathbb{P}(W < \infty) = \sum_{n=1}^{\infty} \mathbb{P}(W = 2n) = 1,$

(c) $\mathbb{E}(W) = \infty.$

*Proof*

(a) Since the event $\{W \geq 2n\}$ is the union of the disjoint events $\{W = 2n\}$ and $\{W \geq 2n + 2\}$, it follows that

$$\mathbb{P}(W = 2n) = \mathbb{P}(W \geq 2n) - \mathbb{P}(W \geq 2n + 2). \qquad (2.44)$$

For $k \geq 2$, the event $\{W \geq 2k\}$ occurs if and only if the initial segment of length $2(k-1)$ of the path is zero-avoiding, i.e., if $S_1 \neq 0, \ldots, S_{2(k-1)} \neq 0$. Thus, the main lemma and Corollary 2.3 provide

$$\mathbb{P}(W \geq 2k) = \dfrac{\binom{2(k-1)}{k-1}}{2^{2(k-1)}} = u_{2(k-1)}. \qquad (2.45)$$

Substituting this result for $k = n+1$ and $k = n$ into (2.44), the assertion follows for the case $n \geq 2$ by direct calculation with binomial coefficients.

> **SAQ 11** Can you perform this calculation?

If $n = 1$, then $\mathbb{P}(W = 2) = \mathbb{P}(X_1 = 1, X_2 = -1) + \mathbb{P}(X_1 = -1, X_2 = 1) = \tfrac{1}{2}$, which, due to $\binom{0}{0} = 1$, also matches the right-hand side of (a).

(b) It follows that $\{W = \infty\} = \bigcap_{k=1}^{\infty} \{W \geq 2k\}$, where the events $\{W \geq 2k\}, k \geq 1$, form a decreasing sequence of sets because $\{W \geq 2(k+1)\} \subset \{W \geq 2k\}$. Since probability measures are continuous from above (cf. (7.2)), it follows with (2.45) and (2.29) that

$$\mathbb{P}(W = \infty) = \lim_{k \to \infty} \mathbb{P}(W \geq 2k) = \lim_{k \to \infty} u_{2(k-1)} = 0,$$

and thus $1 = \mathbb{P}(W < \infty) = \sum_{n=1}^{\infty} \mathbb{P}(W = 2n)$.

(c) With part (a), we get $\mathbb{E}(W) = \sum_{n=1}^{\infty} 2n\,\mathbb{P}(W = 2n) = \sum_{n=1}^{\infty} u_{2(n-1)}$. Due to (2.29), there exists an $n_0$ with the property $u_{2n} \geq 1/(2\sqrt{\pi n})$ for each $n \geq n_0$. It follows that

$$\sum_{n=1}^{\infty} 2n\,\mathbb{P}(W = 2n) \geq \frac{1}{2\sqrt{\pi}} \sum_{n=n_0}^{\infty} \frac{1}{\sqrt{n}},$$

and thus the assertion, since the series $\sum_{n=1}^{\infty} n^{-1/2}$ diverges. ∎

Figure 2.22 shows the bar chart of the distribution of $W$. The probabilities $\mathbb{P}(W = 2n)$ satisfy the recursion formula

$$\frac{\mathbb{P}(W = 2(n+1))}{\mathbb{P}(W = 2n)} = \frac{2n-1}{2n+2}, \quad n \geq 1,$$

and the initial condition $\mathbb{P}(W = 2) = \frac{1}{2}$.

The equation $\mathbb{P}(W < \infty) = 1$ states that an infinitely long random walk starting at 0 will eventually (and thus infinitely often) return to 0 with probability one (the waiting time until the $k$-th return to 0 will be examined at the end of Sect. 2.9). Due to symmetry, this property also holds for any other starting point. This means that every integer $k$ is *recurrent* in the sense that $\mathbb{P}((\cup_{j=1}^{\infty}\{S_j = k\})|S_0 = k) = 1$. It should already be emphasized here that this *recurrence property*, examined more generally in [PO], is lost and the random walk on $\mathbb{Z}$ becomes *transient* if upward and downward steps are not equally likely (see Sect. 4.1). The following consideration shows that a symmetric random walk will reach the point 0 (and thus any other given point) in finite time with probability one from any starting point $k$ (in Sect. 2.9, we will study the distribution of the number of steps needed for this in more detail).

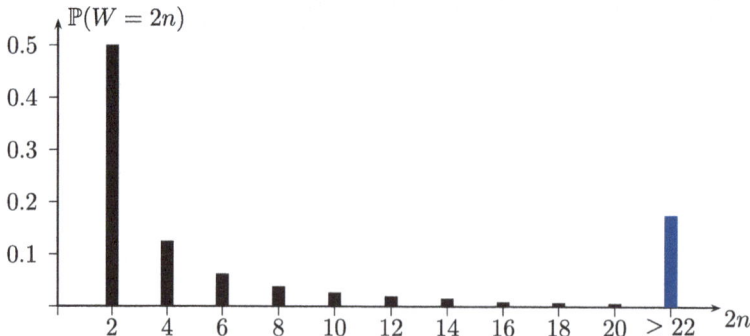

**Fig. 2.22** Bar chart of the distribution of the first return time to zero (note the large probability $\mathbb{P}(W \geq 22)$)

## 2.5 First Return to Zero and Recurrence

Setting

$$p(k) := \mathbb{P}\left(\left(\bigcup_{j=1}^{\infty}\{S_j = 0\}\right)\bigg| S_0 = k\right), \quad k \in \mathbb{Z},$$

then, according to Theorem 2.9 (b), $p(0) = 1$. Decomposing the event $\bigcup_{j=1}^{\infty}\{S_j = 0\}$ according to the two cases $X_1 = 1$ and $X_1 = -1$, the law of total probability (7.4) provides the difference equation

$$p(k) = \frac{1}{2}p(k+1) + \frac{1}{2}p(k-1), \quad k \in \mathbb{Z} \qquad (2.46)$$

(Exercise 2.8). Equation (2.46) indicates that the points $(k, p(k))$, where $k \in \mathbb{Z}$, lie on a straight line passing through $(0, 1)$. Since this line is parallel to the $x$-axis, it follows that $p(k) = 1$ for every $k$. Otherwise, $p(k)$ would be less than 0 for some integer $k$, which is impossible.

The fact that the first return time has an infinite expectation is perhaps surprising at first glance but fits seamlessly with the previous results. If $\mathbb{E}(W)$ were finite, one would expect that the number of visits to zero of the random walk would increase proportionally to the elapsed time, which, according to Theorems 2.6 (c) and 2.7, is not the case. This rather intuitive argument can be made more concrete as follows: If $Z_1, Z_2, \ldots, Z_k$ are independent random variables with the same distribution as $W$, then $Z_1 + \ldots + Z_k$ models the time of the $k$-th return to zero of the random walk. After each first return, due to the independence of the upward and downward steps $X_1, X_2, \ldots$, the calculation of the time span until the next first return starts again from the beginning (see Fig. 2.23).

Since at least $k$ returns to zero occur by time $2n$ if and only if the $k$-th return time to zero is at most $2n$, it follows that

$$\mathbb{P}(N_{2n} \geq k) = \mathbb{P}\left(\sum_{j=1}^{k} Z_j \leq 2n\right).$$

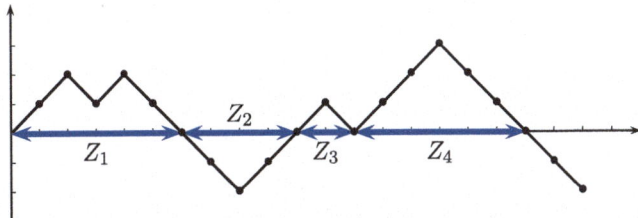

**Fig. 2.23** The time until the fourth return to zero as the sum of independent first return times

Now,

$$\mathbb{P}(N_{2n} \geq k) = \sum_{j=k}^{n} \mathbb{P}(N_{2n} = j) \leq \sum_{j=k}^{n} \frac{j}{k} \mathbb{P}(N_{2n} = j) \leq \frac{1}{k} \mathbb{E}(N_{2n}).$$

According to Theorem 2.6 (c), there exists a constant $C > 0$ such that $\mathbb{E}(N_{2n}) \leq C\sqrt{n}$ for each $n \geq 1$. Thus,

$$\mathbb{P}\left(\frac{1}{k}\sum_{j=1}^{k} Z_j \leq 2 \cdot \frac{n}{k}\right) = \mathbb{P}\left(\sum_{j=1}^{k} Z_j \leq 2n\right) \leq C \frac{\sqrt{n}}{k}. \tag{2.47}$$

If $\mathbb{E}(W) =: a \in (0, \infty)$ were a finite value, then, according to the law of large numbers, $\frac{1}{k}\sum_{j=1}^{k} Z_j$ would converge in probability (even almost surely) to $a$ as $k \to \infty$. Specifically, if we choose $n := \lfloor k^{3/2} \rfloor$, then $\frac{n}{k}$ tends to infinity as $k \to \infty$, and the probability in (2.47) converges to 1. On the other hand, for this choice of $n$, the upper bound in (2.47) tends to 0, which is a contradiction. Therefore, $\mathbb{E}(W)$ must be infinite.

The probabilities $\mathbb{P}(W = 2n)$ are often abbreviated as

$$f_{2n} := \mathbb{P}(W = 2n) = \frac{\binom{2(n-1)}{n-1}}{2^{2(n-1)}} \cdot \frac{1}{2n}. \tag{2.48}$$

If we decompose the event $\{S_{2n} = 0\}$ according to the possible values $2, 4, \ldots, 2n$ of $W$, it follows that

$$\mathbb{P}(S_{2n} = 0) = \sum_{r=1}^{n} \mathbb{P}(W = 2r) \mathbb{P}(S_{2n} = 0 | W = 2r).$$

Given the condition $W = 2r$, a random walk starting at $(2r, 0)$ and ending at $(2n, 0)$ is a $(2n - 2r)$-bridge. Therefore, using the main lemma and $u_{2k} = \binom{2k}{k} 2^{-2k}$ (cf. (2.15)), we get

$$u_{2n} = \sum_{r=1}^{n} f_{2r} u_{2n-2r}, \quad n = 1, 2, \ldots \tag{2.49}$$

Because $u_0 = 1$, summing these equations yields

$$\sum_{n=1}^{k} u_{2n} \leq \sum_{r=1}^{k} f_{2r} \left(1 + \sum_{n=1}^{k} u_{2n}\right) \tag{2.50}$$

for each $k \geq 1$ (Exercise 2.9), and thus

$$\sum_{n=1}^{k} u_{2n} \left(1 + \sum_{n=1}^{k} u_{2n}\right)^{-1} \leq \sum_{r=1}^{k} f_{2r} \leq 1, \qquad k \geq 1.$$

Since $\sum_{n=1}^{\infty} u_{2n} = \infty$, this provides an alternative proof of the relationship $\sum_{r=1}^{\infty} f_{2r} = \mathbb{P}(W < \infty) = 1$.

After extensively dealing with the *first* return time to zero, one can naturally ask the more general question: what is the probability that a symmetric random walk starting at zero has its $k$-th return to zero at time $2n$? Here, $k$ is an arbitrary positive integer. In generalization of Theorem 2.9, Theorem 2.28 (b) provides an answer to this question.

We conclude this section with a surprising phenomenon that is directly related to the time $W$ of the first return to zero. Let $k$ be any non-zero integer, and let

$$B_k := \sum_{j=1}^{\infty} \mathbf{1}\{S_j = k \text{ and } W > j\} \tag{2.51}$$

denote the number of stays of the random walk at height $k$ *before it first returns to zero*. In Sect. 4.3, we will see that $\mathbb{E}(B_k) = 1$, regardless of $k$. So, every symmetric random walk reaches any height $k \neq 0$ on average exactly once before it returns to the starting point 0 for the first time!

## 2.6 Sojourn Times

In the theory of random walks, a *sojourn time* refers to the amount of time that a random walk spends in a given set of states. In this section, we focus on the amount of time that a simple symmetric random walk of length $2n$ starting at 0 spends above the $x$-axis. As before, let $X_j$ denote the direction of the $j$-th step of the random walk, and let $S_0 := 0$ and $S_k := X_1 + \ldots + X_k$ for $k \geq 1$. Then the special sojourn time we are interested in can be expressed in the form

$$O_{2n} := \sum_{k=1}^{2n} \mathbf{1}\{S_k \geq 0 \text{ and } S_{k-1} \geq 0\}. \tag{2.52}$$

Figure 2.24 shows a random walk of length 20 that spends 12 time steps above the $x$-axis. The right endpoints of these time steps, each of one unit, are marked by arrows. The arrowheads point to those $k$ for which the indicator function in (2.52) takes the value 1.

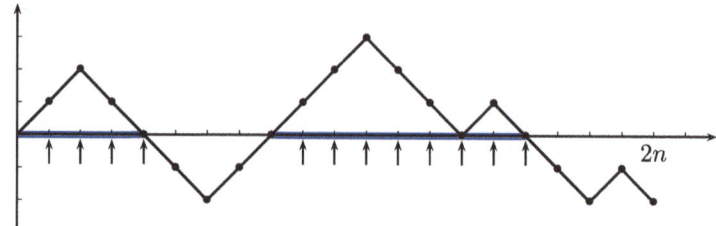

**Fig. 2.24** Random walk of length 20 that spends 12 time steps above the x-axis

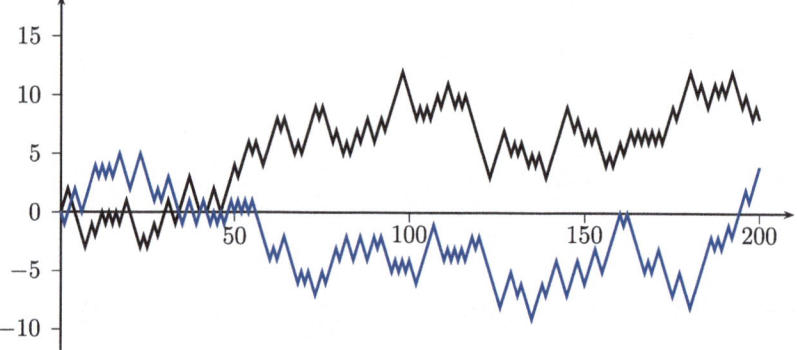

**Fig. 2.25** Sojourn times tend towards extreme values

Apparently, $O_{2n}$ can only take the even values $0, 2, \ldots, 2n$. The extreme cases $O_{2n} = 0$ or $O_{2n} = 2n$ mean that the $2n$-path is non-positive or non-negative, respectively. According to the main lemma and Corollary 2.3,

$$\mathbb{P}(O_{2n} = 0) = \mathbb{P}(S_1 \leq 0, \ldots, S_{2n} \leq 0) = u_{2n}, \qquad (2.53)$$

$$\mathbb{P}(O_{2n} = 2n) = \mathbb{P}(S_1 \geq 0, \ldots, S_{2n} \geq 0) = u_{2n}, \qquad (2.54)$$

with $u_{2n}$ as in (2.15). Figure 2.25 shows two random walks of length 200 whose sojourn times above the x-axis are 50 and 174, respectively, constituting 25 and 87% of the total time. The random walk depicted in Fig. 2.21 is even more extreme, spending 182 time units, or more than 90% of the total time, above the x-axis. In the interpretation of the random walk as a game between two players, one must therefore expect that one of the two players leads most of the time.

Since the sojourn time

$$U_{2n} := \left| \{ k \in \{1, \ldots, 2n\} : S_k \leq 0 \text{ and } S_{k-1} \leq 0 \} \right|$$

below the x-axis has the same distribution as $O_{2n}$, it follows that $\mathbb{E}(O_{2n}) = n$.

## 2.6 Sojourn Times

**SAQ 12** Why do $O_{2n}$ and $U_{2n}$ have the same distribution?

A random walk should therefore, "on average," spend half of its time above the $x$-axis. However, anyone who thinks that the probability distribution of the sojourn time $O_{2n}$ should be bell-shaped near the expected value is greatly mistaken. As the following famous result shows, the opposite is true; the sojourn time $O_{2n}$ tends towards extremely large and small values. The illustrations 2.21 and 2.25 are not exceptions, but the rule!

**Theorem 2.10 (Distribution of the Sojourn Time (cf. [CF]))** *Let $O_{2n}$ be the time that a simple symmetric random walk of length $2n$, starting at $0$, spends above the $x$-axis. Then*

$$\mathbb{P}(O_{2n} = 2k) = u_{2k}u_{2(n-k)} = \frac{\binom{2k}{k}\binom{2(n-k)}{n-k}}{2^{2n}}, \qquad k = 0, 1, \ldots, n. \tag{2.55}$$

*Proof* We set

$$b_{2k,2n} := \mathbb{P}(O_{2n} = 2k). \tag{2.56}$$

Because $u_0 = 1$, the cases $k = 0$ and $k = n$ are obviously handled with (2.53) and (2.54). Therefore, we assume $1 \leq k \leq n-1$ in what follows. This means that the random walk spends both time above and below the $x$-axis. The first return time $W$ therefore takes one of the values $2r$, where $r \in \{1, 2, \ldots, n-1\}$. We decompose the event $\{O_{2n} = 2k\}$ according to the possible values of $W$ and whether $S_1$ is positive or negative, i.e., whether the first step is upward or downward. If $S_1 = -1$, the random walk must visit zero at the latest by time $2n - 2k$; otherwise fewer than $2k$ time steps would be available for staying above the $x$-axis. When $S_1 = 1$, the first visit to zero necessarily occurs at the latest at time $2k$; otherwise $O_{2n} > 2k$ would hold. These considerations provide the representation

$$\{O_{2n} = 2k\} = \biguplus_{r=1}^{k} \{O_{2n} = 2k, W = 2r, S_1 = 1\} \uplus$$

$$\biguplus_{r=1}^{n-k} \{O_{2n} = 2k, W = 2r, S_1 = -1\},$$

where the events on the right-hand side are pairwise disjoint. With the abbreviation $f_{2r} = \mathbb{P}(W = 2r)$, introduced in (2.48), it follows that

$$b_{2k,2n} = \mathbb{P}(O_{2n} = 2k)$$

$$= \sum_{r=1}^{k} \mathbb{P}(S_1 = 1) \mathbb{P}(O_{2n} = 2k, W = 2r | S_1 = 1)$$

$$+ \sum_{r=1}^{n-k} \mathbb{P}(S_1 = -1) \mathbb{P}(O_{2n} = 2k, W = 2r | S_1 = -1)$$

$$= \frac{1}{2} \sum_{r=1}^{k} \mathbb{P}(O_{2n} = 2k, W = 2r | S_1 = 1) + \frac{1}{2} \sum_{r=1}^{n-k} \mathbb{P}(O_{2n} = 2k, W = 2r | S_1 = -1)$$

$$= \frac{1}{2} \sum_{r=1}^{k} f_{2r} \, \mathbb{P}(O_{2n} = 2k | W = 2r, S_1 = 1) \tag{2.57}$$

$$+ \frac{1}{2} \sum_{r=1}^{n-k} f_{2r} \, \mathbb{P}(O_{2n} = 2k | W = 2r, S_1 = -1). \tag{2.58}$$

The last step used the fact that, due to symmetry,

$$\mathbb{P}(W = 2r) = \mathbb{P}(W = 2r, S_1 = 1) + \mathbb{P}(W = 2r, S_1 = -1)$$
$$= 2\mathbb{P}(W = 2r, S_1 = 1) = 2\mathbb{P}(W = 2r, S_1 = -1).$$

Since $\mathbb{P}(S_1 = 1) = \mathbb{P}(S_1 = -1) = \frac{1}{2}$, we obtain

$$f_{2r} = \mathbb{P}(W = 2r) = \mathbb{P}(W = 2r | S_1 = 1) = \mathbb{P}(W = 2r | S_1 = -1).$$

Under the conditions $W = 2r$ and $S_1 = 1$, the random walk stays above the $x$-axis until time $2r$. For the event $\{O_{2n} = 2k\}$ to occur, it must therefore spend another $2k - 2r$ time steps above the $x$-axis, starting from the point $(2r, 0)$. According to the definition of $b_{2k,2n}$ in (2.56), it follows that

$$\mathbb{P}(O_{2n} = 2k | W = 2r, S_1 = 1) = b_{2k-2r, 2n-2r}. \tag{2.59}$$

Under the conditions $W = 2r$ and $S_1 = -1$, the random walk remains below the $x$-axis until time $2r$. For the event $\{O_{2n} = 2k\}$ to occur, this means that, starting from the point $(2r, 0)$, it has to spend $2k$ out of a remaining total of $2n - 2r$ time steps above the $x$-axis. Hence,

$$\mathbb{P}(O_{2n} = 2k | W = 2r, S_1 = -1) = b_{2k, 2n-2r}. \tag{2.60}$$

## 2.6 Sojourn Times

Inserting (2.59) and (2.60) into (2.57) and (2.58), we get

$$b_{2k,2n} = \frac{1}{2}\sum_{r=1}^{k} f_{2r}\, b_{2k-2r,2n-2r} + \frac{1}{2}\sum_{r=1}^{n-k} f_{2r}\, b_{2k,2n-2r}, \quad 1 \leq k \leq n-1. \quad (2.61)$$

We now prove (2.55) by induction on $n$. Since (2.55) holds for $k = 0$ and $k = n$, the base case $n = 1$ is already shown. The induction step $n - 1 \mapsto n$ uses representation (2.61) and the fact that the induction hypothesis is applicable to the terms $b_{2k-2r,2n-2r}$ and $b_{2k,2n-2r}$ appearing on the right-hand side of this equation. Therefore,

$$b_{2k-2r,2n-2r} = u_{2k-2r}\, u_{2n-2r-(2k-2r)}, \quad b_{2k,2n-2r} = u_{2k}\, u_{2n-2r-2k}.$$

Inserting the terms on the above right-hand sides into (2.61), it follows that

$$b_{2k,2n} = \frac{1}{2} \cdot u_{2n-2k} \sum_{r=1}^{k} f_{2r}\, u_{2k-2r} + \frac{1}{2} \cdot u_{2k} \sum_{r=1}^{n-k} f_{2r}\, u_{2n-2k-2r}.$$

According to (2.49), the first sum appearing here is equal to $u_{2k}$, and the second sum equals $u_{2n-2k}$, which demonstrates that $b_{2k,2n} = u_{2k} u_{2n-2k}$ and completes the proof. ∎

In comparison with Theorem 2.4, it is evident that $O_{2n}$ surprisingly has the same distribution as the time $L_{2n}$ of the last visit to zero. Therefore, Fig. 2.12 also shows the U-shaped bar chart of the distribution of $O_{40}$. It can be seen that with approximately a 25% probability, the sojourn time above the $x$-axis is equal to 0 or 40. Thus, every fourth random walk of length 40 stays entirely above or entirely below the $x$-axis!

Since $O_{2n}$ and $L_{2n}$ have the same distribution, Theorem 2.5 also provides the following limit theorem, first stated by P. Lévy ([LEV], Corollaire 2, pp. 303–304), for the distribution of the *proportion of time* $O_{2n}/(2n)$ that the random walk spends above the $x$-axis.

**Theorem 2.11 (Arcsine Law for the Sojourn Time)** *For each $x$ with $0 \leq x \leq 1$,*

$$\lim_{n \to \infty} \mathbb{P}\left(\frac{O_{2n}}{2n} \leq x\right) = \int_0^x \frac{1}{\pi\sqrt{t(1-t)}}\, dt = \frac{2}{\pi} \arcsin\sqrt{x}. \quad (2.62)$$

The implications of this result for long random walks are significant: Referring to Table 2.1 for the distribution function $A(x)$ on the right-hand side of (2.62), we find that $A(0.15) = 0.253$ and $A(0.85) = 0.747$. This indicates that about a quarter of all long symmetric random walks spend at least 85% of the total time above the $x$-axis, while another quarter spend there only at most 15% of their total time. Interpreted as a game consisting of a long series of coin tosses between two players, this result implies that in about half of such games, one player is in the lead for at least 85% of the game's duration!

## 2.7 Maximum and Minimum

In this section, we examine the extent to which a symmetric random walk of length $n$, starting at 0, can deviate from zero upwards or downwards. To this end, we consider the maximum

$$M_n := \max\{S_0, S_1, \ldots, S_n\}$$

and the minimum

$$m_n := \min\{S_0, S_1, \ldots, S_n\}.$$

Interpreting the random walk as a game between two players, A and B, $M_n$ represents the "maximum intermediate loss" in euros for player B during a game based on $n$ coin tosses, where player A bets on heads, and heads corresponds to an upward step. Similarly, $-m_n$ stands for the maximum intermediate loss for player A.

Since, by symmetry, the random variables $m_n$ and $-M_n$ have the same distribution, we will focus on the case of the maximum. Obviously, $M_n$ can take any value from 0 to $n$. The case $M_n = 0$ occurs when the $n$-path is non-positive. The other extreme case, $M_n = n$, happens only when the path completes $n$ upward steps. Surprisingly, we will see that the probabilities $\mathbb{P}(M_n = k)$ depend only on the distribution of $S_n$, i.e., on the stochastic behavior of the random walk at the end of the period. To explore this, we consider any non-negative integer $k$ with $k \leq n$ and decompose the event $\{M_n \geq k\}$ according to whether $S_n > k$, $S_n < k$, or $S_n = k$. We thus obtain

$$\mathbb{P}(M_n \geq k) = \mathbb{P}(M_n \geq k, S_n > k) + \mathbb{P}(M_n \geq k, S_n < k)$$
$$+ \mathbb{P}(M_n \geq k, S_n = k).$$

The crucial point is that the first two probabilities on the right-hand side are equal. For every $n$-path with $M_n \geq k$ and $S_n > k$, there is a unique corresponding $n$-path with $M_n \geq k$ and $S_n < k$ according to the reflection principle: we simply reflect the $n$-path with the first property from the first time it reaches the height $k$ across the horizontal line $y = k$. In this way, a path with the second property is created, and this mapping of paths with $M_n \geq k$ and $S_n > k$ onto paths with $M_n \geq k$ and $S_n < k$ is clearly bijective (see Fig. 2.26).

## 2.7 Maximum and Minimum

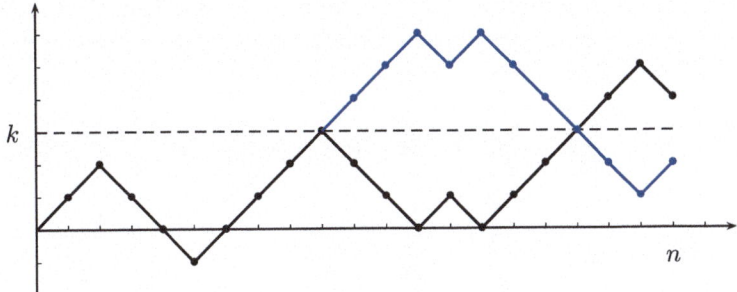

**Fig. 2.26** A path with $M_n \geq k$ is reflected across the line $y = k$ from the first time it reaches the height $k$. This maps paths with the additional property $S_n > k$ or $S_n < k$ bijectively onto each other

Since both $S_n \geq k$ and $S_n = k$ imply $M_n \geq k$, it follows that

$$\begin{aligned} \mathbb{P}(M_n \geq k) &= 2\,\mathbb{P}(M_n \geq k, S_n > k) + \mathbb{P}(M_n \geq k, S_n = k) \\ &= 2\,\mathbb{P}(M_n \geq k, S_n \geq k) - \mathbb{P}(M_n \geq k, S_n = k) \\ &= 2\,\mathbb{P}(S_n \geq k) - \mathbb{P}(S_n = k), \end{aligned} \qquad (2.63)$$

and thus

$$\mathbb{P}(M_n \geq k) = \mathbb{P}(S_n \geq k) + \mathbb{P}(S_n \geq k+1), \qquad k = 0, \ldots, n. \qquad (2.64)$$

Replacing $k$ by $k+1$ and considering the complementary event gives

$$\mathbb{P}(M_n \leq k) = \mathbb{P}(S_n \leq k) + \mathbb{P}(S_n \leq k+1) - 1, \qquad k = 0, \ldots, n. \qquad (2.65)$$

Taking into account (2.64) and

$$\mathbb{P}(M_n = k) = \mathbb{P}(M_n \geq k) - \mathbb{P}(M_n \geq k+1) = \mathbb{P}(S_n \geq k) - \mathbb{P}(S_n \geq k+2),$$

we further obtain

$$\mathbb{P}(M_n = k) = \mathbb{P}(S_n = k) + \mathbb{P}(S_n = k+1), \qquad k = 0, \ldots, n. \qquad (2.66)$$

Since $\mathbb{P}(S_n = k)$ can only be positive if $n$ and $k$ have the same parity, one of the summands in (2.66) vanishes. With the convention $\binom{n}{\ell} := 0$ if $\ell \notin \mathbb{Z}$, and using (2.7), it follows that

$$\mathbb{P}(M_n = k) = \frac{1}{2^n} \cdot \max\left(\binom{n}{\frac{n+k}{2}}, \binom{n}{\frac{n+k+1}{2}}\right)$$

(see [FE1], Theorem 1, p. 89). Using the floor function $\lfloor x \rfloor = \max\{k \in \mathbb{Z} : k \leq x\}$, this result can be stated even more compactly (see, e.g., [KP2]).

**Theorem 2.12 (Distribution of the Maximum of a Random Walk)** *For the maximum $M_n$ of a simple symmetric random walk of length n, starting at 0, the following holds:*

(a) $\mathbb{P}(M_n = k) = \dfrac{1}{2^n} \binom{n}{\lfloor \frac{n+k+1}{2} \rfloor}$,  $n \geq 1;\; k = 0, 1, \ldots, n.$

(b) $\mathbb{E}(M_{2n}) = \left(2n + \dfrac{1}{2}\right) \dfrac{\binom{2n}{n}}{2^{2n}} - \dfrac{1}{2}$,   $\mathbb{E}(M_{2n+1}) = (2n+1) \dfrac{\binom{2n}{n}}{2^{2n}} - \dfrac{1}{2}.$

(c) $\mathbb{V}(M_{2n}) = 2n + \dfrac{1}{4} - \left(\left(2n + \dfrac{1}{2}\right) \dfrac{\binom{2n}{n}}{2^{2n}}\right)^2,$

$\mathbb{V}(M_{2n+1}) = 2n + \dfrac{5}{4} - \left((2n+1) \dfrac{\binom{2n}{n}}{2^{2n}}\right)^2.$

*Proof* Only parts (b) and (c) need to be shown. We first prove the formula for $\mathbb{E}(M_{2n})$; using (7.26), the second equation in (b) follows analogously. Due to part (a), we obtain

$$\mathbb{E}(M_{2n}) = \sum_{k=1}^{n} 2k \frac{\binom{2n}{n+k}}{2^{2n}} + \sum_{k=0}^{n-1} (2k+1) \frac{\binom{2n}{n+k+1}}{2^{2n}}$$

$$= \frac{1}{2^{2n}} \left( \sum_{k=1}^{n} 2k \binom{2n}{n+k} + \sum_{j=1}^{n} (2j-1) \binom{2n}{n+j} \right)$$

$$= \frac{1}{2^{2n}} \sum_{k=1}^{n} (4k-1) \binom{2n}{n+k}. \tag{2.67}$$

Using (7.25) and

$$\sum_{k=1}^{n} \binom{2n}{n+k} = \frac{1}{2}\left(2^{2n} - \binom{2n}{n}\right), \tag{2.68}$$

the assertion now follows by direct calculation (Exercise 2.11). □

**SAQ 13** Why does Eq. (2.68) hold?

## 2.7 Maximum and Minimum

To prove part (c), we proceed similarly. First, note that

$$\mathbb{E}\left(M_{2n}^2\right) = \sum_{k=1}^{n}(2k)^2\mathbb{P}(M_{2n}=2k) + \sum_{k=0}^{n-1}(2k+1)^2\mathbb{P}(M_{2n}=2k+1). \tag{2.69}$$

Invoking (7.25) and (7.27), with some patience, we get

$$\mathbb{E}\left(M_{2n}^2\right) = 2n + \frac{1}{2} - \left(2n+\frac{1}{2}\right)\frac{\binom{2n}{n}}{2^{2n}}$$

and thus the formula for $\mathbb{V}(M_{2n})$ (Exercise 2.12). Using (7.26) and (7.28), the formula for $\mathbb{V}(M_{2n+1})$ follows by direct calculation. ∎

The fact that one of the summands in (2.66) vanishes is illustrated in the bar chart of the distribution of $M_{40}$ (see Fig. 2.27). This bar chart results from the bar chart of the binomial distribution $\text{Bin}(40, \frac{1}{2})$, by plotting the maximum probability in 0 and then the probabilities sorted in descending order to the right at 1, 2, etc. Due to the symmetry of the bar chart of the distribution $\text{Bin}(40, \frac{1}{2})$, each pair of consecutive probabilities is of equal size.

Due to the close connection between the distribution of $M_n$ and that of $S_n$, as given in (2.65), one can readily obtain the limit distribution of $M_n$ as $n \to \infty$.

**Theorem 2.13 (Limit Distribution of the Maximum)** *For the maximum $M_n$ of a simple symmetric random walk of length n,*

$$\lim_{n\to\infty} \mathbb{P}\left(\frac{M_n}{\sqrt{n}} \leq x\right) = 2\Phi(x) - 1, \qquad x \geq 0, \tag{2.70}$$

*with $\Phi(x)$ as in (2.4).*

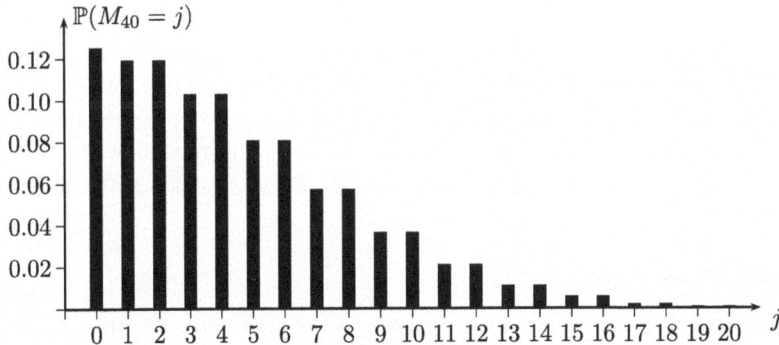

**Fig. 2.27** Bar chart of the distribution of $M_{40}$

***Proof*** The idea of the proof is to use (2.65) and the de Moivre–Laplace central limit theorem (see Sect. 7.4). First, it is clear that the assertion holds for $x = 0$, since then both sides of (2.70) vanish. For fixed $x > 0$, we choose $k_n := \lfloor x\sqrt{n} \rfloor$ as the largest integer less than or equal to $x\sqrt{n}$. Since $M_n$ is integer-valued, $\{M_n \leq k_n\} = \{M_n \leq x\sqrt{n}\}$. Thus, setting $z_n := k_n/\sqrt{n}$, (2.65) yields

$$\begin{aligned}
\mathbb{P}\left(\frac{M_n}{\sqrt{n}} \leq x\right) &= \mathbb{P}(M_n \leq x\sqrt{n}) = \mathbb{P}(M_n \leq k_n) \\
&= \mathbb{P}(S_n \leq k_n) + \mathbb{P}(S_n \leq k_n + 1) - 1 \\
&= 2\,\mathbb{P}(S_n \leq k_n) - 1 + \mathbb{P}(S_n = k_n + 1) \\
&= 2\,\mathbb{P}\left(\frac{S_n}{\sqrt{n}} \leq z_n\right) - 1 + \mathbb{P}(S_n = k_n + 1).
\end{aligned} \qquad (2.71)$$

According to (2.6), $S_n/\sqrt{n}$ converges in distribution to a standard normal distribution. In view of $\lim_{n\to\infty} z_n = x$ and (7.8), the first term in (2.71) converges to $2\Phi(x)$ as $n \to \infty$. The last term is bounded from above by

$$C_n := \max_{k=0,1,\ldots,n} \binom{n}{k} 2^{-n}.$$

Now, $C_{2n} = \binom{2n}{n} 2^{-2n} = u_{2n}$ and $C_{2n+1} = \binom{2n+1}{n+1} 2^{-(2n+1)} = u_{2n}(2n+1)/(2n+2) \leq u_{2n}$. According to (2.29), $\lim_{n\to\infty} u_{2n} = 0$, so the last term in (2.71) converges to zero. Hence, (2.70) follows. ∎

**Remark 2.14** Since $\mathbb{P}(M_n/\sqrt{n} \leq x)$, as a function of $x$, is the distribution function of $M_n/\sqrt{n}$, Theorem 2.13 states that the sequence of distribution functions of $M_n/\sqrt{n}$ converges pointwise to the distribution function of the half-normal distribution. According to Sect. 7.3, we can thus write the statement of Theorem 2.13 in the compact form

$$\frac{M_n}{\sqrt{n}} \xrightarrow{\mathcal{D}} |Z|, \qquad Z \sim N(0,1), \qquad \text{as } n \to \infty.$$

If we set $\Phi^*(x) = 2\Phi(x) - 1$ as before, Table 2.2 gives the values $\Phi^*(0.1) = 0.08$, $\Phi^*(0.672) = 0.5$, and $\Phi^*(1.96) = 0.95$. From Theorem 2.13, it follows, among other things, that in symmetric random walks of length 10,000, the maximum is at most 68 in about half of all cases. With probability 0.08, the random walk reaches a maximum height of at most 10, and only in about 5% of all cases, is the maximum greater than 197.

**Maxima, Minima, and Visits to Zero**

The bar charts of the distributions of $N_{40}$ (Fig. 2.18) and $M_{40}$ (Fig. 2.27) are strikingly similar. It is also striking that, according to Theorems 2.7 and 2.13,

## 2.7 Maximum and Minimum

the number $N_{2n}$ of visits to zero and the maximum $M_{2n}$ of a simple symmetric random walk of length $2n$, after division by $\sqrt{2n}$, have the same limit distribution as $n \to \infty$; both distributions, normalized in the same manner, approach the half-normal distribution, since

$$\lim_{n \to \infty} \mathbb{P}\left(\frac{N_{2n}}{\sqrt{2n}} \leq x\right) = \lim_{n \to \infty} \mathbb{P}\left(\frac{M_{2n}}{\sqrt{2n}} \leq x\right) = 2\int_0^x \varphi(t)\,dt, \qquad x > 0.$$

Moreover, the expectations of $N_{2n}$ and $M_{2n}$ differ in absolute value by at most $\frac{1}{4}$ (Exercise 2.14).

Is there a conceptual connection between the number of visits to zero and the maximum of a random walk? The answer is: Yes, and we have already given this in the proof of Theorem 2.6. There, we saw that a $2n$-path with exactly $j$ visits to zero, which is non-negative up to the $j$-th such visit, can be uniquely associated with a path leading from $(0, 0)$ to $(2n - j, -j)$. Since the path length for the considerations at that time (and also the following ones) does not have to be an even number, we start with an $n$-path for any positive integer $n$ that visits the point 0 at least $j$ times and is non-negative up to the $j$-th such visit. By omitting the first upward step as well as each upward step that directly follows the first $j - 1$ visits to zero, we obtain a path of length $n - j$ whose minimum is less than or equal to $-j$. This mapping is evidently injective. Figures 2.28 and 2.29 illustrate this mapping for the case $j = 4$; the fact that the path can proceed arbitrarily after the $j$-th visit to zero or after the first hit of the height $-j$ is highlighted by a color-coded corridor.

The inverse of this mapping (and thus the proof of its bijectivity) proceeds as in the proof of Theorem 2.6: We start with a $(n - j)$-path whose minimum is less than or equal to $-j$ (see Fig. 2.29 for the case $j = 4$). Such a path reaches the height $-1$ for the first time at some point in its temporal progression (first sub-path), then hits the height $-2$ for the first time (second sub-path), and so on. Eventually, it reaches the height $-j$ for the first time ($j$-th sub-path). By preceding each of these $j$ sub-paths with an upward step and joining these sub-paths together starting at the origin, we obtain a non-negative path that visits zero $j$ times. The last sub-path after the first time hitting the height $-j$ (which somehow runs within the marked

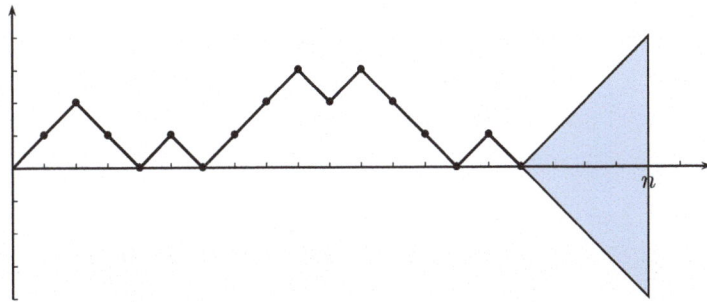

**Fig. 2.28** $n$-path with at least four visits to zero, which is non-negative up to the fourth such visit

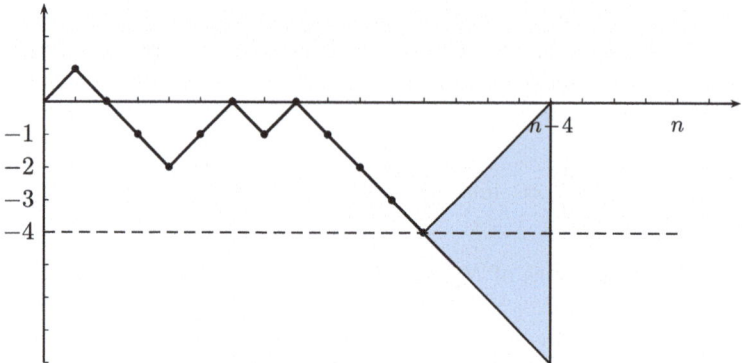

**Fig. 2.29** $(n-4)$-path, whose minimum is less than or equal to $-4$

corridor) is attached unchanged to the $j$-th visit to zero; it can contribute to further visits to zero. In this way, we obtain a path with at least $j$ visits to zero, which is non-negative up to the $j$-th such visit.

By reflection across the $x$-axis, each $(n-j)$-path whose minimum is less than or equal to $-j$ corresponds in a one-to-one manner to a $(n-j)$-path whose maximum is at least $j$. Denoting by $\{N_n \geq j, \geq 0\}$ the set of all $n$-paths with at least $j$ visits to zero which are non-negative up to the $j$-th such visit, it thus follows that

$$\mathbb{P}(N_n \geq j) = \frac{1}{2^n} |\{N_n \geq j\}| = \frac{1}{2^n} \cdot 2^j |\{N_n \geq j, \geq 0\}|$$

$$= \frac{1}{2^n} \cdot 2^j |\{M_{n-j} \geq j\}| = \frac{1}{2^{n-j}} |\{M_{n-j} \geq j\}|.$$

Hence,

$$\mathbb{P}(N_n \geq j) = \mathbb{P}(M_{n-j} \geq j), \quad j = 0, 1, \ldots, \left\lfloor \frac{n}{2} \right\rfloor. \qquad (2.72)$$

This result allows the following very short proof of Theorem 2.7 using Theorem 2.13 (which only uses the de Moivre–Laplace central limit theorem): For $x > 0$, let $j_n = \lceil x\sqrt{2n} \rceil$ be the smallest integer greater than or equal to $x\sqrt{2n}$. Thus, in particular,

$$\lim_{n \to \infty} \frac{j_n}{\sqrt{2n - j_n}} = x. \qquad (2.73)$$

## 2.8 Number and Positions of Maximizers

Since $N_{2n}$ is integer-valued, and using (2.72), we obtain

$$\mathbb{P}\left(\frac{N_{2n}}{\sqrt{2n}} \geq x\right) = \mathbb{P}\left(N_{2n} \geq x\sqrt{2n}\right) = \mathbb{P}(N_{2n} \geq j_n)$$
$$= \mathbb{P}(M_{2n-j_n} \geq j_n)$$
$$= \mathbb{P}\left(\frac{M_{2n-j_n}}{\sqrt{2n-j_n}} \geq \frac{j_n}{\sqrt{2n-j_n}}\right).$$

Since in (2.70) the less-than sign " $<$ " can replace " $\leq$," and $x$ can be replaced by $x_n$, where $(x_n)$ is any sequence converging to $x$ (see Sect. 7.3), (2.70) with $x_n := j_n/\sqrt{2n-j_n}$ and (2.73) yields

$$\lim_{n\to\infty} \mathbb{P}\left(\frac{M_{2n-j_n}}{\sqrt{2n-j_n}} \geq \frac{j_n}{\sqrt{2n-j_n}}\right) = 1 - (2\Phi(x) - 1),$$

and thus,

$$\lim_{n\to\infty} \mathbb{P}\left(\frac{N_{2n}}{\sqrt{2n}} \leq x\right) = 2\Phi(x) - 1,$$

which is the statement of Theorem 2.7.

**Remark 2.15** With greater technical effort, it is even possible to assess the joint distribution of $m_n$, $M_n$ and $S_n$: Let $a, b$, and $v$ be integers with $a \leq 0 \leq b$, $a < b$, $a \leq v \leq b$, and $v \neq a$ and $v \neq b$. Also, let $n$ and $v$ have the same parity. Then the following holds:

$$\mathbb{P}(a < m_n \leq M_n < b, S_n = v)$$
$$= \sum_{k=-\infty}^{\infty} \mathbb{P}\big(S_n = v+2k(b-a)\big) - \sum_{k=-\infty}^{\infty} \mathbb{P}\big(S_n = 2b-v+2k(b-a)\big) \quad (2.74)$$

(see [BI], pp. 95–96). Note that each of the sums only runs over finitely many values of $k$, since $b - a > 0$. A proof of (2.74) can be carried out by induction on $n$. Exercise 2.15 concerns the base case of the induction.

## 2.8 Number and Positions of Maximizers

A symmetric random walk of length $n$ can attain its maximum $M_n = \max_{0 \leq j \leq n} S_j$ at several points in time, but it has at most $\lfloor \frac{n}{2} \rfloor + 1$ maximizers.

**SAQ 14** Why is the number of maximizers at most $\lfloor \frac{n}{2} \rfloor + 1$?

We define

$$D_n := \{k \in \{0, 1, \ldots, n\} : S_k = M_n\}$$

as the set of the *maximizers* of the random walk, and we denote by

$$Q_n := |D_n| = \sum_{k=0}^{n} \mathbf{1}\{S_k = M_n\} \qquad (2.75)$$

the *number of maximizers*. The random variable $Q_n$ can take each of the values from 1 to $\lfloor \frac{n}{2} \rfloor + 1$. In this section, we determine the distribution of $Q_n$. As the following result shows, this distribution is directly related to the distribution of the maximum of random walks (cf. [RE], p. 152).

**Theorem 2.16 (Maximizers and Maximum)** *Let $Q_n$ be the number of maximizers of a symmetric random walk of length n, starting at 0. Then, for each $n = 1, 2, \ldots$ and each $j \in \{0, 1, \ldots, \lfloor \frac{n}{2} \rfloor\}$,*

$$\mathbb{P}(Q_n \geq j+1) = \frac{1}{2^j} \mathbb{P}(M_{n-j} \geq j). \qquad (2.76)$$

*Proof* We do not follow the proof given in [RE], but consider that two sets $A_1$ and $A_2$ of paths that all start at the origin have the same cardinality. The set $A_1$ denotes all paths of length $n - j$ whose maximum is greater than or equal to $j$, and the set $A_2$ stands for all paths of length $n$, where the maximum is attained at least $j + 1$ times. Therefore, $Q_n \geq j + 1$. If $|A_1| = |A_2|$ were shown, the assertion would follow from

$$\mathbb{P}(Q_n \geq j+1) = \frac{|A_2|}{2^n} = \frac{1}{2^j} \frac{|A_1|}{2^{n-j}} = \frac{1}{2^j} \mathbb{P}(M_{n-j} \geq j).$$

Since, in the case $j = 0$, both sides of (2.76) are equal to 1, let $j \geq 1$ in what follows. We show that there is a bijective mapping from $A_1$ onto $A_2$, which yields $|A_1| = |A_2|$.

To construct such a mapping, we start with any $(n - j)$-path with the property $M_{n-j} \geq j$. This path reaches, for each $s = 0, 1, \ldots, j - 1$, the height $M_{n-j} - j + s$ for the first time at some point. We add steps to this path, specifically for each $s \in \{0, \ldots, j - 1\}$ we add a downward step directly after the first time it reaches the height $M_{n-j} - j + s$. In this way, we obtain a path of length $n$ that attains its maximum, which is now $M_{n-j} - j$, at least $j + 1$ times.

## 2.8 Number and Positions of Maximizers

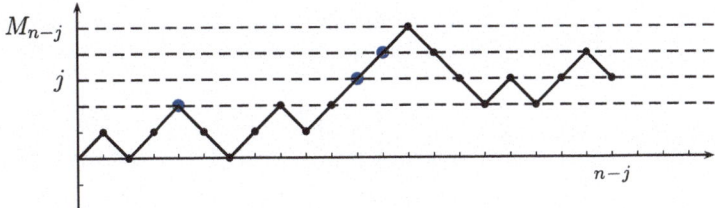

**Fig. 2.30** $(n-j)$-path, whose maximum is greater or equal to $j$

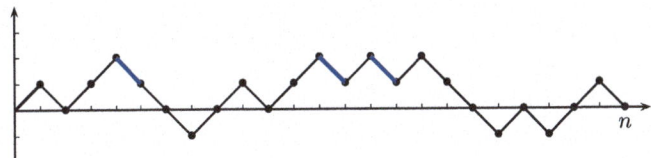

**Fig. 2.31** The $(n-j)$-path from Fig. 2.30 has been transformed into an $n$-path which attains its maximum at least $j+1$ times

As an example, consider the path shown in Fig. 2.30, for which $n = 24$ and $j = 3$. Those points where the height $M_{n-j} - j + s$ is reached for the first time for $s = 0, 1, \ldots, j-1$ are highlighted and marked in blue. If we insert a downward step after each of these points, we obtain the path shown in Fig. 2.31, which attains its maximum at least $j$ times. As a result of the transformation of a path from $A_1$, we obtain a path from the set $A_2$, and it is evident that this mapping from paths in $A_1$ to paths in $A_2$ is injective.

The inverse of this mapping is obtained by starting with any $n$-path that attains its maximum at least $j+1$ times, and leaving the initial segment up to the first occurrence of the maximum unchanged. From the remaining part of the path, which begins with a downward step, remove this downward step as well as the next $j-1$ downward steps that immediately follow maxima, and join the rest together. In Fig. 2.31, for the case $n = 24$ and $j = 3$, the downward steps to be removed are marked in blue. Removing these and joining the rest together yields the path in Fig. 2.30. ∎

The following theorem provides information about the distribution of $Q_n$ (see, e.g., [KP2], pp. 160–161):

**Theorem 2.17 (Distribution of the Number of Maximizers)** *For the number $Q_n$ of maximizers of a random walk of length $n$, the following holds:*

(a) $\mathbb{P}(Q_n \geq k+1) = \dfrac{1}{2^k} - \dfrac{1}{2^n} \sum_{j=0}^{k-1} \binom{n-k}{\lfloor \frac{n-k+j+1}{2} \rfloor}, \quad n \geq 1, \quad 0 \leq k \leq \lfloor \frac{n}{2} \rfloor.$

(b) *For every $n \geq 1$, $Q_{2n}$ and $Q_{2n+1}$ have the same distribution.*

(c) $\mathbb{P}(Q_{2n} = 2k) = \dfrac{1}{2^{2k}} - \dfrac{1}{2^{2n}} \sum_{j=1}^{k-1} \binom{2n-2k+1}{n-j}$, $1 \le k \le \lfloor \dfrac{n+1}{2} \rfloor$,

$\mathbb{P}(Q_{2n} = 2k+1) = \dfrac{1}{2^{2k+1}} - \dfrac{1}{2^{2n}} \left( \sum_{j=1}^{k} \binom{2n-2k}{n-j} - \binom{2n-2k-1}{n-k} \right)$,

$0 \le k \le \lfloor \dfrac{n}{2} \rfloor$.

(d) Setting $u_{2(n+1)} = \binom{2(n+1)}{n+1}/2^{2(n+1)}$, for every $n \ge 1$:

$$\mathbb{E}(Q_{2n}) = 2 - 2\, u_{2(n+1)},$$

$$\mathbb{V}(Q_{2n}) = 2 - \dfrac{2(3n+2)}{n+2} \cdot u_{2(n+1)} - 4\, u_{2(n+1)}^2.$$

**Proof**

(a) follows directly from Theorems 2.16 and 2.12 (a).

(b) According to Theorem 2.16 and (2.72), for every $j \in \{0, 1, \ldots, n\}$:

$$\mathbb{P}(Q_{2n} \ge j+1) = 2^{-j}\, \mathbb{P}(M_{2n-j} \ge j) = 2^{-j}\, \mathbb{P}(N_{2n} \ge j),$$

$$\mathbb{P}(Q_{2n+1} \ge j+1) = 2^{-j}\, \mathbb{P}(M_{2n+1-j} \ge j) = 2^{-j}\, \mathbb{P}(N_{2n+1} \ge j).$$

Since visits to zero only occur at even times, the right-hand sides of these equations are identical, which proves (b).

(c) Of the two statements, we only show the first; the second follows similarly. Due to $\mathbb{P}(Q_{2n} = 2k) = \mathbb{P}(Q_{2n} \ge 2k) - \mathbb{P}(Q_{2n} \ge 2k+1)$, part (a) with $2n$ instead of $n$ and $2k-1$ as well as $2k$ instead of $k$ yields

$$\mathbb{P}(Q_{2n} = 2k) = \dfrac{1}{2^{2k}} - \dfrac{1}{2^{2n}} \left( \sum_{j=0}^{2k-2} \binom{2n-2k+1}{\lfloor n-k+1+\frac{j}{2} \rfloor} \right.$$

$$\left. - \sum_{j=0}^{2k-1} \binom{2n-2k}{\lfloor n-k+\frac{j+1}{2} \rfloor} \right).$$

The sums over binomial coefficients that appear here are equal to

$$\sum_{j=0}^{2k-2} \binom{2n-2k+1}{\lfloor n-k+1+\frac{j}{2} \rfloor} = 2 \sum_{v=0}^{k-2} \binom{2n-2k+1}{n-k+1+v} + \binom{2n-2k+1}{n},$$

$$\sum_{j=0}^{2k-1} \binom{2n-2k}{\lfloor n-k+\frac{j+1}{2} \rfloor} = \binom{2n-2k}{n-k} + 2 \sum_{v=1}^{k-1} \binom{2n-2k}{n-k+v} + \binom{2n-2k}{n}.$$

## 2.8 Number and Positions of Maximizers

Using the recursion formula $\binom{m+1}{\ell} = \binom{m}{\ell} + \binom{m}{\ell-1}$, we get

$$\sum_{v=0}^{k-2}\binom{2n-2k+1}{n-k+1+v} = \sum_{v=0}^{k-2}\binom{2n-2k}{n-k+1+v} + \sum_{v=0}^{k-2}\binom{2n-2k}{n-k+v},$$

and therefore for the difference of the two right-hand sides, as claimed, the expression

$$2\sum_{v=0}^{k-2}\binom{2n-2k}{n-k+v+1} + \binom{2n-2k}{n-k} - \binom{2n-2k}{n-1}$$

$$= \binom{2n-2k}{n-1} + 2\sum_{v=0}^{k-3}\binom{2n-2k}{n-k+1+v} + \binom{2n-2k}{n-k}$$

$$= \sum_{v=0}^{k-2}\binom{2n-2k}{n-k+1+v} + \sum_{v=-1}^{k-3}\binom{2n-2k}{n-k+1+v}$$

$$= \sum_{j=1}^{k-1}\left[\binom{2n-2k}{n-k+j} + \binom{2n-2k}{n-k+j-1}\right]$$

$$= \sum_{j=1}^{k-1}\binom{2n-2k+1}{n-j}.$$

(d) We use the representation (2.75). Due to the additivity of expectation,

$$\mathbb{E}(Q_{2n}) = \sum_{j=0}^{2n} \mathbb{P}(S_j = M_{2n})$$

$$= \mathbb{P}(S_0 = M_{2n}) + \mathbb{P}(S_{2n} = M_{2n}) + \sum_{j=1}^{2n-1}\mathbb{P}(S_j = M_{2n}).$$

According to Corollary 2.3 and (2.15), $\mathbb{P}(S_0 = M_{2n}) = \mathbb{P}(S_1 \leq 0, \ldots, S_{2n} \leq 0) = u_{2n}$. Furthermore,

$$\mathbb{P}(S_{2n} = M_{2n}) = \mathbb{P}(S_0 \leq S_{2n}, S_1 \leq S_{2n}, \ldots, S_{2n-1} \leq S_{2n})$$
$$= \mathbb{P}(X_1 + \ldots + X_{2n} \geq 0, X_2 + \ldots + X_{2n} \geq 0, \ldots, X_{2n} \geq 0)$$
$$= \mathbb{P}(X_1 \geq 0, X_1 + X_2 \geq 0, \ldots, X_1 + \ldots + X_{2n} \geq 0)$$
$$= \mathbb{P}(S_1 \geq 0, S_2 \geq 0, \ldots, S_{2n} \geq 0)$$
$$= u_{2n}.$$

The third equality indicates that $(X_1, \ldots, X_{2n})$ and $(X_{2n}, \ldots, X_1)$ have the same distribution. This so-called *duality principle*, explained further in Sect. 2.14, will be used several times in the following. More generally, the random vector $(X_{\pi(1)}, \ldots, X_{\pi(2n)})$ has the same distribution as $(X_1, \ldots, X_{2n})$ for any permutation $\pi = (\pi(1), \ldots, \pi(2n))$ of $(1, \ldots, 2n)$. By splitting according to even and odd values of $j$, it follows that

$$\sum_{j=1}^{2n-1} \mathbb{P}(S_j = M_{2n}) = \sum_{k=1}^{n-1} \mathbb{P}(S_{2k} = M_{2n}) + \sum_{k=0}^{n-1} \mathbb{P}(S_{2k+1} = M_{2n}).$$

Furthermore, $\mathbb{P}(S_{2k} = M_{2n}) = \mathbb{P}(A_k \cap B_k)$, where $A_k = \{S_0 \leq S_{2k}, \ldots, S_{2k-1} \leq S_{2k}\}$ and $B_k = \{S_{2k+1} \leq S_{2k}, \ldots, S_{2n} \leq S_{2k}\}$. The events $A_k$ and $B_k$ are independent, because $A_k = \{X_1 + \ldots + X_{2k} \geq 0, \ldots, X_{2k-1} + X_{2k} \geq 0, X_{2k} \geq 0\}$ depends only on $X_1, \ldots, X_{2k}$, and $B_k = \{X_{2k+1} \leq 0, X_{2k+1} + X_{2k+2} \leq 0, \ldots, X_{2k+1} + \ldots + X_{2n} \leq 0\}$ is determined only by $X_{2k+1}, \ldots, X_{2n}$. Using Corollary 2.3 and (2.15), it follows that $\mathbb{P}(A_k) = u_{2k}$. According to the above duality principle,

$$\mathbb{P}(B_k) = \mathbb{P}(X_1 \leq 0, X_1 + X_2 \leq 0, \ldots, X_1 + \ldots + X_{2n-2k} \leq 0) = u_{2(n-k)}.$$

Therefore, $\mathbb{P}(S_{2k} = M_{2k}) = u_{2k} u_{2(n-k)}$. Since $S_{2k+1} \geq 0$ automatically implies $S_{2k+2} \geq 0$, we similarly obtain $\mathbb{P}(S_{2k+1} = M_{2n}) = u_{2(k+1)} u_{2(n-k)}$. Therefore,

$$\sum_{j=1}^{2n-1} \mathbb{P}(S_j = M_{2n}) = \sum_{k=1}^{n-1} u_{2k} u_{2(n-k)} + \sum_{k=0}^{n-1} u_{2(k+1)} u_{2(n-k)}.$$

Due to Theorem 2.4 (a) and $u_0 = 1$, it follows that

$$\sum_{k=1}^{n-1} u_{2k} u_{2(n-k)} = \sum_{k=0}^{n} u_{2k} u_{2(n-k)} - 2u_{2n} = 1 - 2u_{2n}.$$

Similarly, an index shift yields

$$\sum_{k=0}^{n-1} u_{2(k+1)} u_{2(n-k)} = \sum_{v=1}^{n} u_{2v} u_{2(n+1-v)} - 2u_{2(n+1)} = 1 - 2u_{2(n+1)}.$$

Overall, as claimed, we get

$$\mathbb{E}(Q_{2n}) = 2u_{2n} + 1 - 2u_{2n} + 1 - 2u_{2(n+1)} = 2 - 2u_{2(n+1)}.$$

We determine the variance via the representation $\mathbb{V}(Q_{2n}) = \mathbb{E}(Q_{2n}^2) - (\mathbb{E}Q_{2n})^2$ and write $Q_{2n}^2$ as the double sum $\sum_{j=0}^{2n} \sum_{k=0}^{2n} \mathbf{1}\{S_j = M_{2n}\} \mathbf{1}\{S_k = M_{2n}\}$. By

## 2.8 Number and Positions of Maximizers

decomposing this double sum into the cases $j = k$ and $j \neq k$, and using the relations $\mathbf{1}\{A \cap B\} = \mathbf{1}\{A\}\mathbf{1}\{B\}$ and $\mathbf{1}\{A\}^2 = \mathbf{1}\{A\}$ for indicator functions of events, the additivity of expectation gives

$$\mathbb{E}(Q_{2n}^2) = \sum_{j=0}^{2n} \mathbb{P}(S_j = M_{2n}) + 2 \sum_{k=1}^{2n} \sum_{j=0}^{k-1} \mathbb{P}(S_j = M_{2n}, S_k = M_{2n}).$$

Here, the first sum is equal to $\mathbb{E}(Q_{2n})$. Since $j$ and $k$ in the double sum must have the same parity for $\mathbb{P}(S_j = M_{2n}, S_k = M_{2n})$ to be positive, it follows that

$$\sum_{k=1}^{2n} \sum_{j=0}^{k-1} \mathbb{P}(S_j = M_{2n}, S_k = M_{2n}) = a_n + b_n,$$

where

$$a_n := \sum_{k=1}^{n} \sum_{j=0}^{k-1} \mathbb{P}(S_{2j} = M_{2n} = S_{2k}), \quad b_n := \sum_{k=1}^{n-1} \sum_{j=0}^{k-1} \mathbb{P}(S_{2j+1} = M_{2n} = S_{2k+1}).$$

Setting

$$A_j := \{S_0 \leq S_{2j}, S_1 \leq S_{2j}, \ldots, S_{2j-1} \leq S_{2j}\},$$
$$B_{j,k} := \{S_{2j+1} \leq S_{2j}, \ldots, S_{2k-1} \leq S_{2j}, S_{2k} = S_{2j}\},$$
$$C_k := \{S_{2k+1} \leq S_{2k}, S_{2k+2} \leq S_{2k}, \ldots, S_{2n} \leq S_{2k}\},$$

we have $\mathbb{P}(S_{2j} = M_{2n} = S_{2k}) = \mathbb{P}(A_j \cap B_{j,k} \cap C_k)$. Since the events $A_j$, $B_{j,k}$ and $C_k$ are independent, it follows that $\mathbb{P}(S_{2j} = M_{2n} = S_{2k}) = \mathbb{P}(A_j)\mathbb{P}(B_{j,k})\mathbb{P}(C_k)$, where $\mathbb{P}(A_j) = u_{2j}$ and $\mathbb{P}(C_k) = u_{2(n-k)}$ (cf. the derivation of $\mathbb{E}(Q_{2n})$).

**SAQ 15** Why are $A_j$, $B_{j,k}$, and $C_k$ independent?

For symmetry reasons, it also holds that

$$\mathbb{P}(B_{j,k}) = \mathbb{P}(S_1 \leq 0, S_2 \leq 0, \ldots, S_{2(k-j)-1} \leq 0, S_{2(k-j)} = 0)$$
$$= \mathbb{P}(S_{2(k-j)} = 0)\,\mathbb{P}(S_1 \leq 0, S_2 \leq 0, \ldots, S_{2(k-j)-1} \leq 0 | S_{2(k-j)} = 0).$$

Here, the first factor is equal to $u_{2(k-j)}$ and the second is the probability that a purely random bridge of length $2(k-j)$ is non-positive. After a short consideration, this results in $\frac{1}{k-j+1}$, if you take into account all paths from

$(1,-1)$ to $(2(k-j),0)$ (these are $\binom{2(k-j)-1}{k-j}$ in number) and subtract those that hit the axis $y=1$. According to the reflection principle, the latter set of paths is the $\binom{2(k-j)}{k-j+1}$-element set of all paths from $(1,-1)$ to $(2(k-j),2)$. With $f_{2n}$ defined in (2.48), it thus holds that $\mathbb{P}(B_{j,k}) = u_{2(k-j)}/(k-j+1) = 2f_{2(k-j+1)}$, and hence

$$a_n = 2\sum_{k=1}^{n}\left(\sum_{j=0}^{k-1} u_{2j} f_{2(k-j+1)}\right) u_{2(n-k)} = 2\sum_{k=1}^{n}\left(\sum_{r=2}^{k+1} f_{2r} u_{2(k+1-r)}\right) u_{2(n-k)}.$$

Due to (2.49), the inner sum over $r$ is equal to $u_{2(k+1)} - f_2 u_{2k} = u_{2(k+1)} - u_{2k}/2$, and Theorem 2.4 (a), along with $u_0 = 1$ and $u_2 = \frac{1}{2}$, yields

$$a_n = 2\sum_{k=1}^{n}\left(u_{2(k+1)} - \frac{u_{2k}}{2}\right) u_{2(n-k)} = 2\sum_{v=2}^{n+1} u_{2v} u_{2(n+1-v)} - \sum_{k=1}^{n} u_{2k} u_{2(n-k)}$$

$$= 2\left(1 - u_{2(k+1)} - \frac{u_{2n}}{2}\right) - (1 - u_{2n})$$

$$= 1 - 2u_{2(n+1)}.$$

With the same considerations and a little perseverance, we obtain

$$b_n = 2\sum_{k=1}^{n-1}\sum_{j=0}^{k-1} u_{2(j+1)} f_{2(k-j+1)} u_{2(n-k)} = 1 - 6u_{2(n+2)} + \left(2 + \frac{1}{n+2}\right) u_{2(n+1)}.$$

Overall, we get

$$\mathbb{E}(Q_{2n}^2) = \mathbb{E}(Q_{2n}) + 2a_n + 2b_n = 6 - 12u_{2(n+2)} + \left(\frac{2}{n+2} - 2\right) u_{2(n+1)} \quad (2.77)$$

and thus the representation for $\mathbb{V}(Q_{2n})$ by direct calculation (Exercise 2.16).

∎

After this somewhat lengthy proof, it is time to comment on the results obtained. Figure 2.32 shows a bar chart of the distribution of $Q_{30}$. It is striking that the probabilities $\mathbb{P}(Q_{30} = k)$ decrease rapidly with increasing $k$ (in fact, $\mathbb{P}(Q_{30} = k) < 0.0001$ for $k \geq 8$), and that the probability for exactly one maximizer is greater than $\frac{1}{2}$. Additionally, $\mathbb{P}(Q_{30} = 2)$ seems to be very close to $\frac{1}{4}$. Indeed, Theorem 2.17 yields

$$\mathbb{P}(Q_{2n} = 1) = \mathbb{P}(Q_{2n+1} = 1) = \frac{1}{2} + \frac{\binom{2n-1}{n}}{2^{2n}} = \frac{1}{2} + \frac{u_{2n}}{2} > \frac{1}{2},$$

## 2.8 Number and Positions of Maximizers

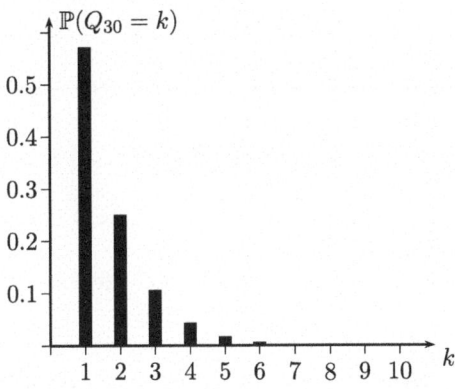

**Fig. 2.32** Bar chart of the distribution of $Q_{30}$

as well as the surprising result

$$\mathbb{P}(Q_{2n} = 2) = \mathbb{P}(Q_{2n+1} = 2) = \frac{1}{4}.$$

Thus, completely independent of their path length, every fourth symmetric random walk has exactly two maximizers. Based on Theorem 2.17 (or Theorem 2.16), one can observe that the distribution of $Q_n$ converges to a *geometric distribution*, shifted by 1, as $n \to \infty$. We note this result as follows:

**Theorem 2.18 (Limit Distribution of the Number of Maximizers)** *For the number $Q_n$ of maximizers of a simple symmetric random walk of length n,*

$$\lim_{n \to \infty} \mathbb{P}(Q_n = k) = \left(\frac{1}{2}\right)^k, \quad k \in \mathbb{N}.$$

We now ask *where* the maximizers of a random walk lie, and for simplicity, we consider random walks of even length $2n$. More precisely, we ask about the distribution of the position of the *first* maximizer

$$E_{2n} := \min\left\{k \in \{0, 1, \ldots, 2n\} : S_k = \max_{0 \le j \le 2n} S_j\right\}. \tag{2.78}$$

The random variable $E_{2n}$ can take any of the values $0, 1, \ldots, 2n$. The extreme case $E_{2n} = 0$ occurs if and only if the path is non-positive, that is, $S_1 \le 0, S_2 \le 0, \ldots, S_{2n} \le 0$. According to Corollary 2.3, the probability for this is equal to $u_{2n} = \binom{2n}{n}/2^{2n}$, so $\mathbb{P}(E_{2n} = 0) = u_{2n}$. The other extreme case $E_{2n} = 2n$ happens when $S_0 < S_{2n}, S_1 < S_{2n}, \ldots, S_{2n-1} < S_{2n}$, that is, for each $j = 1, \ldots, 2n - 1$, the sum $X_j + \ldots + X_{2n}$ is greater than 0. Since $X_1, \ldots, X_{2n}$ are independent and identically distributed, this event has the same probability as the event $\{X_1 > 0, X_1 + X_2 > 0, \ldots, X_1 + \ldots + X_{2n} > 0\}$. With the definition of $S_1, \ldots, S_{2n}$, we thus have

$$\mathbb{P}(E_{2n} = 2n) = \mathbb{P}(S_1 > 0, S_2 > 0, \ldots, S_{2n} > 0), \tag{2.79}$$

which, with Corollary 2.3, means $\mathbb{P}(E_{2n} = 2n) = \frac{1}{2} u_{2n}$. To determine $\mathbb{P}(E_{2n} = k)$, we therefore assume the case $1 \leq k \leq 2n - 1$. Let us briefly set $A_k := \{S_0 < S_k, \ldots, S_{k-1} < S_k\}$ and $B_k := \{S_{k+1} \leq S_k, \ldots, S_{2n} \leq S_k\}$, so that

$$\mathbb{P}(E_{2n} = k) = \mathbb{P}(A_k \cap B_k).$$

Since the events $A_k$ and $B_k$ are independent, $\mathbb{P}(E_{2n} = k) = \mathbb{P}(A_k)\mathbb{P}(B_k)$.

**SAQ 16** Why are $A_k$ and $B_k$ independent?

With the same considerations that led to (2.79), one obtains $\mathbb{P}(A_k) = \mathbb{P}(S_1 > 0, \ldots, S_k > 0)$ and $\mathbb{P}(B_k) = \mathbb{P}(S_1 \leq 0, S_2 \leq 0, \ldots, S_{2n-k} \leq 0)$, which yields the intermediate result

$$\mathbb{P}(E_{2n} = k) = \mathbb{P}(S_1 > 0, \ldots, S_k > 0)\, \mathbb{P}(S_1 \leq 0, \ldots, S_{2n-k} \leq 0). \tag{2.80}$$

We now distinguish between the cases that $k$ is even or odd, i.e., $k = 2\ell$ for an $\ell \in \{1, \ldots, n-1\}$ or $k = 2\ell + 1$ for an $\ell \in \{0, \ldots, n-1\}$. In the first case, the probabilities in (2.80) become

$$\mathbb{P}(S_1 > 0, \ldots, S_{2\ell} > 0) = \frac{1}{2} \mathbb{P}(S_1 \neq 0, \ldots, S_{2\ell} \neq 0) = \frac{1}{2} \cdot u_{2\ell},$$

$$\mathbb{P}(S_1 \leq 0, \ldots, S_{2n-2\ell} \leq 0) = u_{2(n-\ell)}.$$

In the second case, we get

$$\mathbb{P}(S_1 > 0, \ldots, S_{2\ell+1} > 0) = \frac{1}{2} \mathbb{P}(S_2 > 0, \ldots, S_{2\ell+1} > 0 | X_1 = 1) = \frac{1}{2} \cdot u_{2\ell}$$

$$\mathbb{P}(S_1 \leq 0, \ldots, S_{2n-2\ell-1} \leq 0) = \mathbb{P}(S_1 \leq 0, \ldots, S_{2n-2\ell} \leq 0) = u_{2(n-\ell)}.$$

Due to $u_0 = 1$ we thus obtain the following result (cf. [FE1], p. 94):

**Theorem 2.19 (Distribution of the First Maximizer)** *For the first maximizer $E_{2n}$, the following holds:*

$$\mathbb{P}(E_{2n} = 0) = u_{2n},$$

$$\mathbb{P}(E_{2n} = 2\ell) = \frac{1}{2} \cdot u_{2\ell} u_{2(n-\ell)}, \qquad \ell = 1, \ldots, n,$$

$$\mathbb{P}(E_{2n} = 2\ell + 1) = \frac{1}{2} \cdot u_{2\ell} u_{2(n-\ell)}, \qquad \ell = 0, \ldots, n-1.$$

## 2.8 Number and Positions of Maximizers

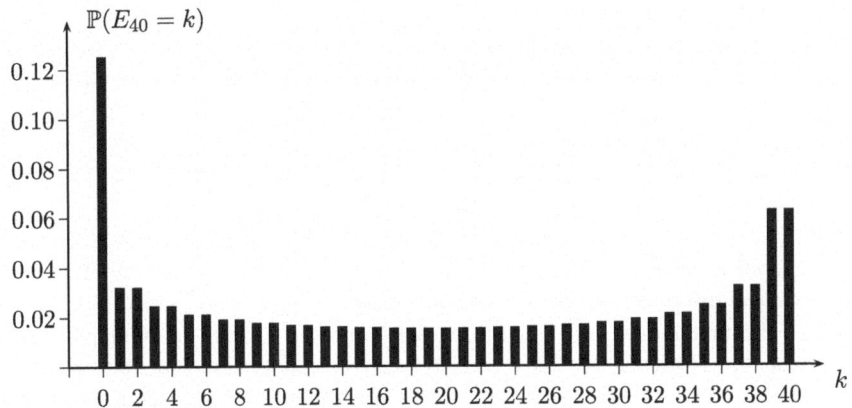

**Fig. 2.33** Bar chart of the distribution of $E_{40}$

Surprisingly, regardless of the length of the path, it is most likely that the maximum is attained right at the beginning.

A comparison with the discrete arcsine distribution of the time $L_{2n}$ of the last visit to zero, given in Theorem 2.4, shows that the distribution of $E_{2n}$ is obtained from that of $L_{2n}$ by keeping the probability at zero unchanged and splitting the probability $\mathbb{P}(L_{2n} = 2k)$ for each $k = 1, 2, \ldots, n$ equally between the points $2k-1$ and $2k$. Fig. 2.33 illustrates this fact in comparison with Fig. 2.12 for the example $2n = 40$.

For each $k = 0, 1, \ldots, n$, it follows that $\mathbb{P}(L_{2n} \leq 2k) = \mathbb{P}(E_{2n} \leq 2k)$, and for each $k = 1, \ldots, n$, $\mathbb{P}(L_{2n} \leq 2k - 1)$ and $\mathbb{P}(E_{2n} \leq 2k - 1)$ differ by the probability $\mathbb{P}(E_{2n} = 2k - 1)$, which is bounded from above by $u_{2n}$, that converges to 0 as $n \to \infty$. These considerations make it clear that the arcsine law from Theorem 2.5 also applies to $E_{2n}$. We record this insight as follows:

**Theorem 2.20 (Arcsine Law for the first maximizer)** *For each x with $0 \leq x \leq 1$,*

$$\lim_{n \to \infty} \mathbb{P}\left(\frac{E_{2n}}{2n} \leq x\right) = \int_0^x \frac{1}{\pi \sqrt{t(1-t)}} \, dt = \frac{2}{\pi} \arcsin\sqrt{x}.$$

In the context of a repeatedly played fair game, Theorems 2.20 and 2.5 state that, over a long series of games, the proportion of time until the first occurrence of the maximum interim gain for person A, relative to the total duration of the game, follows the same distribution as the proportion of time until the last game tie. It should be clear that the distribution of the last maximizer, obtained by replacing "min" with "max" in (2.78), is represented by a bar chart that reflects the bar chart of the distribution of $E_{2n}$ across the axis $x = n$. For this last maximizer, the same asymptotic arcsine law applies as for $E_{2n}$.

## 2.9 First-Passage Times

In this section, we ask when a random walk starting at 0 first reaches the height $k$. Here, $k$ is a given non-zero integer. The random time at which this happens is denoted by $V_k$. Formally,

$$V_k := \inf\{n \geq 0 : S_n = k\}.$$

The random variable $V_k$ represents the *first-passage time* through the height $k$, and it will henceforth be termed the *$k$-th passage time*.

Like the first return time $W$, the random variable $V_k$ can, in principle, take the value $\infty$. However, we have already seen in connection with the proof of Theorem 2.9 (and will also show in another way) that this case occurs only with probability zero. Since, by symmetry, $V_{-k}$ has the same distribution as $V_k$, we assume $k \geq 1$ in what follows. Note that $V_k$ can only take even or odd values, depending on whether $k$ is even or odd.

For each $k \geq 2$, the difference $D_k := V_k - V_{k-1}$ is called a *first-passage increment*. It measures the additional number of steps required to reach the (next higher) integer $k$ for the first time, starting from the time the random walk first reaches the (previous) height $k - 1$. Furthermore, we set $D_1 := V_1$. These concepts are illustrated in Fig. 2.34.

We first examine the distribution of the first-passage time $V_1$.

**Theorem 2.21 (Distribution of the First-passage Time, cf. [FE1], p. 89)** *For the first-passage time $V_1$, the following hold:*

(a) $\mathbb{P}(V_1 = 2n+1) = \dfrac{\binom{2n}{n}}{2^{2n}} \cdot \dfrac{1}{2(n+1)} = \dfrac{u_{2n}}{2(n+1)}, \quad n \geq 0,$

(b) *The random variable $V_1 + 1$ and the first return time $W$, defined in (2.43), have the same distribution, i.e., $V_1 + 1 \sim W$,*

(c) $\mathbb{P}(V_1 < \infty) = 1,$

(d) $\mathbb{E}(V_1) = \infty.$

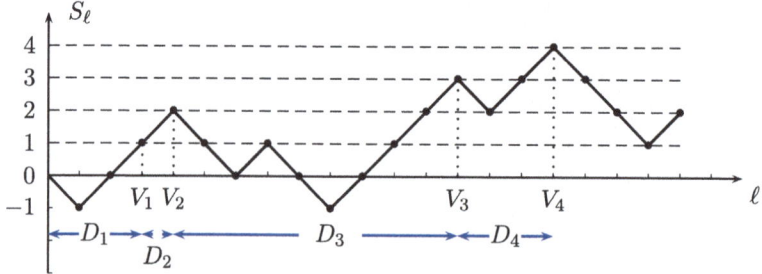

**Fig. 2.34** Illustrating first-passage times and first-passage increments

## 2.9 First-Passage Times

*Proof*

(a) Since $V_1 \geq 2n+1$ if and only if $S_1 \leq 0, \ldots, S_{2n} \leq 0$, Corollary 2.3 implies

$$\mathbb{P}(V_1 \geq 2n+1) = \mathbb{P}(S_1 \leq 0, \ldots, S_{2n} \leq 0) = \frac{\binom{2n}{n}}{2^{2n}} = u_{2n}.$$

Thus, for each $n \geq 0$,

$$\mathbb{P}(V_1 = 2n+1) = \mathbb{P}(V_1 \geq 2n+1) - \mathbb{P}(V_1 \geq 2n+3) = u_{2n} - u_{2(n+1)}$$
$$= \frac{u_{2n}}{2(n+1)}.$$

(b) According to part (a) and Theorem 2.9 (a), $\mathbb{P}(V_1 + 1 = 2n) = \mathbb{P}(V_1 = 2(n-1)+1) = u_{2(n-1)}/(2n) = \mathbb{P}(W = 2n)$, $n \in \mathbb{N}$. The assertions (c) and (d) follow directly from part (b) and parts (b) and (c) of Theorem 2.9.

∎

Whereas the equality in distribution in (b) is immediately obvious (the random walk must take a step up or down for the first return to 0 and then descend or ascend by one step, respectively), the result in (d) contradicts intuition. Although the random walk takes a step up or down at the beginning, the expected time to reach the height 1 is infinite. In the context of a random walk interpreted as a coin toss game, this means that, on average, one waits infinitely long until player A takes the lead for the first time.

As the following result shows, the first-passage times $V_1, V_2, \ldots$ of a random walk are closely linked to the maxima $M_1, M_2, \ldots$, examined in Sect. 2.7.

**Lemma 2.22 (First-passage Times and Maxima)** *If $k$ and $n$ are positive integers with $k \leq n$, then*

$$\{V_k \leq n\} = \{M_n \geq k\}.$$

**SAQ 17** Why does this statement hold?

Together with Theorem 2.13, Lemma 2.22 provides some important conclusions:

**Corollary 2.23** *For a symmetric random walk, the following hold:*

(a) $\mathbb{P}(V_k < \infty) = 1$ *for every $k \geq 1$,*

(b) $\mathbb{P}\left(\bigcap_{k=1}^{\infty}\{V_k < \infty\}\right) = 1,$

(c) $\mathbb{P}\left(\limsup_{n\to\infty} S_n = \infty\right) = 1.$

*Proof*

(a) According to Lemma 2.22, for $n \geq k$,

$$\mathbb{P}(V_k > n) = \mathbb{P}(M_n < k) = \mathbb{P}(M_n \leq k-1) = \mathbb{P}\left(\frac{M_n}{\sqrt{n}} \leq \frac{k-1}{\sqrt{n}}\right).$$

Using Theorem 2.13 and $\{V_k = \infty\} = \bigcap_{n=1}^{\infty}\{V_k > n\}$, it follows that

$$\mathbb{P}(V_k = \infty) = \lim_{n\to\infty} \mathbb{P}(V_k > n) = 2\Phi(0) - 1 = 0,$$

and thus $\mathbb{P}(V_k < \infty) = 1$.

(b) Part (b) follows from (a) and

$$\mathbb{P}\left(\bigcap_{k=1}^{\infty}\{V_k < \infty\}\right) = \lim_{k\to\infty} \mathbb{P}(V_k < \infty),$$

since $\mathbb{P}$, as a probability measure, is continuous from above (see (7.1)). According to part (b), in the probability space $(\Omega, \mathcal{A}, \mathbb{P})$ for infinitely long random walks introduced in Sect. 7.2, there exists a set $\Omega_0 \in \mathcal{A}$ with $\mathbb{P}(\Omega_0) = 1$, such that

$$V_k(\omega) < \infty \quad \text{for every } \omega \in \Omega_0 \text{ and every } k \geq 1.$$

For every $\omega \in \Omega_0$ and every integer $k$, there is thus an integer $n \geq k$ with $S_n(\omega) \geq k$, which means $\limsup_{n\to\infty} S_n(\omega) = \infty$.

■

As a consequence, a symmetric random walk eventually reaches any given height $k$ with probability one (statement (a)), see also the discussion following the proof of Theorem 2.9). Furthermore, with probability one, it will eventually reach every positive integer (statement (b)). Since $V_k \geq V_1$ and $\mathbb{E}(V_1) = \infty$, it follows that $\mathbb{E}(V_k) = \infty$ for every $k \geq 1$. By symmetry, all previous results also apply to first-passage times $V_k$ with $k < 0$. In particular,

$$\mathbb{P}\left(\liminf_{n\to\infty} S_n = -\infty\right) = 1.$$

## 2.9 First-Passage Times

This, together with Corollary 2.23 (c), shows that the symmetric random walk *oscillates*: with probability one, it does not tend to infinity in either direction (positive or negative), but instead continues to return to any given integer infinitely often over time.

From Lemma 2.22, we obtain the distribution of $V_k$ for general $k$ (see, e.g., [FE1], p. 89).

**Theorem 2.24 (Distribution of the $k$-th Passage Time)** *If $k$ and $n$ are positive integers having the same parity, then*

$$\mathbb{P}(V_k = n) = \frac{k}{n2^n} \cdot \binom{n}{\frac{n+k}{2}}, \qquad n \geq k.$$

*Proof* Let $k$ and $n$ be as above. According to Lemma 2.22 and (2.63),

$$\mathbb{P}(V_k = n) = \mathbb{P}(V_k \leq n) - \mathbb{P}(V_k \leq n-1)$$
$$= \mathbb{P}(M_n \geq k) - \mathbb{P}(M_{n-1} \geq k)$$
$$= 2\mathbb{P}(S_n \geq k) - \mathbb{P}(S_n = k) - (2\mathbb{P}(S_{n-1} \geq k) - \mathbb{P}(S_{n-1} = k))$$
$$= 2(\mathbb{P}(S_n \geq k) - \mathbb{P}(S_{n-1} \geq k)) - \mathbb{P}(S_n = k) + \mathbb{P}(S_{n-1} = k).$$

Now,

$$\mathbb{P}(S_n \geq k) = \mathbb{P}(S_n \geq k, X_n = 1) + \mathbb{P}(S_n \geq k, X_n = -1)$$
$$= \mathbb{P}(X_n = 1)\mathbb{P}(S_n \geq k | X_n = 1) + \mathbb{P}(X_n = -1)\mathbb{P}(S_n \geq k | X_n = -1)$$
$$= \frac{1}{2}\mathbb{P}(S_{n-1} \geq k - 1) + \frac{1}{2}\mathbb{P}(S_{n-1} \geq k + 1).$$

Since $n-1$ and $k$ have different parity, it follows that $\mathbb{P}(S_{n-1} = k) = 0$. Furthermore,

$$\mathbb{P}(S_n = k) = \mathbb{P}(S_n = k | X_n = 1) \cdot \frac{1}{2} + \mathbb{P}(S_n = k | X_n = -1) \cdot \frac{1}{2}$$
$$= \frac{1}{2} \cdot \left(\mathbb{P}(S_{n-1} = k - 1) + \mathbb{P}(S_{n-1} = k + 1)\right),$$

which results in

$$\mathbb{P}(V_k = n) = \frac{1}{2}\left(\mathbb{P}(S_{n-1} = k - 1) - \mathbb{P}(S_{n-1} = k + 1)\right).$$

The assertion now follows from (2.7) by substituting the binomial probabilities and combining. ∎

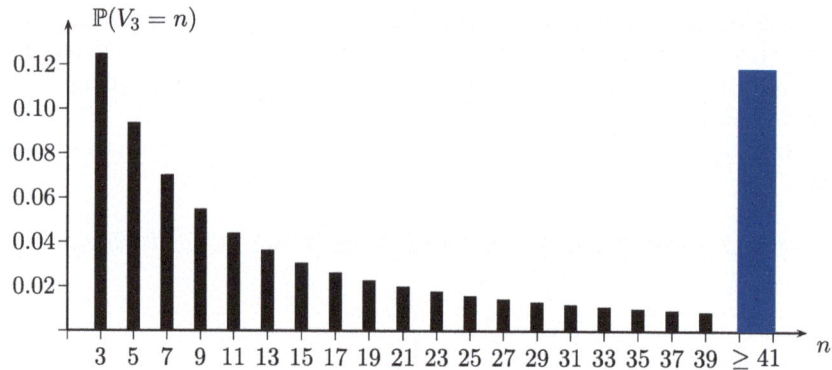

**Fig. 2.35** Bar chart of the distribution of $V_3$ (note the high probability $\mathbb{P}(V_3 \geq 41)$)

**SAQ 18** Can you do this calculation?

The probabilities $\mathbb{P}(V_k = n)$ satisfy the recursion formula

$$\mathbb{P}(V_k = n+2) = \frac{n(n+1)}{(n+2+k)(n+2-k)} \mathbb{P}(V_k = n), \qquad n \geq k, \quad (2.81)$$

and the initial condition $\mathbb{P}(V_k = k) = 2^{-k}$ (Exercise 2.18). Equation (2.81) shows that $\mathbb{P}(V_k = n)$ converges to zero as $n \to \infty$, albeit at a very slow rate. This effect is also evident in the bar chart of the distribution of $V_3$ sketched in Fig. 2.35. The probability that one must wait at least 41 time steps for a symmetric random walk to reach the height 3 for the first time is 0.356, which is represented in Fig. 2.35 by a colored bar with triple thickness and a height of $\frac{0.356}{3}$. The probability that even 1000 time steps are not enough to reach the height 3 is still a considerable 7.55%!

We now consider the first-passage increments $D_k = V_k - V_{k-1}$ for $k \geq 1$, i.e., the random time spans that the random walk needs to ascend from height $k - 1$ to height $k$. Due to the property of being memoryless and the probability of taking another step up or down being independent of the height of the random walk, it can be conjectured that the first-passage increments $D_1, D_2, D_3, \ldots$ are independent random variables, all having the same distribution as $V_1$. The following result shows that this assumption is correct.

**Theorem 2.25 (First-passage Increments Are i.i.d. Random Variables)** *The first-passage increments $D_k = V_k - V_{k-1}$, $k \geq 1$, are i.i.d. random variables, all having the same distribution as $V_1$.*

***Proof*** We use the fact that the heights $S_k = X_1 + \ldots + X_k$ for $k \geq 1$ are partial sums of i.i.d. random variables $X_1, X_2, \ldots$, and first prove the independence of $D_1$

## 2.9 First-Passage Times

and $D_2$ by showing

$$\mathbb{P}(D_2 = r | D_1 = n) = \mathbb{P}(D_2 = r) \qquad (2.82)$$

for any odd integers $r$ and $n$. Under the condition $D_1 = n$, the event $D_2 = r$ occurs if and only if for every $j$ with $n + 1 \leq j \leq n + r - 1$ the inequality $S_j \leq 1$ holds, and $S_{n+r} = 2$. Because $S_n = 1$ (which follows from $D_1 = n$) and $S_j = S_n + (S_j - S_n)$, this is equivalent to $1 + \sum_{i=n+1}^{j} X_i \leq 1$ for $j = n+1, \ldots, n+r-1$ and $1 + \sum_{i=n+1}^{n+r} X_i = 2$, and thus to

$$\sum_{i=n+1}^{j} X_i \leq 0, \quad j = n+1, \ldots, n+r-1, \quad \sum_{i=n+1}^{n+r} X_i = 1. \qquad (2.83)$$

Therefore, the left-hand side of (2.82) is the conditional probability of the event in (2.83) under the condition $D_1 = n$. Since this condition is an event formed by the random variables $X_1, \ldots, X_n$, and (2.83) involves $X_{n+1}, \ldots, X_{n+r}$, we can omit the condition $D_1 = n$ because of the independence of all $X_j$ and obtain

$$\mathbb{P}(D_2 = r | D_1 = n) = \mathbb{P}\left(\sum_{i=n+1}^{j} X_i \leq 0, \ j = n+1, \ldots, n+r-1; \ \sum_{i=n+1}^{n+r} X_i = 1\right).$$

Due to the independence and identical distribution of all $X_j$, an index shift $i \to i - n$ can be performed on the right-hand side without changing the probability. After this shift, the event becomes $\{S_1 \leq 0, \ldots, S_{r-1} \leq 0, S_r = 1\}$, which is $\{D_1 = r\}$, and we obtain

$$\mathbb{P}(D_2 = r | D_1 = n) = \mathbb{P}(D_2 = r) = \mathbb{P}(D_1 = r).$$

The random variables $D_1$ and $D_2$ are thus i.i.d. as $V_1$. The rest of the proof does not require any new idea. The simplest approach is to proceed inductively by showing the equality

$$\mathbb{P}(D_{k+1} = n_{k+1} | D_1 = n_1, \ldots, D_k = n_k) = \mathbb{P}(D_{k+1} = n_{k+1}) = \mathbb{P}(D_1 = n_{k+1})$$

for any odd integers $n_1, \ldots, n_{k+1}$. ∎

According to Theorem 2.25, the $k$-th passage time

$$V_k = D_1 + D_2 + \ldots + D_k \qquad (2.84)$$

is a sum of i.i.d. random variables. For such sums, a law of large numbers would hold in the sense that, with probability one, $\frac{1}{k} V_k$ converges as $k \to \infty$ if the expectation of the summands existed. Since this condition does not hold in our case, this law does not apply to $\frac{1}{k}V_k$. The following result states that $V_k$ should not be divided by $k$, but by the much larger $k^2$, and then in the limit $k \to \infty$, a limiting distribution is obtained (see, e.g., [FE1], p. 90).

**Theorem 2.26 (Limit Distribution of the $k$-th Passage Time)** *For the $k$-th passage time,*

$$\lim_{k \to \infty} \mathbb{P}\left(\frac{V_k}{k^2} \leq x\right) = 2\left(1 - \Phi\left(\frac{1}{\sqrt{x}}\right)\right) = 2\int_{x^{-1/2}}^{\infty} \varphi(t)\,dt, \qquad x > 0.$$

*Proof* Fix any $x > 0$, and set $n_k := \lfloor x \cdot k^2 \rfloor$. Since $V_k$ is integer-valued, Lemma 2.22 yields

$$\begin{aligned} \mathbb{P}\left(\frac{1}{k^2} V_k \leq x\right) &= \mathbb{P}(V_k \leq xk^2) = \mathbb{P}(V_k \leq n_k) \\ &= \mathbb{P}(M_{n_k} \geq k) \\ &= \mathbb{P}\left(\frac{M_{n_k}}{\sqrt{n_k}} \geq \frac{k}{\sqrt{n_k}}\right). \end{aligned}$$

Given that $\lim_{k \to \infty} k/\sqrt{n_k} = 1/\sqrt{x}$, Theorem 2.13 implies

$$\lim_{k \to \infty} \mathbb{P}\left(\frac{M_{n_k}}{\sqrt{n_k}} \geq \frac{k}{\sqrt{n_k}}\right) = 1 - \left(2\Phi\left(\frac{1}{\sqrt{x}}\right) - 1\right),$$

which was to be shown. ∎

**Remark 2.27** If $Z$ is a standard normally distributed random variable, then for every $x > 0$,

$$\mathbb{P}\left(\frac{1}{Z^2} \leq x\right) = \mathbb{P}\left(Z^2 \geq \frac{1}{x}\right) = \mathbb{P}\left(|Z| \geq \frac{1}{\sqrt{x}}\right)$$
$$= 2\left(1 - \Phi\left(\frac{1}{\sqrt{x}}\right)\right).$$

Since $\mathbb{P}(V_k/k^2 \leq x)$, as a function of $x$, is the distribution function of $V_k/k^2$, Theorem 2.26 states that, as $k \to \infty$, the sequence of distribution functions of $V_k/k^2$ converges pointwise to the distribution function $L : \mathbb{R} \to \mathbb{R}$, defined by

$$L(x) := 2\left(1 - \Phi\left(\frac{1}{\sqrt{x}}\right)\right) \quad \text{for } x > 0, \tag{2.85}$$

and $L(x) := 0$ for $x \leq 0$. The distribution associated with $L$ is called the (standard) *Lévy distribution*. This distribution belongs to the class of *stable distributions*, which

## 2.9 First-Passage Times

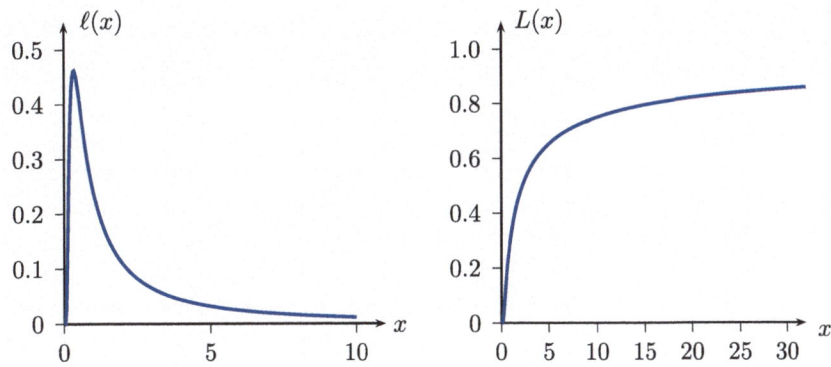

**Fig. 2.36** Density (left) and distribution function (right) of the Lévy distribution

includes the normal and Cauchy distributions (see, e.g., [NO], pp. 2–3). As was done in (2.31) and (2.41), we can therefore state the result of Theorem 2.26 in the compact form

$$\frac{V_k}{k^2} \xrightarrow{\mathcal{D}} Y, \qquad Y \sim L, \quad \text{as } k \to \infty.$$

Here, $Y \sim 1/Z^2$ with $Z \sim N(0, 1)$. The distribution function of the Lévy distribution is sketched on the right in Fig. 2.36. The left-hand illustration shows the density of the Lévy distribution, denoted by

$$\ell(x) := \frac{1}{(2\pi x^3)^{1/2}} \exp\left(-\frac{1}{2x}\right), \quad x > 0, \qquad \ell(x) := 0, \ x \leq 0.$$

Table 2.3 provides some values of the distribution function $L$. Apparently, the values $L(x)$ converge to 1 remarkably slowly as $x \to \infty$. The median of $L$ is 2.2, the 0.95-quantile is 254, and the 0.99-quantile is an impressive 6366! The consequences for the $k$-th passage time $V_k$ are striking, as according to Theorem 2.26 for large $k$,

$$\mathbb{P}(V_k \leq 2.2 \cdot k^2) \approx 0.5, \quad \mathbb{P}(V_k \leq 254 \cdot k^2) \approx 0.95, \quad \mathbb{P}(V_k \leq 6{,}366 \cdot k^2) \approx 0.99.$$

Thus, every second random walk requires more than 22,000 steps to reach the height of 100 for the first time, and for every twentieth random walk, even two and a half million steps are not sufficient!

For each positive integer $k$, let

$$W_k := \inf\left\{2j : j \geq 1, \ \sum_{\ell=1}^{j} \mathbf{1}\{S_{2\ell} = 0\} = k\right\}$$

**Table 2.3** Values of the distribution function $L(x) := 2(1 - \Phi(1/\sqrt{x}))$ of the Lévy distribution

| $x$ | $L(x)$ | $x$ | $L(x)$ | $x$ | $L(x)$ |
|---|---|---|---|---|---|
| 0.1 | 0.0016 | 2.5 | 0.527 | 20 | 0.823 |
| 0.2 | 0.025 | 3 | 0.564 | 30 | 0.855 |
| 0.3 | 0.068 | 4 | 0.617 | 40 | 0.874 |
| 0.4 | 0.114 | 5 | 0.655 | 50 | 0.888 |
| 0.5 | 0.157 | 6 | 0.683 | 75 | 0.908 |
| 0.7 | 0.232 | 7 | 0.706 | 100 | 0.920 |
| 1.0 | 0.317 | 8 | 0.724 | 200 | 0.944 |
| 1.5 | 0.414 | 9 | 0.734 | 500 | 0.964 |
| 2.0 | 0.480 | 10 | 0.752 | 1000 | 0.975 |
| 2.2 | 0.500 | 15 | 0.796 | 5000 | 0.989 |

denote the *kth return time*, which is the number of steps the random walk needs to return to the starting point 0 for the $k$-th time. With the insights gained on first-passage times, we obtain the following results for the random variable $W_k$ (cf. [FE1], pp. 90–91):

**Theorem 2.28 ((Limit-)Distribution of the $k$-th Return Time)** *For the $k$-th return time $W_k$ of a symmetric random walk, the following holds:*

(a) $W_k \sim V_k + k$,

(b) $\mathbb{P}(W_k = 2n) = \dfrac{k}{2n-k} \dfrac{\binom{2n-k}{n}}{2^{2n-k}}, \quad n \geq k$,

(c) $\lim\limits_{k \to \infty} \mathbb{P}\left(\dfrac{W_k}{k^2} \leq x\right) = L(x), \quad x \geq 0.$

*Proof* Referring to Theorem 2.24, statements (a) and (b) are obviously equivalent. Statement (a) follows conceptually by noting that $W_k$ follows the same distribution as the sum $Z_1 + \ldots + Z_k$, where the $Z_j$ are independent and identically distributed like the first return time $W$ (cf. Fig. 2.23 and the preceding discussion). According to Theorem 2.21 (b), each $Z_j$ is distributed like a first-passage time, increased by 1. From this, the stated equality in distribution follows.

An alternative proof demonstrates (b) (and thus part (a)) using the reflection principle. To this end, consider a *non-positive* path that has its $k$-th visit to zero at time $2n$. If the downward step at the beginning as well as the downward step after the $j$-th visit to 0 for each $j = 1, \ldots, k-1$, is omitted, we transform this path into a path whose $k$-th passage time $V_k$ is equal to $2n-k$. This mapping is injective and can be inverted by inserting an additional downward step at the beginning and for each $j = 1, \ldots, k - 1$ directly after hitting the height $j$. Since each given non-positive path with its $k$-th visit to zero equal to $2n$ leads to $2^k$ different paths with the same visits to zero, part (b) follows. Part (c) is a consequence of (a) and Theorem 2.26 together with (7.8), noting that

$$\mathbb{P}\left(\frac{W_k}{k^2} \leq x\right) = \mathbb{P}\left(\frac{V_k + k}{k^2} \leq x\right) = \mathbb{P}\left(\frac{V_k}{k^2} \leq x_k\right)$$

with $x_k = x - \frac{1}{k}$ and $x_k \to x$. ∎

## 2.10 Collisions of Random Walks

Two symmetric random walks start simultaneously at different heights and proceed independently of each other. When do they meet (or "collide") for the first time? In other words, when do their corresponding paths first intersect at integer coordinates? How long does it take on average until this first meeting occurs? Figure 2.37 illustrates this situation with two random walks starting at heights 0 and 4, respectively, meeting for the first time after 16 steps.

Due to parity reasons, a meeting is only possible if the initial heights of the random walks are either both even or both odd. Since the questions posed depend only on the height differences of the random walks over time, we can further assume without loss of generality that one of the random walks starts at zero and the other at height $2k$. Here, $k$ is any fixed positive integer.

We model this situation using independent random variables $X_1, X_1^*, X_2, X_2^*, \ldots$ with $\mathbb{P}(X_j = 1) = \mathbb{P}(X_j^* = 1) = \frac{1}{2}$ and $\mathbb{P}(X_j = -1) = \mathbb{P}(X_j^* = -1) = \frac{1}{2}$ for $j = 1, 2, \ldots$. We set $S_0 := 0$, $S_n := X_1 + \ldots + X_n$, $S_0^* := 2k$, and $S_n^* := 2k + X_1^* + \ldots + X_n^*$ for $n \geq 1$. The random variables $S_n$ and $S_n^*$ represent the height of the random walk starting at zero and the random walk starting at height $2k$ after $n$ steps, respectively. The random time at which both random walks meet for the first time is

$$T_k := \inf\{n \geq 1 : S_n^* = S_n\}.$$

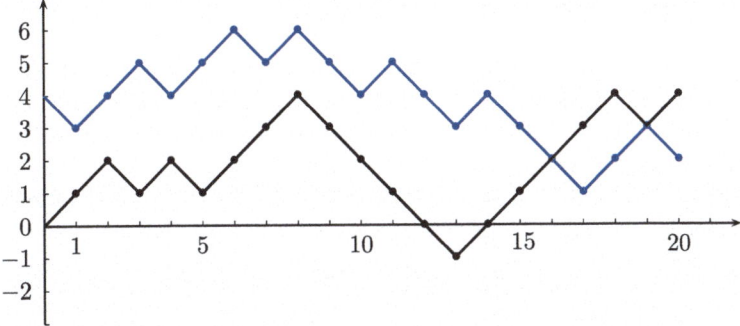

**Fig. 2.37** Two random walks that meet for the first time after 16 steps

**Table 2.4** Realizations of $X_j$, $X_j^*$, and $\delta_j$ for the random walks in Fig. 2.37

| $X_j$ | 1 | 1 | -1 | 1 | -1 | 1 | 1 | 1 | -1 | -1 | -1 | -1 | -1 | 1 | 1 | 1 | 1 | 1 | -1 | 1 |
|---|---|---|---|---|---|---|---|---|---|---|---|---|---|---|---|---|---|---|---|---|
| $X_j^*$ | -1 | 1 | 1 | -1 | 1 | 1 | -1 | 1 | -1 | -1 | 1 | -1 | -1 | 1 | -1 | -1 | -1 | 1 | 1 | -1 |
| $\delta_j$ | 2 | 0 | -2 | 2 | -2 | 0 | 2 | 0 | 0 | 0 | -2 | 0 | 0 | 0 | 2 | 2 | 2 | 0 | -2 | 2 |

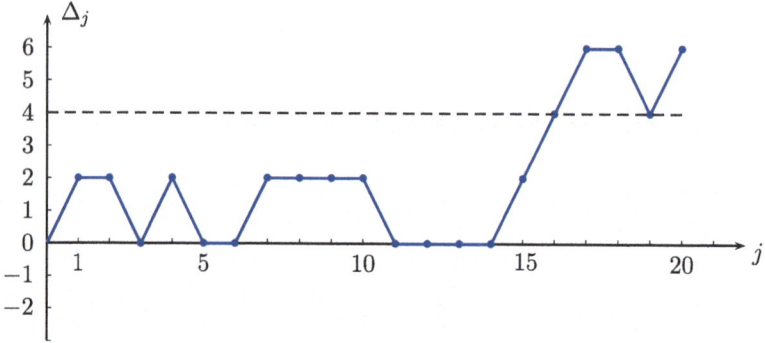

**Fig. 2.38** The random walk $(j, \Delta_j)$, $0 \leq j \leq 20$, for the data in Table 2.4

The usual convention $\inf \emptyset = \infty$ accounts for the fact that such a meeting may never occur (e.g., if each $X_j$ and each $X_j^*$ takes the value 1). The random variable $T_k$ is subsequently be referred to as the *2k-meeting time*, highlighting the initial distance of $2k$ between the two starting points.

According to the definitions of $S_n$ and $S_n^*$, the equation $S_n = S_n^*$ is equivalent to $\sum_{j=1}^{n}(X_j - X_j^*) = 2k$. Therefore, with the abbreviations

$$\delta_j := X_j - X_j^*, \quad j \geq 1, \qquad \Delta_n := \delta_1 + \ldots + \delta_n, \quad n \geq 1,$$

the 2k-meeting time may be written in the form

$$T_k = \inf\{n \geq 1 : \Delta_n = 2k\}. \tag{2.86}$$

Table 2.4 shows the realizations of $X_j$, $X_j^*$, and $\delta_j$ for $j = 1, \ldots, 20$ of the random walks illustrated in Fig. 2.37.

If we set $\Delta_0 := 0$ and connect the points $(j, \Delta_j)$ for $j$ from 0 to 20, we create a polygonal chain, as shown in Fig. 2.38. In this figure, the time of the first meeting of the random walks from Fig. 2.37 is just the number of steps at which the height 4 is reached for the first time.

Generally, the points $(j, \Delta_j)$, $j \geq 0$, model a random walk that starts at zero and, without memory, either takes two steps up, two steps down, or remains at the current

## 2.10 Collisions of Random Walks

height at each step. This is because the random variables $\delta_1, \delta_2, \ldots$ are independent and take the values 2, 0, and $-2$. Since

$$\mathbb{P}(\delta_j = 2) = \mathbb{P}(X_j = 1, X_j^* = -1) = \frac{1}{4},$$

$$\mathbb{P}(\delta_j = -2) = \mathbb{P}(X_j = -1, X_j^* = 1) = \frac{1}{4},$$

and $\mathbb{P}(\delta_j = 0) = \frac{1}{2}$, one can imagine generating this new random walk as follows: At each step, a fair coin is tossed. If the coin shows tails, the random walk is "lazy" and stays at the current height. In the case of heads, the coin is tossed again. If it shows tails, the random walk takes two steps up; otherwise, it takes two steps down.

In the same way as with the first-passage times and the first-passage increments of the symmetric random walk examined in the previous section, one can essentially focus on the case $k = 1$ when studying the random variable $T_k$ in (2.86) because the following result holds:

**Theorem 2.29** *Let $k \geq 2$ be an integer, $E_1 := T_1$, and $E_j := T_j - T_{j-1}$ for $j = 2, \ldots, k$. Then the random variables $E_1, \ldots, E_k$ are independent and identically distributed, with $E_j \sim T_1$ for $j = 1, \ldots, k$.*

***Proof*** The proof follows verbatim the proof of Theorem 2.25. As in that proof, one employs the fact that $E_1, E_2, \ldots, E_k$ are first-passage increments of a random walk, which are based on the partial sums $\Delta_n$ of the independent, identically distributed random variables $X_j - X_j^*$ with expectation 0 (see Fig. 2.39). This random walk first reaches the height 2 (black path segment in Fig. 2.39) at the random time $T_1 = E_1$. Since the random walk evolves without memory, the subsequent path segment (drawn in blue in Fig. 2.39) until reaching the height 4 is completely unaffected by the stochastic behavior of the random walk up to $T_1$, as if the blue path segment

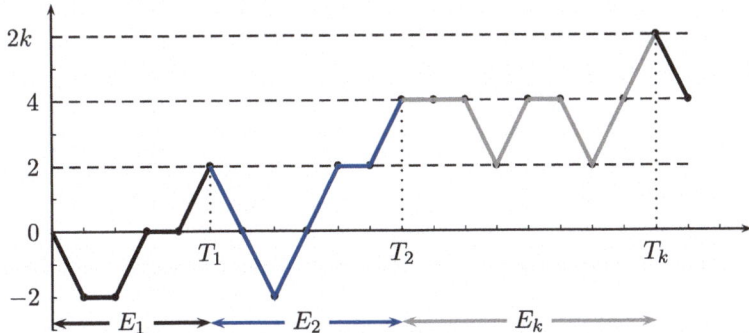

**Fig. 2.39** First-passage times and first-passage increments for the random walk $(j, \Delta_j)$, $j \geq 0$

started at zero and had to reach height 2. The same holds for the subsequent gray path segment until reaching the height $2k$, where in Fig. 2.39, $k = 3$. ∎

To determine the distribution of $T_1$, we derive a closed expression for the generating function $g$ of $T_1$, which is defined by

$$g(t) := \sum_{n=1}^{\infty} \mathbb{P}(T_1 = n) \cdot t^n, \qquad |t| \leq 1$$

(see Sect. 7.9 for the most important properties of generating functions). The plan is to expand the function $g$ into a power series around 0. Since the coefficients of this power series are unique, one can then read off the probabilities $\mathbb{P}(T_1 = n)$.

First we note:

$$\mathbb{P}(T_1 = 1) = \frac{1}{4}, \qquad \mathbb{P}(T_1 = 2) = \frac{1}{8}. \tag{2.87}$$

**SAQ 19** Why do these equations hold?

To obtain the probabilities $\mathbb{P}(T_1 = n)$ for $n \geq 3$, we derive a recursion formula by considering the first step of the random walk. For this step, the mutually exclusive cases $\delta_1 = 2$, $\delta_1 = 0$, and $\delta_1 = -2$ are possible, which occur with probabilities $\frac{1}{4}$, $\frac{1}{2}$, and $\frac{1}{4}$, respectively. According to the law of total probability (7.4), for each $n \geq 3$:

$$\mathbb{P}(T_1 = n) = \frac{1}{4}\mathbb{P}(T_1 = n|\delta_1 = 2) + \frac{1}{2}\mathbb{P}(T_1 = n|\delta_1 = 0) + \frac{1}{4}\mathbb{P}(T_1 = n|\delta_1 = -2).$$

Here, the first summand on the right-hand side vanishes because $\mathbb{P}(T_1 = 1|\delta_1 = 2) = 1$. If $\delta_1 = 0$, the random walk remains on the $x$-axis for one time step, which means that starting from the point $(1, 0)$, it has one less time step available to reach the height 2. Thus,

$$\mathbb{P}(T_1 = n|\delta_1 = 0) = \mathbb{P}(T_1 = n - 1). \tag{2.88}$$

In the remaining case $\delta_1 = -2$, after one time step, we have a random walk starting from the point $(1, -2)$ that needs to reach the height 2. This is equivalent to starting from the origin after one time step and reaching the height 4. According to Theorem 2.29, the time to reach the height 4 has the same distribution as $E_1 + E_2$. Here, $E_1$ and $E_2$ are independent random variables with the same distribution as $T_1$. These considerations lead to the equation

$$\mathbb{P}(T_1 = n|\delta_1 = -2) = \mathbb{P}(E_1 + E_2 = n - 1). \tag{2.89}$$

## 2.10 Collisions of Random Walks

Substituting the right-hand sides of (2.88) and (2.89) into the recursion formula and sum over $n$, we get for each $t$ with $|t| \leq 1$:

$$\sum_{n=3}^{\infty} \mathbb{P}(T_1 = n)\, t^n = \frac{1}{2} \sum_{n=3}^{\infty} \mathbb{P}(T_1 = n-1)\, t^n + \frac{1}{4} \sum_{n=3}^{\infty} \mathbb{P}(E_1 + E_2 = n-1)\, t^n.$$

By the definition of $g(t)$ and display (2.87), the left-hand side is equal to $g(t) - \frac{t}{4} - \frac{t^2}{8}$. The first sum on the right-hand side, after factoring out $t$ and shifting the index, as well as using (2.87), is equal to $t\left(g(t) - \frac{t}{4}\right)$. Since $\mathbb{P}(E_1 + E_2 \geq 2) = 1$, the second sum is equal to $t\, h(t)$, where $h$ is the generating function of $E_1 + E_2$. Due to the independence of $E_1$ and $E_2$ and the equalities in distribution $E_1 \sim T_1$ and $E_2 \sim T_1$, it follows that $h(t) = g(t)^2$ (see Sect. 7.9). Thus,

$$g(t) - \frac{t}{4} - \frac{t^2}{8} = \frac{t}{2}\left(g(t) - \frac{t}{4}\right) + \frac{t}{4} g(t)^2,$$

which yields, for every $t \in [-1, 1]$ with $t \neq 0$, the quadratic equation for $g(t)$:

$$g(t)^2 + 2\left(1 - \frac{2}{t}\right) g(t) + 1 = 0.$$

Considering that $g(t)$ converges to $\mathbb{P}(T_1 = 0) = 0$ as $t \to 0$, the solution is

$$g(t) = \frac{2\left(1 - \sqrt{1-t}\right) - t}{t}, \qquad |t| \leq 1,\ t \neq 0,$$

with the continuous extension $g(0) := \lim_{t \to 0} g(t) = 0$. Using the binomial series

$$\sqrt{1-t} = (1-t)^{1/2} = 1 - 2 \sum_{n=1}^{\infty} \frac{\binom{2n-2}{n-1}}{n\, 2^{2n}}\, t^n, \qquad |t| < 1, \qquad (2.90)$$

(see Sect. 7.11), a direct calculation (Exercise 2.19) yields

$$g(t) = \sum_{n=1}^{\infty} \frac{\binom{2n}{n}}{(n+1)\, 2^{2n}}\, t^n, \qquad |t| < 1.$$

Thus, we obtain the following result (cf. [VH], Theorem 3.14 for the case $k = 1$).

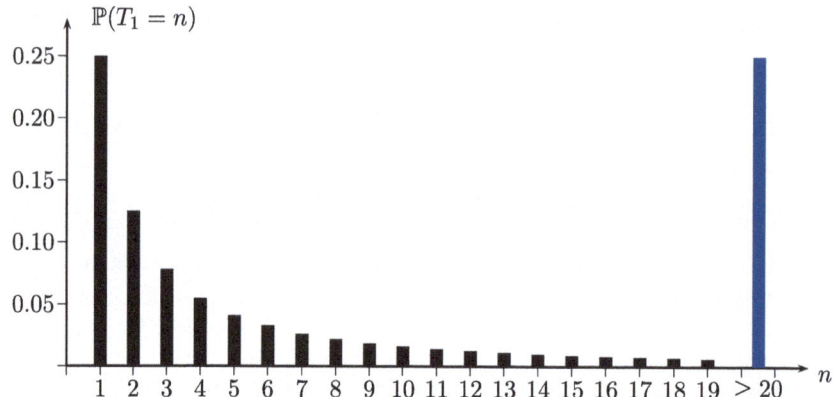

**Fig. 2.40** Bar chart of the distribution of the 2-meeting time (note the large probability $\mathbb{P}(T_1 \geq 20) \approx 0.251$)

**Theorem 2.30 (Distribution of the 2-Meeting Time)** *Let $T_1$ be the time that two independent simple symmetric random walks, starting at heights 0 and 2, first meet. Then, for each $n \geq 1$,*

$$\mathbb{P}(T_1 = n) = \frac{\binom{2n}{n}}{(n+1)\, 2^{2n}}.$$

Figure 2.40 shows a bar chart of the distribution of $T_1$. The probabilities $\mathbb{P}(T_1 = n)$ satisfy the recursion formula

$$\mathbb{P}(T_1 = n+1) = \frac{2n+1}{2n+4} \mathbb{P}(T_1 = n), \qquad n \geq 1,$$

and thus decrease relatively slowly with increasing $n$. For example, $\mathbb{P}(T_1 \geq 20) \approx 0.251$ and $\mathbb{P}(T_1 \geq 127) \approx 0.100$. If you repeatedly start a symmetric random walk at zero and simultaneously a second one at height 2, they will meet, in the long run, in every fourth case not before the twentieth step, and in 10% of all cases, at least 127 steps must occur before the first meeting.

If we compare the result of Theorem 2.30 with the distribution of the second passage time $V_2$ of the symmetric random walk given in Theorem 2.24, it follows that

$$\mathbb{P}(T_1 = n) = \mathbb{P}(V_2 = 2n) = \mathbb{P}\left(\frac{V_2}{2} = n\right), \qquad n \geq 1.$$

Thus, the initially surprising equality in distribution,

## 2.10 Collisions of Random Walks

$$T_1 \sim \frac{V_2}{2}, \qquad (2.91)$$

holds. The random time until the first meeting of two independently evolving symmetric random walks, starting at the origin and at height 2, has the same distribution as half the time it takes for a symmetric random walk starting at the origin to reach height 2 for the first time.

This connection can be conceptually understood as follows: The random variable $T_1$ models the time needed for a random walk starting at the origin, with step heights of 2, 0, or $-2$ (with corresponding probabilities $\frac{1}{4}$, $\frac{1}{2}$ and $\frac{1}{4}$, respectively), to reach height 2 for the first time. If we interpret an upward or downward step of this random walk as *two consecutive upward or downward steps of the symmetric random walk*, and a pause at a reached height for a time step as *two consecutive steps of the symmetric random walk* (one upward and one downward), it becomes clear that certain paths of the symmetric random walk are aggregated and their time duration is halved. For instance, the two paths $(1, -1, 1, 1)$ and $(-1, 1, 1, 1)$ of the symmetric random walk, which make up the event $\{V_2 = 4\}$, are combined into the path $(0, 2)$ of the random walk based on the random variables $\delta_1, \delta_2, \ldots$.

According to (2.84), $V_{2k}$ has the same distribution as the sum $\sum_{j=1}^{2k} D_j$, where $D_1, \ldots, D_{2k}$ are i.i.d. random variables, each distributed like the first-passage time $V_1$ of the symmetric random walk. If we group each pair of these $D_j$'s together by setting $V_{2,j} := D_{2j-1} + D_{2j}$ for $j = 1, \ldots, k$, then $V_{2,1}, \ldots, V_{2,k}$ are i.i.d. random variables, all distributed like $V_2$. Since, according to (2.91), $V_{2,j}/2$ has the same distribution as the random variable $E_j$ appearing in Theorem 2.29, and $T_k = E_1 + \ldots + E_k$, the equality in distribution (2.91) extends to general $k$, i.e.,

$$T_k \sim \frac{V_{2k}}{2}, \qquad k \geq 1. \qquad (2.92)$$

Together with the results for first-passage times in the symmetric random walk, this implies $\mathbb{P}(T_k < \infty) = 1$ and $\mathbb{E}(T_k) = \infty$. Although a meeting of the random walks occurs in finite time with probability one, one waits infinitely long, on average, for such a meeting.

From (2.92) and Theorem 2.24, we directly obtain the distribution of $T_k$ for general $k$ (cf. [VH], Theorem 3.14).

**Theorem 2.31 (Distribution of the $2k$-Meeting Time)** *Let $T_k$ be the time that two independent simple symmetric random walks, starting at heights 0 and $2k$, meet for the first time. Then, for any positive integers $n$ and $k$ with $n \geq k$,*

$$\mathbb{P}(T_k = n) = \binom{2n}{n+k} \frac{k}{n 2^{2n}}.$$

Given the previous considerations, it is not surprising that, in general, one must wait quite a long time for two symmetric random walks starting at different heights

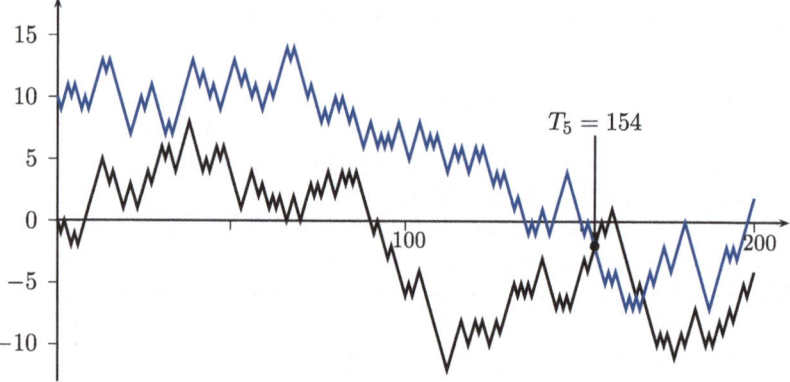

**Fig. 2.41** Two random walks starting at zero and at height 10, meeting for the first time after 154 steps

to meet. Figure 2.41 shows the result of a simulation of two random walks, each of length 200, with one starting at zero and the other at height 10. In this case, the 10-meeting time $T_5$ takes the value 154.

Finally, we aim to investigate the stochastic behavior of $T_k$ for large $k$ to estimate the time of the first meeting of two random walks starting far apart. According to Theorem 2.26,

$$\lim_{k \to \infty} \mathbb{P}\left(\frac{V_k}{k^2} \leq x\right) = L(x), \qquad x > 0,$$

where $L$ is the distribution function of the Lévy distribution defined in (2.85). Replacing $k$ with $2k$ and using the equality in distribution $T_k \sim V_{2k}/2$, we obtain the following limit theorem for the distribution of $T_k$ as $k \to \infty$.

**Theorem 2.32 (Limit Distribution of the $2k$-Meeting Time)** *For each $x > 0$,*

$$\lim_{k \to \infty} \mathbb{P}\left(\frac{T_k}{2k^2} \leq x\right) = L(x).$$

In a shorthand notation that has already been used several times,

$$\frac{T_k}{2k^2} \xrightarrow{\mathcal{D}} Y, \qquad Y \sim L, \quad \text{as } k \to \infty.$$

This means that the sequences $(T_k/(2k^2))$ and $(V_k/k^2)$ have the same limit distribution as $k \to \infty$. Some values of the function $L$ are found in Table 2.3. For large $k$, we thus obtain

$$\mathbb{P}(T_k \leq 2k^2 x) \approx L(x).$$

Since $L(2.2) \approx 0.5$ and $L(254) \approx 0.95$ (see Table 2.3), it particularly holds that

$$\mathbb{P}(T_k \leq 4.4 \cdot k^2) \approx 0.5, \qquad \mathbb{P}(T_k \leq 508 \cdot k^2) \approx 0.95$$

for large $k$. Thus, if one random walk starts at zero and the other at the height $2k = 100$, the probability is approximately $\frac{1}{2}$ that the first meeting occurs by time 11,000. On the other hand, there is a 5% probability that the two random walks will not have met even after 1,270,000 steps!

A final remark: After the first meeting time, the next step of the two symmetric random walks can either be in the same or in opposite directions, with both possibilities being equally likely. In the first case, the random walks meet again after one more time step. In the second case, the height difference is 2. Therefore, the random time interval between the first and the second meeting time has the same distribution as $1 + UT_1$. Here, $U$ is a random variable with $\mathbb{P}(U = 1) = \mathbb{P}(U = 0) = \frac{1}{2}$, which is independent of $T_1$.

## 2.11 Changes of Sign

How often does a symmetric random walk cross the $x$-axis and thus change its sign? How often does the lead change in the context of a random walk as a game between two people? From Sect. 2.4, we know that there are surprisingly few visits to zero and thus few game ties. Since every change of sign in a random walk can only occur after a preceding visit to zero, we must expect significantly fewer changes in leadership than game ties. For example, the random walk depicted in Fig. 2.42 exhibits 8 changes of sign and about twice as many visits to zero, namely 17. In the following, we will specify the term *change of sign* and examine the stochastic behavior of the number of changes of sign in a symmetric random walk. The results are astonishing: regardless of the length of the random walk, it is most likely that no change of sign occurs at all! Furthermore, in a random walk of length 10,000, you can safely bet that at most 34 changes of sign will occur.

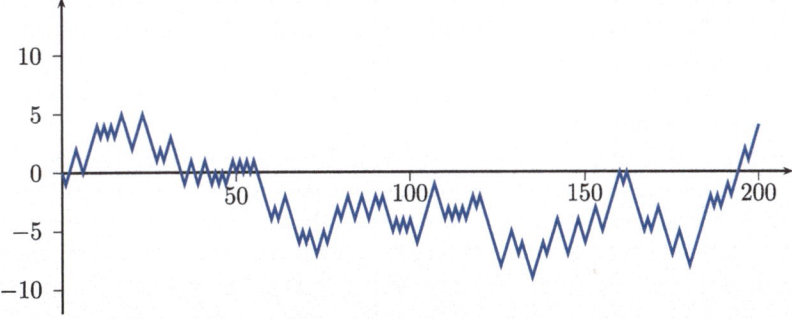

**Fig. 2.42** Random walk with 8 changes of sign and 17 visits to zero

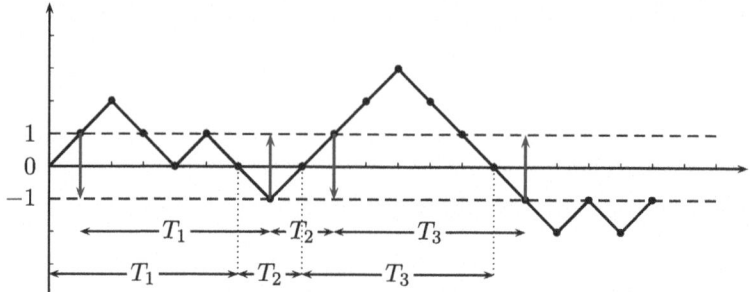

**Fig. 2.43** The time intervals between changes of sign are independent and identically distributed as $V_2$

A random walk has a *change of sign* at time $2k$, where $k \geq 1$, if $S_{2k+1} S_{2k-1} = -1$. Note that changes of sign can only occur at even time points, because at those times $S_{2k} = 0$ necessarily. In the following, let

$$C_{2n+1} := \sum_{k=1}^{n} \mathbf{1}\{S_{2k+1} S_{2k-1} = -1\}$$

denote the number of changes of sign in a symmetric random walk of length $2n+1$. The length was chosen to be odd because, starting from the origin, the first step leads away from 0 and only after that (at most $n$) changes of sign are possible.

What is the probability that exactly $r$ changes of sign occur? The answer to this question is not straightforward, but it can be deduced using the first-passage times introduced in Sect. 2.9: Referring to Fig. 2.43, the random walk initially moves one step up to height 1. For a change of sign to occur subsequently, it must reach the height $-1$, i.e., descend by 2 steps. The time span required for this, denoted by $T_1$, has the same distribution as the first-passage time $V_{-2}$ (in Fig. 2.43, $T_1$ takes the value 6). If the first step of the random walk had been downward, it would have needed to ascend by 2 steps up to height 1 to induce the first change of sign. The random time span for this has the same distribution as the first-passage time $V_2$. Since $V_2$ and $V_{-2}$ have the same distribution, the random time span $T_1$ until the first change of sign occurs is, regardless of whether $S_1 = 1$ or $S_1 = -1$, distributed like the first-passage time $V_2$. Note that $T_1$ counts from time point 1, but includes the last step after reaching zero.

Once the first change of sign has occurred, the random walk is either at height 1 or $-1$. To induce the next change of sign, it must descend or ascend by 2 steps. Due to the memoryless property of the random walk, the random time span required for this, denoted by $T_2$, is independent of $T_1$ and follows the same distribution as $V_2$, and so on (see Fig. 2.43). The blue arrows in Fig. 2.43, pointing alternatively downward and upward, indicate that, from the respective time points, the random walk must descend or ascend by two steps to cause the next change of sign.

## 2.11 Changes of Sign

Let, as above, $T_1$ denote the time until the first change of sign occurs. Furthermore, for each $j \geq 2$, let $T_j$ be the time span between the $(j-1)$-th and the $j$-th change of sign. Then

$$\{C_{2n+1} \geq r\} = \left\{\sum_{j=1}^{r} T_j \leq 2n\right\}. \tag{2.93}$$

**SAQ 20** Why does (2.93) hold?

Since the random variables $T_1, \ldots, T_r$ are independent and each $T_j$ has the same distribution as $V_2$, $T_1 + \ldots + T_r$ has the same distribution as $V_{2r}$. Thus, (2.93) yields

$$\mathbb{P}(C_{2n+1} \geq r) = \mathbb{P}(V_{2r} \leq 2n).$$

Using Lemma 2.22 and (2.63), it follows that

$$\mathbb{P}(C_{2n+1} \geq r) = \mathbb{P}(M_{2n} \geq 2r) \tag{2.94}$$
$$= 2\mathbb{P}(S_{2n} \geq 2r) - \mathbb{P}(S_{2n} = 2r),$$

and thus

$$\mathbb{P}(C_{2n+1} = r) = \mathbb{P}(C_{2n+1} \geq r) - \mathbb{P}(C_{2n+1} \geq r+1)$$
$$= 2\mathbb{P}(S_{2n} \geq 2r) - \mathbb{P}(S_{2n} = 2r)$$
$$\quad - (2\mathbb{P}(S_{2n} \geq 2(r+1)) - \mathbb{P}(S_{2n} = 2(r+1)))$$
$$= \mathbb{P}(S_{2n} = 2r) + \mathbb{P}(S_{2n} = 2r+2).$$

Since

$$\mathbb{P}(S_{2n+1} = 2r+1) = \mathbb{P}(S_{2n} + X_{2n+1} = 2r+1)$$
$$= \mathbb{P}(S_{2n} + 1 = 2r+1 | X_{2n+1} = 1) \cdot \frac{1}{2}$$
$$+ \mathbb{P}(S_{2n} - 1 = 2r+1 | X_{2n+1} = -1) \cdot \frac{1}{2}$$
$$= \frac{1}{2}\mathbb{P}(S_{2n} = 2r) + \frac{1}{2}\mathbb{P}(S_{2n} = 2(r+1)),$$

we get

$$\mathbb{P}(C_{2n+1} = r) = 2\mathbb{P}(S_{2n+1} = 2r+1).$$

Using (2.7), we arrive at the following result (see [FE1], p. 84):

**Theorem 2.33 (Distribution of the Number of Changes of Sign)** *For the number $C_{2n+1}$ of changes of sign in a simple symmetric random walk of length $2n+1$, the following hold:*

(a)
$$\mathbb{P}(C_{2n+1} = r) = \binom{2n+1}{r+n+1} \frac{1}{2^{2n}}, \qquad r = 0, 1, \ldots, n. \tag{2.95}$$

(b) $\mathbb{E}(C_{2n+1}) = (n+1) \dfrac{\binom{2n+1}{n}}{2^{2n+1}} - \dfrac{1}{2}.$

(c) $\mathbb{V}(C_{2n+1}) = \dfrac{n+1}{2} - \dfrac{1}{4} - \left( (n+1) \dfrac{\binom{2n+1}{n}}{2^{2n+1}} \right)^2.$

*Proof* Only parts (b) and (c) need to be shown. Using (7.26), it follows that

$$\mathbb{E}(C_{2n+1}) = \frac{1}{2^{2n}} \sum_{r=1}^{n} r \binom{2n+1}{r+n+1} = \frac{1}{2^{2n}} \left( \frac{n+1}{2} \binom{2n+1}{n} - 2^{2n-1} \right),$$

and thus (b) (see also [FE2], Theorem 4). Using (7.28), we get analogously

$$\mathbb{E}\left(C_{2n+1}^2\right) = \sum_{r=1}^{n} r^2 \binom{2n+1}{r+n+1} \frac{1}{2^{2n}} = \frac{n+1}{2} - \frac{2n+1}{2} \frac{\binom{2n}{n}}{2^{2n}}. \tag{2.96}$$

Together with part (b), the formula for the variance of $C_{2n+1}$ now follows by direct calculation (Exercise 2.20). ∎

The result of Theorem 2.33 is surprising. First of all, it is astonishing that the probabilities $\mathbb{P}(C_{2n+1} = r)$ strictly decrease in $r$, since

$$\frac{\mathbb{P}(C_{2n+1} = r+1)}{\mathbb{P}(C_{2n+1} = r)} = \frac{n-r}{n+r+2}, \qquad r = 0, 1, \ldots, n-1.$$

Thus, regardless of the length of the random walk, it is most likely to observe no changes of sign! The next most likely case is that exactly one change of sign occurs, followed by the next most likely case of exactly two changes of sign, and so on.

Figure 2.44 shows a bar chart of the distribution of $C_{101}$. The probability that in a symmetric random walk of length 101 no change of sign is observed at all is a

## 2.11 Changes of Sign

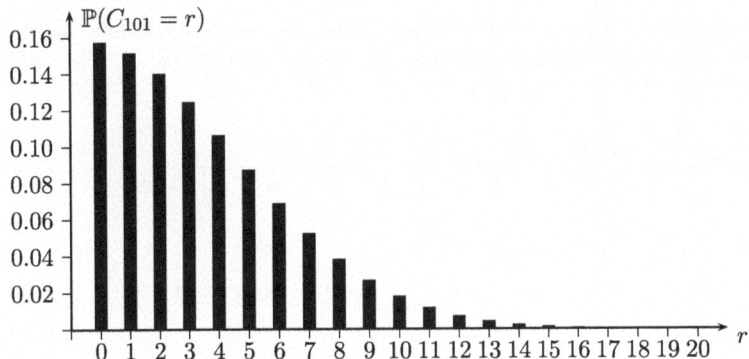

**Fig. 2.44** Bar chart of the distribution of $C_{101}$

remarkable 15.8%, while the probability that at most four changes of sign occur is 68%. At least 10 changes of sign occur only with a probability of 4.6%!

**An Alternative Approach ("Thinning Out Visits to Zero")**

At the beginning of this section, we noted that there can be at most as many changes of sign as there are visits to zero. Does the distribution of the number $N_{2n}$ of visits to zero up to time $2n$, known from Sect. 2.4, help us derive the distribution of $C_{2n+1}$? The answer is: Yes. We just have to realize that, independently of each other, each visit to zero of the random walk causes a change of sign if and only if the two steps before and after hitting zero are both upward or both downward. This happens with probability $\frac{1}{2}$. Therefore, the distribution of $C_{2n+1}$ should result from tossing a fair coin for each visit to zero and declaring it as a change of sign only in the case of heads. In this way, we "stochastically thin out" the number $N_{2n}$ of visits to zero.

To make this idea mathematically precise, let $U_1, U_2, \ldots, U_n$ be independent Bernoulli-distributed random variables with $\mathbb{P}(U_j = 1) = \mathbb{P}(U_j = 0) = \frac{1}{2}$ for each $j = 1, \ldots, n$, which are also independent of $N_{2n}$. The thinning described above then leads to the random variable

$$\widetilde{C}_{2n+1} := \sum_{j=1}^{N_{2n}} U_j, \tag{2.97}$$

which, as we will see shortly, has the same distribution as $C_{2n+1}$. Because the upper summation index $N_{2n}$ in (2.97) is a random variable, the expression on the right-hand side of (2.97) is also referred to as a *randomized sum*. Conditionally on $N_{2n} = k$, this sum takes the value $\sum_{j=1}^{k} U_j$ and thus has a binomial distribution with parameters $k$ and $\frac{1}{2}$. If $N_{2n}$ takes the value 0, we define the sum to be 0.

We determine $\mathbb{P}(\widetilde{C}_{2n+1} = r)$ by decomposing the randomized sum based on the possible values $k$ of $N_{2n}$ and applying the law of total probability. Since $k \geq r$ necessarily, due to the independence of $N_{2n}$ and $U_1, \ldots, U_n$,

$$\mathbb{P}(\widetilde{C}_{2n+1} = r) = \sum_{k=r}^{n} \mathbb{P}\left(\sum_{j=1}^{k} U_j = r \middle| N_{2n} = k\right) \mathbb{P}(N_{2n} = k)$$

$$= \sum_{k=r}^{n} \mathbb{P}\left(\sum_{j=1}^{k} U_j = r\right) \mathbb{P}(N_{2n} = k).$$

Since $\sum_{j=1}^{k} U_j$ has the binomial distribution $\text{Bin}(k, \frac{1}{2})$, we use (2.32) to obtain

$$\mathbb{P}(\widetilde{C}_{2n+1} = r) = \sum_{k=r}^{n} \binom{k}{r} \frac{1}{2^k} \binom{2n-k}{n} \frac{1}{2^{2n-k}}$$

$$= \frac{1}{2^{2n}} \sum_{k=r}^{n} \binom{k}{r} \binom{2n-k}{n}.$$

The comparison with (2.95) shows that we would be done if we could show

$$\binom{2n+1}{r+n+1} = \sum_{k=r}^{n} \binom{k}{r} \binom{2n-k}{n}, \qquad r = 0, 1, \ldots, n, \tag{2.98}$$

because then $C_{2n+1}$ and $\widetilde{C}_{2n+1}$ would have the same distribution.

According to Lemma 2.1, the left-hand side of (2.98) represents the number of all paths leading from $(0, 0)$ to $(2n+1, 2r+1)$. Such a path is sketched in Fig. 2.45. Every path from $(0, 0)$ to $(2n+1, 2r+1)$ must run within the dashed rectangle in Fig. 2.45. We decompose the set of all paths from $(0, 0)$ to $(2n+1, 2r+1)$ based on the point at which they first hit the bold line, which lies on the straight line $y = 2r + 2 - x$. The last step before reaching the barrier must have been an upward step, leaving a point on the line $y = -x + 2r$. The points in question have the coordinates $(k, 2r-k)$, where $k \in \{r, r+1, \ldots, n\}$. According to Lemma 2.1, there are $\binom{k}{r}$ paths from $(0, 0)$ to $(k, 2r-k)$. Every path arriving at the point $(k, 2r-k)$ reaches the point $(k+1, 2r-k+1)$ in the next step. According to (2.11), from there it has $\binom{2n-k}{n}$ ways to reach the point $(2n+1, 2r+1)$. This proves (2.98).

## 2.11 Changes of Sign

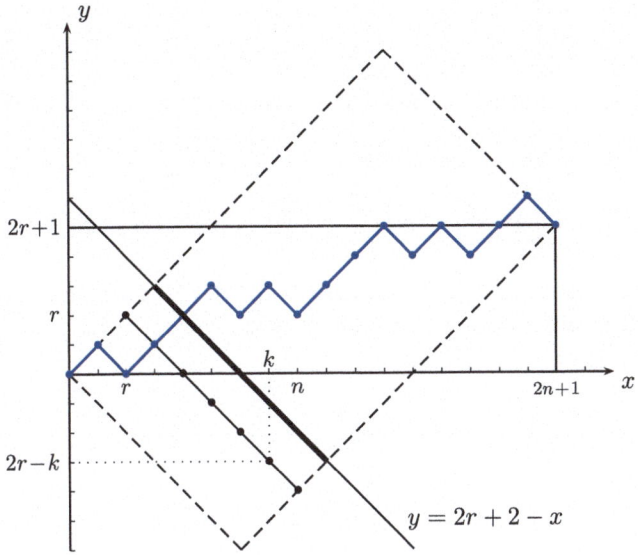

**Fig. 2.45** Illustrating the proof of (2.98) (decomposition of the set of all paths leading from $(0, 0)$ to $(2n + 1, 2r + 1)$)

Using (2.97) and (2.98), as well as the binomial distribution of $\sum_{j=1}^{k} U_j$, we obtain the following alternative derivation of the expectation of $C_{2n+1}$:

$$\mathbb{E}(C_{2n+1}) = \mathbb{E}(\widetilde{C}_{2n+1}) = \sum_{r=0}^{n} r\, \mathbb{P}(\widetilde{C}_{2n+1} = r)$$

$$= \sum_{k=0}^{n} \left[ \sum_{r=0}^{k} r\, \mathbb{P}\left( \sum_{j=1}^{k} U_j = r \right) \right] \mathbb{P}(N_{2n} = k) = \sum_{k=r}^{n} \frac{k}{2} \mathbb{P}(N_{2n} = k)$$

$$= \frac{1}{2} \mathbb{E}(N_{2n})$$

$$= (2n + 1) \binom{2n}{n} 2^{-(2n+1)} - \frac{1}{2},$$

in accordance with part (b) of Theorem 2.33. This result was also expected in view of the thinning process carried out using coin tosses and follows directly from (2.97) using iterated expectation (see, e.g., [RO], Section 7.5): since $N_{2n}$ and $U_1, \ldots, U_n$ are independent and the $U_j$ have the same binomial distribution $\text{Bin}(1, \frac{1}{2})$, it follows that

$$\mathbb{E}\left( \sum_{j=1}^{N_{2n}} U_j \right) = \mathbb{E}\left[ \mathbb{E}\left( \sum_{j=1}^{N_{2n}} U_j \Big| N_{2n} \right) \right] = \mathbb{E}\big( N_{2n}\, \mathbb{E}(U_1) \big) = \frac{1}{2} \mathbb{E}(N_{2n}). \qquad (2.99)$$

The limit behavior of $C_{2n+1}$ as $n \to \infty$ is given by the following result (see [FE1], p. 86):

**Theorem 2.34 (Limit Distribution of the Number of Changes of Sign)** *For every* $x \geq 0$,

$$\lim_{n \to \infty} \mathbb{P}\left(\frac{C_{2n+1}}{\sqrt{2n+1}} \leq x\right) = 2\Phi(2x) - 1 = 2\int_0^{2x} \varphi(t)\,dt. \qquad (2.100)$$

*Proof* We can assume $x > 0$, as otherwise both sides of (2.100) are equal to zero. Let $r_n$ be the smallest integer $r$ such that $r \geq x\sqrt{2n+1}$. Since $C_{2n+1}$ is integer-valued, (2.94) gives

$$\mathbb{P}\left(\frac{C_{2n+1}}{\sqrt{2n+1}} \geq x\right) = \mathbb{P}(C_{2n+1} \geq r_n) = \mathbb{P}(M_{2n} \geq 2r_n)$$

$$= \mathbb{P}\left(\frac{M_{2n}}{\sqrt{2n}} \geq \frac{2r_n}{\sqrt{2n}}\right).$$

In view of $\lim_{n \to \infty} (2r_n/\sqrt{2n}) = 2x$, Theorem 2.13 yields

$$\lim_{n \to \infty} \mathbb{P}\left(\frac{C_{2n+1}}{\sqrt{2n+1}} \geq x\right) = 1 - (2\Phi(2x) - 1),$$

and the claim follows. ∎

**Remark 2.35** The right-hand side of (2.100) is the distribution function of $\frac{|Z|}{2}$, where $Z$ is a standard normally distributed random variable. Therefore, we can write the assertion of Theorem 2.34 in the compact form:

$$\frac{C_{2n+1}}{\sqrt{2n+1}} \xrightarrow{\mathcal{D}} \frac{1}{2}|Z|, \quad Z \sim N(0,1), \quad \text{as } n \to \infty. \qquad (2.101)$$

Compared to the limit distribution of $N_{2n}$ (cf. (2.41)), the factor $\frac{1}{2}$ in (2.101) comes precisely from the thinning process carried out.

Table 2.5 provides some values of the distribution function of $\frac{|Z|}{2}$, denoted by $C$. Since $C$ has a median of 0.337, it follows from Theorem 2.34, among other things, that the probability $\mathbb{P}(C_{2n+1} \leq 0.337\sqrt{2n+1})$ is approximately equal to 0.5 for large $n$. In a random walk of length 10,001, there is therefore a probability of about 0.5 that there will be at most 34 changes of sign. Furthermore, because $C(1.29) \approx 0.99$, more than 130 changes of sign occur only in every hundredth random walk of this length.

**Table 2.5** Values of the distribution function $C(x) := 2\Phi(2x) - 1$ of $\frac{|Z|}{2}$, where $Z \sim N(0, 1)$

| $x$  | $C(x)$ | $x$  | $C(x)$ | $x$  | $C(x)$ |
|------|--------|------|--------|------|--------|
| 0.00 | 0.000  | 0.45 | 0.632  | 0.90 | 0.928  |
| 0.05 | 0.080  | 0.50 | 0.683  | 0.95 | 0.943  |
| 0.10 | 0.159  | 0.55 | 0.729  | 1.0  | 0.955  |
| 0.15 | 0.236  | 0.60 | 0.770  | 1.1  | 0.972  |
| 0.20 | 0.311  | 0.65 | 0.806  | 1.2  | 0.984  |
| 0.25 | 0.383  | 0.70 | 0.839  | 1.3  | 0.991  |
| 0.30 | 0.452  | 0.75 | 0.866  | 1.4  | 0.995  |
| 0.35 | 0.516  | 0.80 | 0.890  | 1.5  | 0.997  |
| 0.40 | 0.576  | 0.85 | 0.911  | 1.6  | 0.998  |

## 2.12 The Maximum Modulus

How likely is it that a symmetric random walk of length $n$ will reach a height of $k$ or $-k$? In other words, using the notations $M_n = \max(S_1, \ldots, S_n)$ and $m_n = \min(S_1, \ldots, S_n)$ from Sect. 2.7, what is the probability that at least one of the events $\{M_n \geq k\}$ or $\{m_n \leq -k\}$ occurs? Conversely, what is the probability that a symmetric random walk of length $n$ stays entirely within a corridor centered on the $x$-axis with a width of $2(k-1)$? These questions lead us to examine the distribution of the random variable

$$|M|_n = \max_{j=1,\ldots,n} |S_j|,$$

which denotes the *maximum modulus* of the random walk up to time $n$.

To this end, we first determine the probability $\mathbb{P}(|M|_n \geq k)$ for $k = 1, \ldots, n$. Here, the cases $k = 1$ and $k = n$ are quickly addressed.

**SAQ 21** Why are these cases quickly addressed?

In what follows, we fix any $k \in \{2, \ldots, n-1\}$ and briefly write $r_n$ for the number of all $n$-paths with the property $|M|_n \geq k$, without making the dependence of $r_n$ on $k$ explicit. We determine $r_n$ and thus $\mathbb{P}(|M|_n \geq k)$ by elementary counting using the inclusion-exclusion principle.

For $s \geq 1$, let $r_{n,s}^+$ be the number of all $n$-paths that reach the height $k$ and subsequently reach the respective opposite height at least $s - 1$ more times, i.e., first $-k$, then $k$, and so on. Likewise, $r_{n,s}^-$ denotes the number of all $n$-paths that reach the height $-k$ and subsequently reach the respective opposite height at least $s - 1$ more times, i.e., first $k$, then $-k$, and so on.

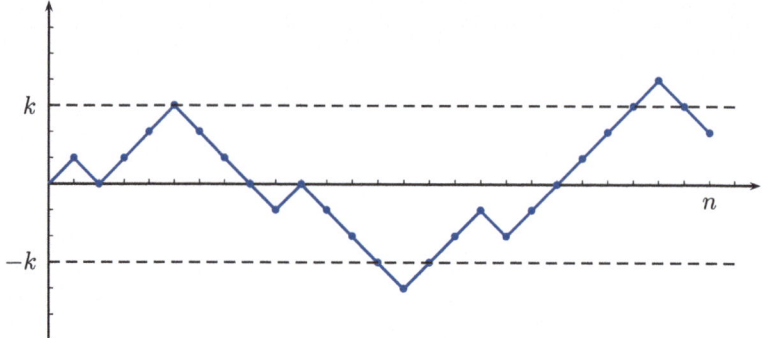

**Fig. 2.46** Path with $|M|_n \geq k$ that first reaches $+k$, then $-k$, and finally $+k$

Since such a path requires at least $k + 2k(s - 1)$ time steps, it follows that $k + 2k(s - 1) \leq n$, which leads to

$$s \leq s_n := s_n(k) := \left\lfloor \frac{n-k}{2k} + 1 \right\rfloor. \tag{2.102}$$

In the sum $r_{n,1}^+ + r_{n,1}^-$, each path that reaches the height $k$ or the height $-k$ is counted at least once. However, paths that reach both the heights $k$ and $-k$, like the one shown in Fig. 2.46, are counted twice. Subtracting the sum $r_{n,2}^+ + r_{n,2}^-$ ensures that paths reaching either the height $k$ or $-k$ (possibly multiple times) or both heights, but subsequently not returning to the first reached height (i.e., $k$ or $-k$), are counted exactly once. The path shown in Fig. 2.46 is no longer counted because it contributes to each of the four terms $r_{n,j}^+$, $r_{n,j}^-$, where $j = 1, 2$. For this reason, we need to add the sum $r_{n,3}^+ + r_{n,3}^-$ (note that the path in Fig. 2.46 contributes to $r_{n,3}^+$).

Continuing in this manner and considering that the inclusion-exclusion process terminates due to (2.102) and, for symmetry (reflection across the $x$-axis), $r_{n,s}^+ = r_{n,s}^-$ for $s \leq s_n$, it follows that

$$r_n = 2 \sum_{s=1}^{s_n} (-1)^{s-1} r_{n,s}^+,$$

and thus

$$\mathbb{P}(|M|_n \geq k) = 2 \sum_{s=1}^{s_n} (-1)^{s-1} \frac{r_{n,s}^+}{2^n}.$$

In this alternating sum (in which $s_n$ and $r_{n,s}^+$ depend on $k$), the quotient denotes the probability that a random walk of length $n$ reaches the height $k$ and then alternates between moving down and up by $2k$ steps at least $s - 1$ more times. The time required

## 2.12 The Maximum Modulus

for this is distributed like the sum of $s-1$ independent $2k$-th passage times. Since a total of $k+2k(s-1)$ steps are necessary, together with the first reaching of the height $k$, the total time required for this is thus distributed as $V_{k+2k(s-1)}$. Since this random time is at most $n$ if and only if the event described above occurs, we have $r_{n,s}^+/2^n = \mathbb{P}(V_{(2s-1)k} \leq n)$. By Lemma 2.22 and Theorem 2.12, we thus arrive at the following result (which may also be obtained from the solution to Problem 17 on p. 369 of [FE1]):

**Theorem 2.36 (Distribution of the Maximum Modulus $|M|_n$)** *For every* $k = 1, \ldots, n$,

$$\mathbb{P}(|M|_n \geq k) = 2 \sum_{s=1}^{\lfloor \frac{n}{2k}+\frac{1}{2} \rfloor} (-1)^{s-1} \mathbb{P}(M_n \geq (2s-1)k) \tag{2.103}$$

$$= 2 \sum_{s=1}^{\lfloor \frac{n}{2k}+\frac{1}{2} \rfloor} (-1)^{s-1} \sum_{j=(2s-1)k}^{n} \frac{1}{2^n} \binom{n}{\lfloor \frac{n+j+1}{2} \rfloor}. \tag{2.104}$$

The probabilities $\mathbb{P}(|M|_n = k)$ can be obtained from this by taking differences. For future purposes, we note that, according to the inclusion-exclusion principle, the partial sums in (2.103) are alternately too large and too small, i.e., the so-called Bonferroni[11] *inequalities* hold:

$$\mathbb{P}(|M|_n \geq k) \leq 2 \sum_{s=1}^{2r+1} (-1)^{s-1} \mathbb{P}(M_n \geq (2s-1)k), \tag{2.105}$$

$$\mathbb{P}(|M|_n \geq k) \geq 2 \sum_{s=1}^{2r} (-1)^{s-1} \mathbb{P}(M_n \geq (2s-1)k). \tag{2.106}$$

These inequalities are valid for any fixed $r$ such that the upper summation limit in (2.105) or (2.106) is at least 1 and at most $s_n$.

Figure 2.47 shows a bar chart of the distribution of $|M|_{40}$. It is noticeable that, for $j \geq 4$, the probabilities $\mathbb{P}(|M|_{40} = 2j)$ and $\mathbb{P}(|M|_{40} = 2j-1)$ are approximately equal. To understand this phenomenon, note that in general

$$\mathbb{P}(|M|_{2n} \geq k) = \mathbb{P}(\{M_{2n} \geq k\} \cup \{m_{2n} \leq -k\})$$

$$= \mathbb{P}(M_{2n} \geq k) + \mathbb{P}(m_{2n} \leq -k) - \mathbb{P}(M_{2n} \geq k, m_{2n} \leq -k)$$

$$= 2\mathbb{P}(M_{2n} \geq k) - \mathbb{P}(M_{2n} \geq k, m_{2n} \leq -k).$$

---

[11] Carlo Emilio Bonferroni (1892–1960), Professor of Financial Mathematics in Bari and Florence. His main areas of work included financial mathematics, probability theory, statistics, and analysis.

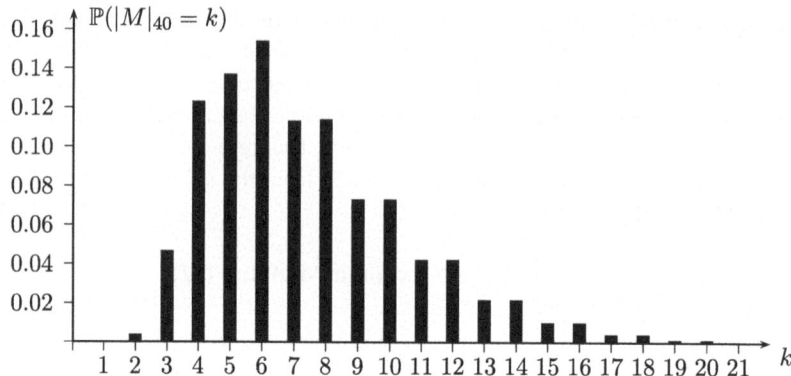

**Fig. 2.47** Bar chart of the distribution of $|M|_{40}$

For $k$ sufficiently large compared to $n$, the probability of the intersection of the events $\{M_{2n} \geq k\}$ and $\{m_{2n} \leq -k\}$ is so small that for such $k$, the approximation $\mathbb{P}(|M|_{2n} \geq k) \approx 2\mathbb{P}(M_{2n} \geq k)$ holds, and thus (by taking differences):

$$\mathbb{P}(|M|_{2n} = k) \approx 2\mathbb{P}(M_{2n} = k). \tag{2.107}$$

For sufficiently large $k$, equality even holds in (2.107).

**SAQ 22** Why does "$=$" hold in (2.107) for sufficiently large $k$?

Comparing Fig. 2.47 with the bar chart of the distribution of $M_{40}$ shown in Fig. 2.27 reveals that the approximation (2.107) is very good for $k \geq 9$.

**Remark 2.37** An alternative expression for the right-hand side of (2.104) in terms of sine and cosine functions is obtained by adding the displays (7.40) and (7.41) in [ET]. In the notation of that book, one has to set $W = L = k$ and $p = q = \frac{1}{2}$.

Regarding the asymptotic behavior of $|M|_n$ as $n \to \infty$, the following limit theorem holds:

**Theorem 2.38 (Limit Distribution of the Maximum Modulus $|M|_n$)** *For each $x > 0$,*

$$\lim_{n \to \infty} \mathbb{P}\left(\frac{|M|_n}{\sqrt{n}} \leq x\right) = R(x),$$

## 2.12 The Maximum Modulus

where

$$R(x) := 1 - 4\sum_{s=1}^{\infty}(-1)^{s-1}\big(1 - \Phi((2s-1)x)\big). \qquad (2.108)$$

**Proof** For $x > 0$, let $k_n = k_n(x)$ be the smallest integer that is greater than or equal to $x\sqrt{n}$. From (2.103) and the fact that $|M|_n$ is integer-valued, it follows that

$$\mathbb{P}\left(\frac{|M|_n}{\sqrt{n}} \geq x\right) = \mathbb{P}(|M|_n \geq k_n)$$

$$= 2\sum_{s=1}^{s_n}(-1)^{s-1}\mathbb{P}\left(\frac{M_n}{\sqrt{n}} \geq (2s-1)\frac{k_n}{\sqrt{n}}\right), \qquad (2.109)$$

with $s_n = s_n(k_n)$ as in (2.102). Since $\lim_{n\to\infty} k_n/\sqrt{n} = x$, Theorem 2.13 yields

$$\lim_{n\to\infty}\mathbb{P}\left(\frac{M_n}{\sqrt{n}} \geq (2s-1)\frac{k_n}{\sqrt{n}}\right) = 1 - \big(2\Phi((2s-1)x) - 1\big)$$

$$= 2\big(1 - \Phi((2s-1)x)\big).$$

Thus, each individual summand in (2.109) converges as $n \to \infty$. The Bonferroni inequalities then provide the bounds

$$\limsup_{n\to\infty}\mathbb{P}\left(\frac{|M|_n}{\sqrt{n}} \geq x\right) \leq 4\sum_{s=1}^{2r+1}(-1)^{s-1}(1 - \Phi((2s-1)x)), \qquad r = 0, 1, \ldots$$

$$\liminf_{n\to\infty}\mathbb{P}\left(\frac{|M|_n}{\sqrt{n}} \geq x\right) \geq 4\sum_{s=1}^{2r}(-1)^{s-1}(1 - \Phi((2s-1)x)), \qquad r = 1, 2, \ldots$$

According to the Leibniz criterion for alternating series, the series $\sum_{s=1}^{m}(-1)^{s-1}(1-\Phi((2s-1)x))$ converges as $m \to \infty$. Using (7.9), the claim follows by taking the limit $r \to \infty$ and complementing. ∎

**Remark 2.39** The function $R : \mathbb{R} \to \mathbb{R}$, defined in (2.108) and extended by $R(x) := 0$ for $x \leq 0$, has the properties of a distribution function; in particular, it is monotonically increasing. This property, which is not immediately apparent from (2.108), results from the fact that, according to Theorem 2.38, $R$ is the limit of a sequence of monotonically increasing functions. The associated distribution is

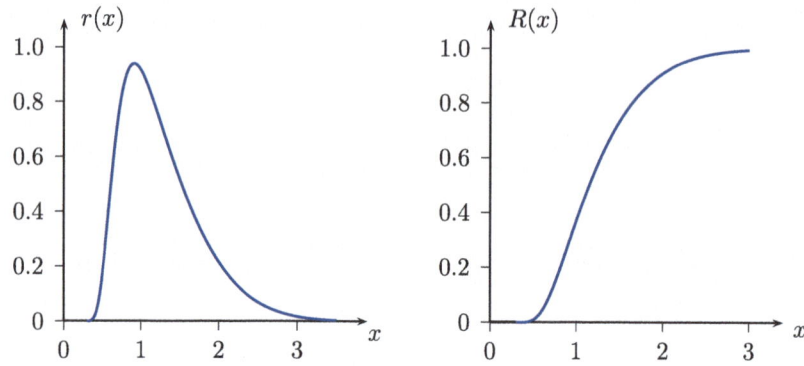

**Fig. 2.48** Density (left) and distribution function (right) of the Rényi distribution

also referred to as the *Rényi*[12] *distribution*. According to Theorem 2.38, the Rényi distribution is the limit distribution of the rescaled maximum modulus of the simple symmetric random walk, scaled by $1/\sqrt{n}$, as $n \to \infty$. If $Z$ is a random variable with the distribution function $R$ (denoted $Z \sim R$), the statement of Theorem 2.38 can be written in the compact form

$$\frac{|M|_n}{\sqrt{n}} \xrightarrow{\mathcal{D}} Z, \qquad Z \sim R, \quad \text{as } n \to \infty.$$

The density and the distribution function of the Rényi distribution are illustrated in Fig. 2.48. Values of the distribution function $R$ can be found in Table 2.6. For the creation of Fig. 2.48 and the preparation of Table 2.6, a significantly faster converging series representation, compared to the representation (2.108), was used (see [RE1]):

$$R(x) = \frac{4}{\pi} \sum_{s=0}^{\infty} \frac{(-1)^s}{2s+1} \exp\left(-(2s+1)^2 \frac{\pi^2}{8x^2}\right), \qquad x > 0. \qquad (2.110)$$

The density $r$ of $R$ results from this by term-by-term differentiation, i.e.,

$$r(x) = \frac{\pi}{x^3} \sum_{s=0}^{\infty} (-1)^s (2s+1) \exp\left(-(2s+1)^2 \frac{\pi^2}{8x^2}\right), \qquad x > 0.$$

---

[12] Alfréd Rényi (1921–1970) was a Hungarian mathematician who, from 1950, served as the director of the Institute for Applied Mathematics of the Hungarian Academy of Sciences, later named after him. Since 1952, he was a professor at the University of Budapest. His main areas of work included number theory, probability theory, combinatorics, graph theory, and analysis.

## 2.13 A Test of Symmetry

**Table 2.6** Values of the distribution function of the Rényi distribution

| $x$ | $R(x)$ | $x$ | $R(x)$ | $x$ | $R(x)$ |
|---|---|---|---|---|---|
| 0.00 | 0.000 | 1.20 | 0.540 | 2.10 | 0.929 |
| 0.40 | 0.001 | 1.30 | 0.613 | 2.20 | 0.944 |
| 0.50 | 0.009 | 1.40 | 0.677 | 2.242 | 0.950 |
| 0.60 | 0.041 | 1.50 | 0.733 | 2.30 | 0.957 |
| 0.70 | 0.103 | 1.60 | 0.781 | 2.40 | 0.967 |
| 0.80 | 0.185 | 1.70 | 0.822 | 2.50 | 0.975 |
| 0.90 | 0.278 | 1.80 | 0.856 | 2.60 | 0.981 |
| 1.00 | 0.371 | 1.90 | 0.885 | 2.70 | 0.986 |
| 1.10 | 0.459 | 1.96 | 0.900 | 2.80 | 0.990 |
| 1.149 | 0.500 | 2.0 | 0.909 | 2.90 | 0.993 |

Because $R(1.149) \approx 0.5$, Theorem 2.38 states, among other things, that in a symmetric random walk of length 10,000, the greatest distance of the path from the $x$-axis is at most 115 with a probability of $\frac{1}{2}$. Interpreting the random walk as a game between two players consisting of 10,000 coin tosses, on average, in every second game of this length, it can be expected that at some point during the game, one of the players will lead by at least 116 tosses.

**Remark 2.40** Theorem 2.38 holds more generally for the maximum modulus, if the underlying i.i.d. random variables $X_1, X_2, \ldots$ have an expectation of 0 and variance of 1 (see Theorem II. of [EK]). There, the limit distribution function of $|M|_n/\sqrt{n}$ as $n \to \infty$ is stated in the form given in (2.110).

## 2.13 A Test of Symmetry

Suppose $n$ people with hypertension receive a blood pressure-lowering medication. Let $Y_j$ and $Z_j$ denote the (random) blood pressure of the $j$-th patient before and after administration of the medication, respectively. Under simplifying model assumptions, the differences $D_j := Y_j - Z_j$ (for $j = 1, \ldots, n$) can be regarded as independent and identically distributed random variables. The unknown distribution function of $D_1$, which is assumed to be continuous, is denoted by $F(x) := \mathbb{P}(D_1 \leq x)$.

If the medication has no effect, the difference $Y_1 - Z_1$ ("before minus after") is stochastically indistinguishable from $Z_1 - Y_1$ ("after minus before"), which means that $D_1$ and $-D_1$ have the same distribution. An equivalent way to state this is that the distribution of $D_1$ is symmetric around 0.

Because $\mathbb{P}(-D_1 \leq x) = \mathbb{P}(D_1 \geq -x) = 1 - F(-x)$, the hypothesis $H_0$ of equality in distribution of $D_1$ and $-D_1$—that is, in the above situation, the ineffectiveness of the medication— can be written as

$$H_0: F(x) + F(-x) - 1 = 0 \quad \text{for each } x \in \mathbb{R}. \tag{2.111}$$

Butler ([BU]) suggested testing $H_0$ using the *empirical distribution function*

$$\widehat{F}_n(x) := \frac{1}{n} \sum_{j=1}^n \mathbf{1}\{D_j \le x\}, \qquad x \in \mathbb{R},$$

of $D_1, \ldots, D_n$, and rejecting the hypothesis for large values of the test statistic

$$Q_n := \sup_{x \le 0} \left| \widehat{F}_n(x) + \widehat{F}_n(-x) - 1 \right|. \tag{2.112}$$

This approach appears reasonable because $\widehat{F}_n(x)$, as the arithmetic mean of i.i.d. random variables with expectation $\mathbb{E}(\mathbf{1}\{D_1 \le x\}) = \mathbb{P}(D_1 \le x) = F(x)$, almost surely converges to $F(x)$ as $n \to \infty$ by the strong law of large numbers. By the Glivenko[13]–Cantelli[14] theorem, it even holds that

$$\lim_{n \to \infty} \sup_{x \in \mathbb{R}} |\widehat{F}_n(x) - F(x)| = 0$$

almost surely (see, e.g., [DU], p. 76). Here, $D_1, D_2, \ldots$ is a sequence of i.i.d. random variables with distribution function $F$, all defined on the same probability space. The existence of such a probability space is ensured by general theorems. Note that the supremum in (2.112) is only taken over $x \le 0$ because the hypothesis $H_0$ holds if we only require the validity of the equation in (2.111) for $x \le 0$.

What does this question have to do with the maximum modulus of a symmetric random walk? For any test, one must first consider the distribution of the test statistic—in our case, $Q_n$—under the assumption that the hypothesis $H_0$ holds. When $H_0$ is true, we want to reject $H_0$ only with a small pre-specified maximum probability, thereby committing a Type I error. We will see that under $H_0$, the random variable $nQ_n$ has the same distribution as the maximum modulus $|M|_n$ of the simple symmetric random walk, regardless of the specific form of the unknown underlying distribution function $F$, as long as $F$ is continuous. To show this, set

$$R_n(x) := n\left(\widehat{F}_n(x) + \widehat{F}_n(-x) - 1\right).$$

Note that, due to $1 = \widehat{F}_n(-x) + \frac{1}{n}\sum_{j=1}^n \mathbf{1}\{D_j > -x\}$,

$$R_n(x) = \sum_{j=1}^n \left( \mathbf{1}\{D_j \le x\} - \mathbf{1}\{D_j > -x\} \right). \tag{2.113}$$

---

[13] Valery Ivanovich Glivenko (1897–1940), Professor at the Pedagogical Institute of the University of Moscow from 1928. His main areas of work were mathematical logic and probability theory.

[14] Francesco Paolo Cantelli (1875–1966), Actuary in the Italian Ministry of Finance from 1903, Professor of Finance and Actuarial Mathematics at various universities from 1925. His main areas of work were financial mathematics and stochastics.

## 2.13 A Test of Symmetry

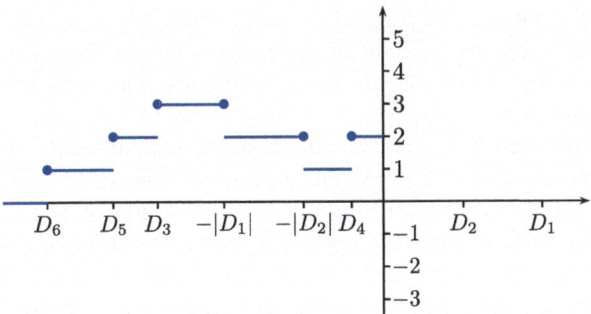

**Fig. 2.49** Graph of the function $R_n(x)$

In the following, we assume that $|D_1|, |D_2|, \ldots, |D_n|$ are pairwise distinct and nonzero, as this event has probability one due to the continuity of $F$. If $x < \min(-|D_1|, \ldots, -|D_n|)$, then $R_n(x) = 0$ because each of the indicator variables in (2.113) equals 0. At the point $x = -|D_j|$, the function $R_n$ jumps either up or down by one unit, depending on whether $D_j < 0$ or $D_j > 0$. It is right-continuous at an upward jump and left-continuous at a downward jump due to the less-than-or-equal and greater-than signs in (2.113). Figure 2.49 shows the graph of $R_n$, with the function value at a jump point highlighted by a small filled circle.

We now show that under the hypothesis $H_0$, the random variables $|D_1|, \ldots, |D_n|$ and $\text{sgn}(D_1), \ldots, \text{sgn}(D_n)$ are independent, and that $\mathbb{P}(\text{sgn}(D_j) = 1) = \mathbb{P}(\text{sgn}(D_j) = -1) = \frac{1}{2}$ for $j = 1, \ldots, n$. Here, as usual, $\text{sgn}(t)$ denotes the sign of a real number $t$.

First, due to the continuity of $F$, under $H_0$, for each $j = 1, \ldots, n$:

$$\frac{1}{2} = F(0) = \mathbb{P}(D_j < 0) = \mathbb{P}(\text{sgn}(D_j) = 1) = \mathbb{P}(\text{sgn}(D_j) = -1),$$

and for $x > 0$,

$$\mathbb{P}(|D_j| \leq x) = \mathbb{P}(-x \leq D_j \leq x) = F(x) - F(-x) = 2F(x) - 1.$$

Furthermore, for $x > 0$,

$$\mathbb{P}(\text{sgn}(D_j) = 1, |D_j| \leq x) = \mathbb{P}(0 < D_j \leq x) = F(x) - F(0)$$
$$= F(x) - \frac{1}{2} = \frac{1}{2}(2F(x) - 1)$$
$$= \mathbb{P}(\text{sgn}(D_j) = 1)\mathbb{P}(|D_j| \leq x).$$

Since $D_1, \ldots, D_n$ are independent, the assertion follows.

With this understanding, it becomes clear that the graph of the function $R_n$, like a simple symmetric random walk of length $n$, takes equally probable upward or downward steps that are independent of each other. Moreover, when taking the supremum in (2.112), the location of the jump points of the function $R_n$ does not matter, as long as they are different from each other and from 0 (this is where the continuity of $F$ plays a role!). Thus, under $H_0$, we can assume without loss of generality that $F$ is a specific distribution function, such as the uniform distribution on the interval $[-\frac{1}{2}, \frac{1}{2}]$. This assumption ensures that the jump points $-|D_1|, \ldots, -|D_n|$ are distinct. Regarding the distribution of $Q_n$, it is therefore equivalent to consider a symmetric random walk starting at the origin with length $n$. Since it is also irrelevant for taking the supremum in (2.112) whether one interpolates linearly between consecutive jump heights or not, we arrive at the following result:

**Theorem 2.41 ($H_0$-Distribution of Butler's Test Statistic)** *Under the hypothesis $H_0$ of symmetry, $nQ_n$ has the same distribution as the maximum modulus $|M|_n$ of a simple symmetric random walk.*

This result allows us to perform the test in a concrete situation. Given an upper bound $\alpha$ for the probability of a Type I error—i.e., incorrectly rejecting the hypothesis $H_0$ when it is actually true—one would reject $H_0$ if and only if the inequality $Q_n \geq k_0/n$ holds. Here, $k_0$ is the smallest integer $k$ such that

$$\mathbb{P}_{H_0}\left(Q_n \geq \frac{k}{n}\right) = \mathbb{P}(|M|_n \geq k) \leq \alpha, \qquad (2.114)$$

and $\mathbb{P}(|M|_n \geq k)$ is given in (2.103). Note that the probability on the left-hand side is indexed with $H_0$ to emphasize that the calculation of the probability of the event $\{Q_n \geq \frac{k}{n}\}$ is made under the assumption that the hypothesis $H_0$ is true. According to the above theorem, equality holds in (2.114).

Table 2.7 shows the diastolic blood pressure of 15 patients immediately before and 2 hours after taking a blood pressure-lowering drug (Source: [HA], p. 56). The last row shows the steps of the corresponding random walk, determined by the signs of the respective differences of the values above.

From the last row, it is evident that the test statistic $Q_{15}$ of Butler takes the value $\frac{11}{15}$ for this dataset. Because $\mathbb{P}_{H_0}(Q_{15} \geq \frac{11}{15}) = \mathbb{P}(|M|_{15} \geq 11) = 0.00836$, the result obtained is so unlikely under $H_0$ that the hypothesis of the ineffectiveness of the drug is rejected even at a maximum allowable probability of 0.01 of making a type I error.

**Table 2.7** Diastolic blood pressure (in mmHg) before (first row) and after (second row) taking a medication in 15 patients as well as steps of the corresponding random walk (last row)

| 130 | 122 | 124 | 104 | 112 | 101 | 121 | 124 | 115 | 102 | 98 | 119 | 106 | 107 | 100 |
|---|---|---|---|---|---|---|---|---|---|---|---|---|---|---|
| 125 | 121 | 121 | 106 | 101 | 85 | 98 | 105 | 103 | 98 | 90 | 98 | 110 | 103 | 82 |
| 1 | 1 | 1 | −1 | 1 | 1 | 1 | 1 | 1 | 1 | 1 | 1 | −1 | 1 | 1 |

According to Theorem 2.38, the limit $H_0$-distribution of $|M|_n/\sqrt{n}$ as $n \to \infty$ is a Rényi distribution. Hence, under $H_0$,

$$\sqrt{n}\, Q_n = \sqrt{n} \sup_{x \leq 0} \left| \widehat{F}_n(x) + \widehat{F}_n(-x) - 1 \right| \xrightarrow{\mathcal{D}} Z, \qquad Z \sim R. \quad (2.115)$$

Because $R(2.242) = 0.95$ (see Table 2.6), one rejects the hypothesis of symmetry of the distribution around 0 for large $n$ in the case $\alpha = 0.05$ if

$$Q_n \geq \frac{2.242}{\sqrt{n}}.$$

This test has asymptotically the level 0.05 as $n \to \infty$, i.e.,

$$\lim_{n \to \infty} \mathbb{P}_{H_0}(Q_n \geq 2.242/\sqrt{n}) = 0.05.$$

Finally, we consider how the Butler test behaves with increasing sample size $n$ when the hypothesis $H_0$ does not hold. In this case, there is at least one $x_0 \leq 0$ with $\delta := |F(x_0) + F(-x_0) - 1| > 0$. If we choose an $\varepsilon > 0$ such that $0 < \delta - \varepsilon$, then, due to the stochastic convergence of $|\widehat{F}_n(x_0) + \widehat{F}_n(-x_0) - 1|$ to $\delta$ and the inequality $Q_n \geq |\widehat{F}_n(x_0) + \widehat{F}_n(-x_0) - 1|$, it follows that $\lim_{n \to \infty} \mathbb{P}(Q_n \geq \delta - \varepsilon) = 1$. We thus obtain

$$\lim_{n \to \infty} \mathbb{P}(\sqrt{n}\, Q_n \geq c) = 1 \quad \text{for every fixed } c > 0.$$

According to (2.115), $\sqrt{n}\, Q_n$ has a limiting distribution under $H_0$ as $n \to \infty$, implying that the sequence of critical values of the test using $\sqrt{n}\, Q_n$ as the test statistic remains bounded. Therefore, the probability of rejecting $H_0$ when it is false converges to 1 as $n \to \infty$. Hence, Butler's test of symmetry is consistent against any fixed alternative to $H_0$.

## 2.14 Duality: New Insights

The stochastic behavior of a simple symmetric random walk is completely characterized by the independence of the random variables $X_1, X_2, \ldots$, which model the step directions, and the assumption $\mathbb{P}(X_j = 1) = \mathbb{P}(X_j = -1) = \frac{1}{2}$ for $j \geq 1$, indicating the equal probability of upward and downward steps. Under these assumptions, all $2^n$ random walks of length $n$ are equally likely. By convention, the height of a random walk at time $k \geq 0$ is given by $S_k = X_1 + \ldots + X_k$, with the empty sum set to 0.

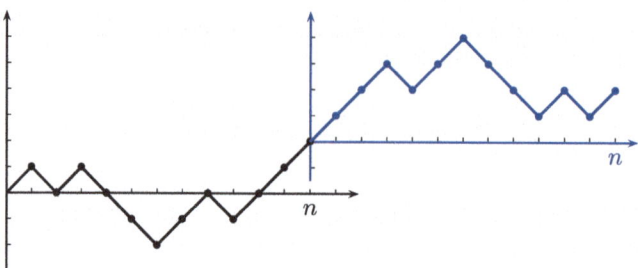

**Fig. 2.50** Random walk and dual random walk (180° rotation about endpoint)

We will now construct a so-called *dual random walk* of length $n$ that exhibits the same stochastic behavior and provides new, surprising insights (see [FE1], p. 91 ff.). To this end, we set

$$X_1^* := X_n, \quad X_2^* := X_{n-1}, \ldots, X_n^* := X_1. \tag{2.116}$$

Since the random vectors $(X_1, \ldots, X_n)$ and $(X_1^*, \ldots, X_n^*)$ have the same distribution (i.e., $\mathbb{P}(X_1 = a_1, \ldots, X_n = a_n) = \mathbb{P}(X_1^* = a_1, \ldots, X_n^* = a_n) = 2^{-n}$ for any choice of $(a_1, \ldots, a_n) \in \{-1, 1\}^n$), functions of these vectors also have the same distribution. For $X_1^*, \ldots, X_n^*$, the partial sums are given by

$$S_0^* := 0, \quad S_k^* := X_1^* + \ldots + X_k^* = S_n - S_{n-k}, \quad k = 1, \ldots, n. \tag{2.117}$$

The dual random walk $\{(k, S_k^*) : k = 0, \ldots, n\}$ is obtained from the original random walk $\{(k, S_k) : k = 0, \ldots, n\}$ by subjecting the latter to a 180° rotation about its endpoint $(n, S_n)$ and starting the resulting random walk at the origin (Fig. 2.50).

Since the above rotation is equivalent to a point reflection of the random walk at its endpoint, the transition to the dual random walk is geometrically a *point reflection principle*. This follows the proverbial saying, "the last will be the first and the first will be last," because definition (2.116) signifies a permutation of the upward and downward steps in this sense.

Since every event that can be formulated using $S_0, S_1, \ldots, S_n$ corresponds to an equally probable *dual event*, which arises from replacing $S_k$ with $S_k^*$ for $k = 0, 1, \ldots, n$, one might wonder whether new insights into the symmetric random walk can be gained through this *duality principle*.

**Record at Time $n$ (First Visit to the Endpoint at the End)**

How likely is it that a symmetric random walk of length $n \geq 2$ reaches the height $S_n$ *for the first time at time $n$*? In other words, what is the probability of the event

$$B_n := \{S_n > \max(S_0, \ldots, S_{n-1})\} \uplus \{S_n < \min(S_0, \ldots, S_{n-1})\}?$$

## 2.14 Duality: New Insights

This question can be easily answered by transitioning to the dual random walk. According to (2.117), the event $\{S_n > \max(S_0, \ldots, S_{n-1})\}$ is equivalent to the dual event

$$\{S_n - S_{n-k} > 0 \text{ for } k = 1, \ldots, n\} = \{S_1^* > 0, \ldots, S_n^* > 0\}.$$

Similarly,

$$\{S_n < \min(S_0, \ldots, S_{n-1})\} = \{S_n - S_{n-k} < 0 \text{ for } k = 1, \ldots, n\}$$
$$= \{S_1^* < 0, \ldots, S_n^* < 0\},$$

and we obtain

$$\mathbb{P}(B_n) = \mathbb{P}(S_1^* \neq 0, \ldots, S_n^* \neq 0) = \mathbb{P}(S_1 \neq 0, \ldots, S_n \neq 0).$$

Since $S_{2j} > 0$ necessarily implies $S_{2j+1} > 0$, and this implication also holds for the case of the less-than sign, it follows from the main lemma and Corollary 2.1 that for every positive integer $j$,

$$\mathbb{P}(B_n) = \frac{\binom{2j}{j}}{2^{2j}}, \quad \text{if} \quad n \in \{2j, 2j+1\}.$$

In the context of a random walk as a repeated coin toss game between two people, according to the main lemma, there are just as many game sequences of even length that end in a tie as there are game sequences where the final score never appears earlier in the game!

**Maxima at Time $n$ and First-passage Times**

In addition to the condition $S_n > \max(S_0, \ldots, S_{n-1})$, if we also specify a value $k$ with $k \in \{1, \ldots, n\}$ for $S_n$, then using the $k$-th passage time $V_k = \inf\{n \geq 1 : S_n = k\}$ introduced in Sect. 2.9, we obtain the event equality

$$\{S_n > \max(S_0, \ldots, S_{n-1}), S_n = k\} = \{V_k = n\}.$$

Since the left-hand event corresponds to the dual event $\{S_1^* > 0, \ldots, S_n^* > 0, S_n^* = k\}$, we get

$$\mathbb{P}(V_k = n) = \mathbb{P}(S_1^* > 0, \ldots, S_n^* > 0, S_n^* = k) \qquad (2.118)$$

(Note that Fig. 2.50 illustrates the case $k = 2$). The favorable $n$-paths for the occurrence of the event $\{S_1^* > 0, \ldots, S_n^* > 0, S_n^* = k\}$ were counted in connection with the ballot problem (this is a ballot problem with $a + b = n$ and $a - b = k$).

From (2.118), we derive the result already obtained by a different method in Theorem 2.24:

$$\mathbb{P}(V_k = n) = \frac{1}{2^n}\left(\binom{n}{\frac{n+k}{2}} - 2\binom{n-1}{\frac{n+k}{2}}\right) = \frac{1}{2^n}\frac{k}{n}\binom{n}{\frac{n+k}{2}}.$$

**Zeros and Final Height Visits**

In Sect. 2.4, we examined the number

$$N_{2n} := \sum_{j=1}^{n} \mathbf{1}\{S_{2j} = 0\}$$

of visits to 0. If we replace $S_{2j}$ here by $S_{2j}^* = S_{2n} - S_{2n-2j}$, we get the random variable

$$N_{2n}^* := \sum_{j=1}^{n} \mathbf{1}\{S_{2n} = S_{2n-2j}\} = \sum_{k=0}^{n-1} \mathbf{1}\{S_{2k} = S_{2n}\},$$

which denotes the number of times before $2n$ that the random walk reaches the final height $S_{2n}$. Thus, the random variable $N_{2n}^*$ has the same distribution as $N_{2n}$, as given in Theorem 2.6.

If we interpret a change of sign as an oscillation around the level 0, then, according to the duality principle, the number $C_{2n+1}$ of changes of sign examined in Sect. 2.11 has the same distribution as the number

$$C_{2n+1}^* = \sum_{j=0}^{n-1} \mathbf{1}\{(S_{2j} - S_{2n+1})(S_{2(j+1)} - S_{2n+1}) = -1\}$$

of oscillations of the random walk around the final height $S_{2n+1}$.

## 2.15 Outlook: Brown–Wiener Process and Law of the Iterated Logarithm

**Brown–Wiener Process and Invariance Principle**

In this chapter, we considered the simple symmetric random walk on $\mathbb{Z}$. To graphically represent random walks of length $n$, the (realizations of the) points $(0, S_0), (1, S_1), \ldots, (n, S_n)$ were plotted in a coordinate system, and consecutive points were connected, creating a polygonal chain called a *path*. Two of the numerous limit theorems, concerning various aspects of random walks as $n \to \infty$, made statements about *time proportions* relative to the total duration of the random walk, namely Theorem 2.5 (elapsed time proportion until the last visit to zero) and Theorem 2.11 (proportion of time the walk spends above the $x$-axis).

## 2.15 Outlook: Brown–Wiener Process and Law of the Iterated Logarithm

If we normalize the duration $n$ of a symmetric random walk of length $n$ to 1 and consider that $S_n/\sqrt{n}$ is asymptotically standard normally distributed as $n \to \infty$ (cf. (2.6)), it is natural to plot not $(0, S_0), (1, S_1), \ldots, (n, S_n)$, but the points

$$\left(\frac{j}{n}, \frac{S_j}{\sqrt{n}}\right), \quad j = 0, 1, \ldots, n,$$

and to connect them linearly. In this way, a (random) continuous function, defined on the unit interval $[0, 1]$, is created, whose (random) value at the point $t \in [0, 1]$ is equal to

$$W_n(t) := \frac{S_j}{\sqrt{n}} + \frac{t - \frac{j}{n}}{1/n} \cdot \frac{X_{j+1}}{\sqrt{n}}, \quad \text{if } t \in \left[\frac{j}{n}, \frac{j+1}{n}\right]$$

($j = 0, 1, \ldots, n-1$), or more compactly written as

$$W_n(t) = \frac{S_{\lfloor nt \rfloor}}{\sqrt{n}} + (nt - \lfloor nt \rfloor) \frac{X_{\lfloor nt \rfloor + 1}}{\sqrt{n}}, \quad 0 \leq t \leq 1. \tag{2.119}$$

The family of random variables $\{W_n(t), 0 \leq t \leq 1\}$ is commonly referred to as the *n-th partial sum process of the sequence* $(X_j)_{j \geq 1}$.

Since $X_1, X_2, \ldots$ and their partial sums $S_k = X_1 + \ldots + X_k$, for $k \geq 1$, are random variables defined on the probability space $(\Omega, \mathcal{A}, \mathbb{P})$ given in Sect. 7.2, $W_n(t)$ is also a random variable on $\Omega$ for each $t$. To emphasize this fact, we should actually write in (2.119) as follows:

$$W_n^\omega(t) := \frac{S_{\lfloor nt \rfloor}(\omega)}{\sqrt{n}} + (nt - \lfloor nt \rfloor) \frac{X_{\lfloor nt \rfloor + 1}(\omega)}{\sqrt{n}}, \quad 0 \leq t \leq 1, \omega \in \Omega. \tag{2.120}$$

Let $C[0, 1]$ denote the set of continuous functions on the interval $[0, 1]$. Then (2.120) shows that for a fixed $\omega \in \Omega$, the function $W_n^\omega$, defined by $[0, 1] \ni t \mapsto W_n^\omega(t)$, represents the so-called *path* of the partial sum process $W_n$ for the realization $\omega$ and is an element of $C[0, 1]$. For the case $n = 100$, three such paths obtained by simulation are sketched in Fig. 2.51. As $n$ increases, the simulated paths appear to stabilize in a certain stochastic manner. In fact, for fixed $t \in (0, 1]$, the second summand in (2.119) converges to zero in probability as $n \to \infty$. The first summand is given by $c_n S_{\lfloor nt \rfloor}/\sqrt{\lfloor nt \rfloor}$, where $c_n = \sqrt{\lfloor nt \rfloor}/\sqrt{n}$. Since $c_n \to \sqrt{t}$ and $S_{\lfloor nt \rfloor}/\sqrt{\lfloor nt \rfloor} \xrightarrow{\mathcal{D}} N(0, 1)$ as $n \to \infty$ (cf. (2.6)), it follows that $W_n(t) \xrightarrow{\mathcal{D}} N(0, t)$ as $n \to \infty$.

Moreover, it can be shown (see, e.g., [BI], Chap. 8), that not only does $W_n(t)$ converge in distribution for fixed $t$, but as $n \to \infty$, there is a convergence in distribution

$$W_n \xrightarrow{\mathcal{D}} W \quad \text{as } n \to \infty$$

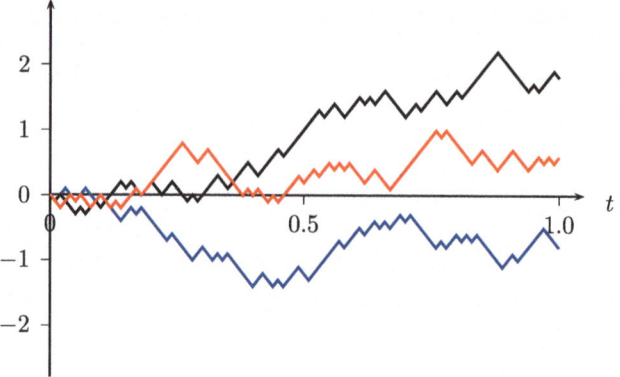

**Fig. 2.51** Three realizations of the partial sum process $W_{100}$

for the entire sequence of partial sum processes $(W_n) = (W_n(t))_{0 \leq t \leq 1}$ to a stochastic process $W = (W(t))_{0 \leq t \leq 1}$. This convergence is defined by the limit relationship

$$\lim_{n \to \infty} \mathbb{E} h(W_n) = \mathbb{E} h(W)$$

for all bounded functions $h : C[0, 1] \to \mathbb{R}$ that are continuous with respect to the supremum norm $\|f\| := \max_{0 \leq t \leq 1} |f(t)|, t \in C[0, 1]$, on $C[0, 1]$. This includes, in particular, the limit relations

$$\lim_{n \to \infty} \mathbb{P}\big(W_n(t_1) \leq x_1, \ldots, W_n(t_k) \leq x_k\big) = \mathbb{P}\big(W(t_1) \leq x_1, \ldots, W(t_k) \leq x_k\big)$$

for any choice of $k \geq 1$, $t_1, \ldots, t_k \in [0, 1]$ with $0 \leq t_1 < t_2 < \ldots < t_k \leq 1$, and real numbers $x_1, \ldots, x_k$. The stochastic process $W(t), 0 \leq t \leq 1$, whose realizations (paths) $W^\omega$ are continuous functions on $[0, 1]$, is called the *Brown*[15]– *Wiener*[16] *process*. It forms the starting point for many other stochastic processes and is characterized by the following properties:

(a) $\mathbb{P}(W(0) = 0) = 1$ (the process starts at 0),
(b) $W$ has *independent increments*, i.e., for any choice of $k \geq 2$ and any choice of $t_1, \ldots, t_k$ with $0 < t_1 < \ldots < t_k \leq 1$, the random variables $W(t_1), W(t_2) - W(t_1), \ldots, W(t_k) - W(t_{k-1})$ are independent,
(c) For $s$ and $t$ with $0 \leq s < t$, $W(t) - W(s) \sim N(0, t - s)$.

---

[15] Robert Brown (1773–1858) was a Scottish physician and botanist. He became a Fellow of the Royal Society in 1810 and served as a Fellow and President of the Linnean Society from 1849 to 1853.

[16] Norbert Wiener (1894–1964) was an American mathematician and the founder of cybernetics. His main fields of work included analysis, probability theory, and neurophysiology.

## 2.15 Outlook: Brown–Wiener Process and Law of the Iterated Logarithm

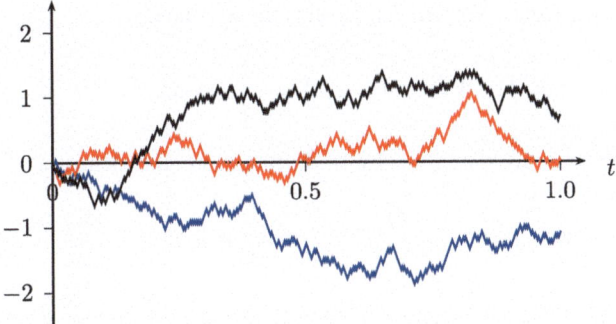

**Fig. 2.52** Three realizations of the partial sum process $W_{1000}$

Figure 2.52 shows three paths of $W_n$ for the case $n = 1000$. A further increase in $n$ would not convey a different impression. In this respect, this figure presents some approximate realizations of the Brown–Wiener process.

The convergence in distribution $W_n \xrightarrow{\mathcal{D}} W$ as $n \to \infty$ implies the convergence in distribution of certain real-valued functionals $h(W_n)$ to the distribution of $h(W)$. In particular, the following hold (see, e.g., [BI], p. 94 ff.):

$$\sup\{t \in [0,1] : W_n(t) = 0\} \xrightarrow{\mathcal{D}} \sup\{t \in [0,1] : W(t) = 0\}, \quad (2.121)$$

$$\lambda^1(\{t \in [0,1] : W_n(t) > 0\}) \xrightarrow{\mathcal{D}} \lambda^1(\{t \in [0,1] : W(t) > 0\}), \quad (2.122)$$

$$\max_{0 \le t \le 1} W_n(t) \xrightarrow{\mathcal{D}} \max_{0 \le t \le 1} W(t), \quad (2.123)$$

$$\max_{0 \le t \le 1} |W_n(t)| \xrightarrow{\mathcal{D}} \max_{0 \le t \le 1} |W(t)|. \quad (2.124)$$

Here, $\lambda^1$ denotes the Lebesgue measure in $\mathbb{R}^1$. Due to

$$\frac{L_{2n}}{2n} = \sup\{t \in [0,1] : W_n(t) = 0\},$$

(2.121) and Theorem 2.5 provide the arcsine law

$$\mathbb{P}(\sup\{t \in [0,1] : W(t) = 0\} \le x) = \frac{2}{\pi} \arcsin\sqrt{x}, \quad 0 \le x \le 1,$$

for the distribution of the time of the last visit to zero of the Brown–Wiener process $W$. Similarly,

$$\frac{O_{2n}}{2n} = \lambda^1(\{t \in [0,1] : W_n(t) > 0\}),$$

along with (2.122) and Theorem 2.11, yields the arcsine law

$$\mathbb{P}\left(\lambda^1(\{t \in [0,1] : W(t) > 0\} \leq x\right) = \frac{2}{\pi} \arcsin \sqrt{x}, \quad 0 \leq x \leq 1,$$

for the time spent above the $x$-axis by $W$. Furthermore, the identities

$$\max_{0 \leq t \leq 1} W_n(t) = \frac{M_n}{\sqrt{n}}, \qquad \max_{0 \leq t \leq 1} |W_n(t)| = \frac{|M|_n}{\sqrt{n}},$$

together with (2.123), (2.124) and Theorems 2.13 and 2.38, provide the distributions of the maximum and the absolute maximum of $W$. It holds that:

$$\mathbb{P}\left(\max_{0 \leq t \leq 1} W(t) \leq x\right) = 2\Phi(x) - 1, \qquad x \geq 0,$$

$$\mathbb{P}\left(\max_{0 \leq t \leq 1} |W(t)| \leq x\right) = R(x), \qquad x > 0.$$

If $X_1, X_2, \ldots$ is a sequence of i.i.d. random variables with $\mathbb{E}(X_1) = 0$ and $\mathbb{V}(X_1) = 1$, and if $S_n := X_1 + \ldots + X_n$ denotes the $n$-th partial sum of the $X_j$, then the Lindeberg–Lévy central limit theorem states that $S_n/\sqrt{n}$ converges in distribution to a standard normal distribution as $n \to \infty$. (More generally, if $\mathbb{E}(X_1) = a$ and $\mathbb{V}(X_1) = \sigma^2 > 0$, standardization is required; thus $X_j$ should be replaced by $(X_j - a)/\sigma$.) The limiting distribution of $S_n/\sqrt{n}$ is therefore *invariant* with respect to the original distribution of the $X_j$, as long as the latter is centered ($\mathbb{E}(X_1) = 0$) and rescaled ($\mathbb{V}(X_1) = 1$).

A famous result, the so-called *Donsker's*[17] *invariance principle* (see, e.g., [BI], Theorem 8.2), states that this invariance property continues to hold for the partial sum process $W_n$ in (2.119): the sequence $(W_n)$ also converges in this significantly more general situation to the Brown–Wiener process $W$, compared to the symmetric random walk (i.e., $\mathbb{P}(X_1 = -1) = \mathbb{P}(X_1 = 1) = \frac{1}{2}$). Figure 2.53 shows three realizations of $W_{100}$ for the case where $X_1$ has the distribution function $F(t) = 1 - \exp(-(t+1))$, $t \geq -1$ ($F(t) := 0$, if $t < -1$), and is thus distributed like a standard exponential random variable, reduced by 1. The realizations of $W_n$ shown in Fig. 2.54, each based on $n = 1000$ random variables with the same initial distribution, are qualitatively indistinguishable from those in Fig. 2.52.

Since the convergence in distribution of $W_n$ to $W$ extends to the convergence in distribution of certain real-valued functionals $h(W_n)$ to $h(W)$, the statements (2.121)–(2.124) also hold for partial sum processes with a general (centered and rescaled) distribution of the underlying random variables $X_1, X_2, \ldots$. For a game between two players over $n$ rounds, where the game score (in terms of the net

---

[17] Monroe David Donsker (1925–1991) was an American mathematician and professor at New York University. His main field of work was probability theory.

## 2.15 Outlook: Brown–Wiener Process and Law of the Iterated Logarithm

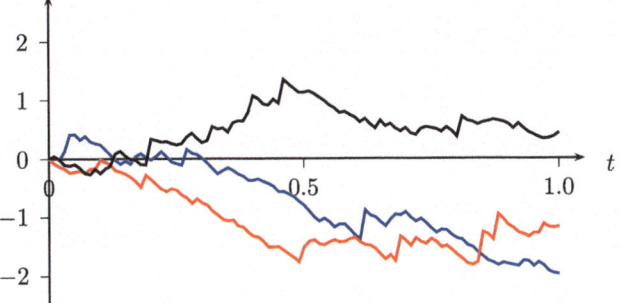

**Fig. 2.53** Realizations of $W_{100}$ with (centered)-exponentially distributed $X_j$

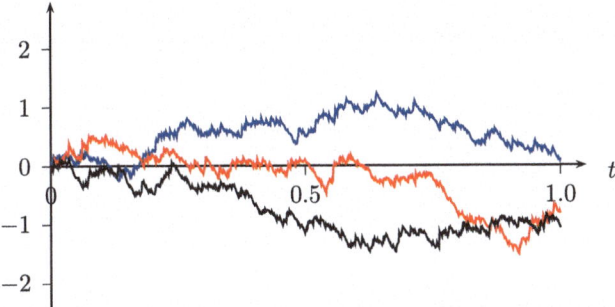

**Fig. 2.54** Realizations of $W_{1000}$ with (centered)-exponentially distributed $X_j$

amount of money flowed to player A) after $k$ rounds is given by the sum $X_1 + \ldots + X_k$ and interpolated between the individual rounds by forming the partial sum process $W_n$ in (2.119), this means that for large $n$, the elapsed fraction of time until the last tie is approximately given by the arcsine distribution, i.e.,

$$\mathbb{P}(\sup\{t \in [0,1] : W_n(t) = 0\} \leq x) \approx \frac{2}{\pi} \arcsin \sqrt{x}, \qquad 0 \leq x \leq 1.$$

Because the distribution of the $X_j$ is arbitrary, as long as $\mathbb{E}(X_j) = 0$ and $\mathbb{V}(X_j) = 1$, this is a truly universal statement!

### The Law of the Iterated Logarithm

The height $S_n$ of a simple symmetric random walk on $\mathbb{Z}$, after division by $\sqrt{n}$, is asymptotically standard normally distributed as $n \to \infty$ (see (2.6)). From this result, one can infer the intervals in which $S_n$ is likely to lie for a *fixed* large $n$. A natural question concerns the *almost sure fluctuation behavior* of the sequence

$(S_n)_{n\geq 1}$. Is there a positive increasing function $\lambda : \mathbb{N} \to \mathbb{R}$ such that for any fixed positive $\varepsilon$ the following holds:

$$\mathbb{P}(S_n \geq (1+\varepsilon)\lambda(n) \text{ for infinitely many } n) = 0, \tag{2.125}$$

$$\mathbb{P}(S_n \geq (1-\varepsilon)\lambda(n) \text{ for infinitely many } n) = 1? \tag{2.126}$$

Since the average of countably infinitely many events of probability 1 also has probability 1, and the union of countably many sets of probability 0 also has probability 0, it follows from (2.125) and (2.126) that, if we set

$$A_\varepsilon := \limsup_{n\to\infty} \left\{ \frac{S_n}{\lambda(n)} \geq 1+\varepsilon \right\}, \qquad B_\varepsilon := \limsup_{n\to\infty} \left\{ \frac{S_n}{\lambda(n)} \geq 1-\varepsilon \right\},$$

and consider the definition of the limit superior of a sequence of sets in Sect. 7.13, we get

$$\mathbb{P}\left( \bigcap_{k=1}^{\infty} A_{1/k} \setminus \left( \bigcup_{k=1}^{\infty} B_{1/k} \right) \right) = 1.$$

As a subset in the canonical probability space for infinitely long random walks from Sect. 7.2, the event occurring here consists of all sequences $\omega = (\omega_j)_{j\geq 1}$ for which the limit superior of the real sequence $S_n(\omega)/\lambda(n)$ is equal to 1. That such a function $\lambda$ exists, was first proved by A. Khintchin[18] [KH] for the situation of the simple symmetric random walk. The form of the function $\lambda$ gives the following famous result its name.

**Theorem 2.42 (Law of the Iterated Logarithm)** *Let $X_1, X_2, \ldots$ be independent random variables with the same distribution, where $\mathbb{E}(X_1) = 0$ and $\mathbb{V}(X_1) = 1$. Then, for the sequence $(S_n)$ of partial sums $S_n = X_1 + \ldots + X_n$,*

$$\mathbb{P}\left( \limsup_{n\to\infty} \frac{S_n}{\sqrt{2n \log \log n}} = 1 \right) = 1.$$

A proof of this theorem can be found in [BI2], p. 153 ff. If one applies the above result to the random variables $-X_1, -X_2, \ldots$, it follows that

$$\mathbb{P}\left( \liminf_{n\to\infty} \frac{S_n}{\sqrt{2n \log \log n}} = -1 \right) = 1.$$

---

[18] Alexander Khintchin (1894–1959) was a Russian mathematician and a highly influential member of the Russian school of mathematics. His main areas of work included analysis, probability theory, and statistical physics.

## 2.15 Outlook: Brown–Wiener Process and Law of the Iterated Logarithm

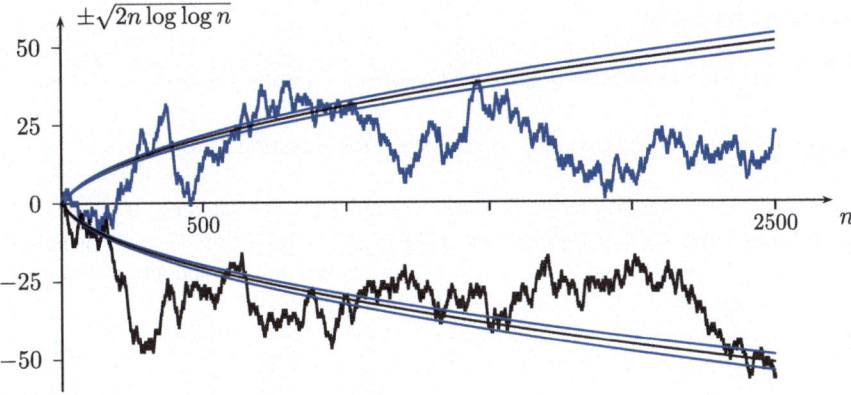

**Fig. 2.55** Illustrating the law of the iterated logarithm

Figure 2.55 shows graphs of the functions $n \mapsto \sqrt{2n \log \log n}$ and $n \mapsto -\sqrt{2n \log \log n}$ along with three simulated random walks of length 2500. Additionally, the graphs of the functions $n \mapsto \pm(1 \pm \varepsilon)\sqrt{2n \log \log n}$ for $\varepsilon = 0.05$ are also plotted in blue. For these, (2.125) and (2.126) hold with $\lambda(n) = \sqrt{2n \log \log n}$.

### Answers to the Self-Assessment Questions

**Answer 1** The event $\{S_7 = 3\}$ occurs if and only if five of the random variables $X_1, \ldots, X_7$ take the value $+1$ and two take the value $-1$. It thus follows that

$$\mathbb{P}(S_7 = 3) = \binom{7}{5}\left(\frac{1}{2}\right)^7 = \frac{21}{128} \approx 0.164.$$

**Answer 2** Because $S_{2n+1} = X_{2n+1} + S_{2n}$ and $|X_{2n+1}| = 1$, $\mathbb{E}|S_{2n+1}| \leq 1 + \mathbb{E}|S_{2n}|$. Furthermore, $S_{2n} = -X_{2n+1} + S_{2n+1}$, and thus $\mathbb{E}|S_{2n}| \leq 1 + \mathbb{E}|S_{2n+1}|$. The assertion follows.

**Answer 3** The left-hand side is equal to $\mathbb{P}\left(\sum_{j=1}^{n} X_j = \ell - a\right)$, and the right-hand side equals $\mathbb{P}\left(\sum_{j=m+1}^{m+n} X_j = \ell - a\right)$. Since $X_1, \ldots, X_{m+n}$ are independent and have the same distribution, the assertion follows.

**Answer 4** Yes. The binomial coefficient $\binom{n}{k}$ can be regarded as the number of binary $n$-tuples $(a_1, \ldots, a_n) \in \{0, 1\}^n$ with exactly $k$ ones. If you flip the ones and zeroes, i.e., if you switch to the tuple $(1 - a_1, \ldots, 1 - a_n)$, you get a binary $n$-tuple with exactly $n - k$ ones. The flipping corresponds to a reflection across the diagonal $\{(x, \ldots, x) : x \in \mathbb{R}\}$ in $\mathbb{R}^n$.

**Answer 5** Equation (2.18) follows from the Euler's formula $e^{it} = \cos t + i \sin t$, $t \in \mathbb{R}$, because, due to $e^{iy} e^{-iy} = e^0 = 1$, the right-hand side of (2.18), after expanding, is equal to

$$1 - (x\cos y + i\sin y) - x\bigl(\cos(-y) + i\sin(-y)\bigr) + x^2.$$

Since $\cos(-y) = \cos y$ and $\sin(-y) = -\sin y$, the assertion follows.

**Answer 6** According to (2.19), $F(1,0) = 0$ and $F(1,1) = 2$, since $0! = 1$. With the defining Eq. (2.21), it follows that $f(1) = F(1,0) + F(1,1) = 2$, which is identical to (2.23). The base case $n = 1$ is thus shown. For the induction step $n \mapsto n+1$ we must prove

$$f(n+1) = 2^{2(n+1)-3}(n+1)\bigl(3(n+1)+1\bigr). \tag{2.127}$$

The recursion formula (2.22) and the induction hypothesis yield

$$f(n+1) = 4 \cdot \frac{n+1}{n} \cdot \frac{3n+4}{3n+1} \cdot 2^{2n-3} n(3n+1),$$

which is the same as (2.127).

**Answer 7** If $x = 1$ then the left-hand side in (2.24) equals 1, but also the right-hand side, since the improper integral $\int_0^1 g(t)\,dt$ exists and equals 1.

**Answer 8** Since the density $a$ is symmetric about $\frac{1}{2}$, a random variable $Y$ with the distribution function $A$ has the same distribution as $1 - Y$. It follows for every $x \in (0,1)$ that

$$A(x) = \mathbb{P}(Y \leq x) = \mathbb{P}(1 - Y \leq x) = \mathbb{P}(Y \geq 1 - x)$$
$$= 1 - \mathbb{P}(Y \leq 1 - x) = 1 - A(1 - x).$$

**Answer 9** Such a replacement is possible, because according to the main lemma, there is a bijective mapping from the set of zero-avoiding paths of a certain length to the set of bridge paths of the same length.

**Answer 10** According to (2.42), $(\mathbb{E}|Z|)^2 = \frac{2}{\pi}$, and $\mathbb{E}(Z^2) = 1$, since $Z$ has a standard normal distribution.

**Answer 11** With $u_{2n} = \binom{2n}{n}/2^{2n}$, it follows that

$$u_{2n} = \frac{(2n)!}{n!^2 2^{2n}} = \frac{2n(2n-1)(2(n-1))!}{4n^2(n-1)!^2 2^{2(n-1)}} = \frac{2n-1}{2n} \cdot u_{2(n-1)}.$$

Hence, $u_{2(n-1)} - u_{2n} = u_{2(n-1)}/(2n)$, which was to be shown.

## 2.15 Outlook: Brown–Wiener Process and Law of the Iterated Logarithm

**Answer 12** The equality in distribution $U_{2n} \sim O_{2n}$ holds for reasons of symmetry, because $(X_1, \ldots, X_{2n})$ has the same distribution as $(-X_1, \ldots, -X_{2n})$, and $U_{2n}$ is the same function of $(X_1, \ldots, X_{2n})$ as $O_{2n}$ is of $(-X_1, \ldots, -X_{2n})$ (reflection across the $x$-axis!)

**Answer 13** Equation (2.68) only uses the symmetry $\binom{2n}{n+k} = \binom{2n}{n-k}$, $k = 1, \ldots, n$, of the binomial coefficients, as well as $\sum_{k=0}^{2n} \binom{2n}{k} = 2^{2n}$.

**Answer 14** The largest possible number $\lfloor n/2 \rfloor + 1$ of maximizers occurs if a downward step is taken at the beginning and then alternately an upward and a downward step.

**Answer 15** Because

$$A_j = \{X_1 + \ldots + X_{2j} \geq 0, X_2 + \ldots + X_{2j} \geq 0, \ldots, X_{2j} \geq 0\},$$
$$B_{j,k} = \{X_{2j+1} \leq 0, \ldots, X_{2j+1} + \ldots + X_{2k-1} \leq 0, X_{2j+1} + \ldots + X_{2k} = 0\},$$
$$C_k = \{X_{2k+1} \leq 0, \ldots, X_{2k+1} + \ldots, +X_{2n} \leq 0\},$$

the events $A_j$, $B_{j,k}$ and $C_k$ are formed by pairwise disjoint blocks of the random variables $X_1, \ldots, X_{2n}$.

**Answer 16** $A_k$ depends only on $X_1, \ldots, X_k$ and $B_k$ only on $X_{k+1}, \ldots, X_{2n}$.

**Answer 17** The event $\{V_k \leq n\}$ means that the random walk reaches the height $k$ at the latest at time $n$. This is equivalent to the fact that the maximum $M_n$ is greater than or equal to $k$.

**Answer 18** According to (2.7),

$$\mathbb{P}(S_{n-1} = k - 1) = \frac{\binom{n-1}{\frac{n-1+k-1}{2}}}{2^{n-1}}, \quad \mathbb{P}(S_{n-1} = k + 1) = \frac{\binom{n}{\frac{n-1+k+1}{2}}}{2^{n-1}}.$$

Thus, $\frac{1}{2}\big(\mathbb{P}(S_{n-1} = k-1) - \mathbb{P}(S_{n-1} = k+1)\big)$ becomes

$$\frac{1}{n 2^n} \left( \frac{n+k}{2} \binom{n}{\frac{n+k}{2}} - \left(n - \frac{n+k}{2}\right)\binom{n}{\frac{n+k}{2}} \right) = \frac{k}{n 2^n} \binom{n}{\frac{n+k}{2}}.$$

**Answer 19** Note that $\{T_1 = 1\} = \{\delta_1 = 2\}$ and $\{T_1 = 2\} = \{\delta_1 = 0, \delta_2 = 2\}$, and the probabilities of these events are $\frac{1}{4}$ and $\frac{1}{2} \cdot \frac{1}{4} = \frac{1}{8}$, respectively.

**Answer 20** At least $r$ changes of sign take place by the time $2n + 1$ at the latest, if and only if the time defined by $T_1 + \ldots + T_r$ until the $r$-th change of sign is at most equal to $2n$.

**Answer 21** Since $S_1$ can only take the values 1 and $-1$, $\mathbb{P}(|M|_n \geq 1) = 1$. Because there are exactly two paths for which $|M|_n$ takes the maximum possible value $n$ (namely the steadily ascending or descending path), $\mathbb{P}(|M|_n \geq n) = 2/2^n$.

**Answer 22** If both of the events $\{M_{2n} = k\}$ and $\{m_{2n} = -k\}$ occur then necessarily $k + 2k \leq 2n$, because if the random walk is to reach both the height $k$ and the height $-k$, it must perform at least $3k$ steps.

### Exercises

**Exercise 2.1** Let $X_1, \ldots, X_{2n}$ be independent random variables, where $\mathbb{P}(X_j = 1) = \mathbb{P}(X_j = -1) = \frac{1}{2}$ for $j = 1, \ldots, 2n$. Furthermore, let $S_{2n} := X_1 + \ldots + X_{2n}$. Prove the following:

(a) $\mathbb{E}|S_{2n}| = 2n \cdot \dfrac{\binom{2n}{n}}{2^{2n}}$.

(b) $\mathbb{E}|S_{2n}| \sim \sqrt{\dfrac{2}{\pi}} \cdot \sqrt{2n}$ as $n \to \infty$.

**Hint for (a):** Note (7.25).

**Exercise 2.2** On a $(m \times n)$-grid with coordinates $(i, j)$, $i \in \{0, \ldots, m\}$, $j \in \{0, \ldots, n\}$ (see the following illustration for the case $m = 8, n = 6$), a robot starts at point $(0, 0)$. As shown, it can only move to the right or upwards per step.

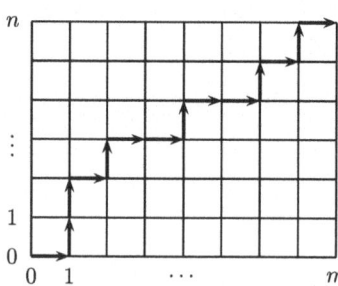

(a) How many such paths are there in total?
(b) How many of these paths pass through the point $(a, b)$, which is different from $(0, 0)$ and $(m, n)$?

## 2.15 Outlook: Brown–Wiener Process and Law of the Iterated Logarithm

(c) By a suitable division of cases for paths, show the *Vandermonde*[19] *identity*

$$\binom{m+n}{m} = \sum_{k=0}^{n}\binom{n}{k}\binom{m}{m-k}.$$

**Exercise 2.3** In the context of the ballot problem, let $E$ be the event that $A$ has *always at least as many* votes as $B$ during the entire vote count. Show that

$$\mathbb{P}(E) = \frac{a-b+1}{a+1}.$$

**Exercise 2.4** Prove part (b) of Theorem 2.6 by induction with the use of (2.35).

**Exercise 2.5** Derive the representation (2.37) using (2.36).

**Exercise 2.6** Prove (2.38).

**Exercise 2.7** Let $W$ be the time of the first return to zero of the simple symmetric random walk on $\mathbb{Z}$ and

$$\mathbb{E}[W|W \leq 2n] = \sum_{k=1}^{n} 2k\mathbb{P}(W = 2k|W \leq 2n)$$

be the conditional expectation of $W$ given $W \leq 2n$, where $n \geq 1$. Further, let $u_{2k} := \binom{2k}{k}/2^{2k}$, where $k \geq 1$. Prove:

(a) $\mathbb{E}[W|W \leq 2n] = \dfrac{(2n-1)u_{2(n-1)}}{1-u_{2n}}$,

(b) $\mathbb{E}[W|W \leq 2n] \sim \dfrac{2}{\pi}\sqrt{n}$ as $n \to \infty$.

Recall the general notation $a_n \sim b_n :\Longleftrightarrow a_n/b_n \to 1$ as $n \to \infty$.

**Exercise 2.8** Derive the difference Eq. (2.46).

**Exercise 2.9** Prove the inequality (2.50).

**Exercise 2.10** Let $W$ be the time of the first return to zero of the simple symmetric random walk on $\mathbb{Z}$. Show that

$$\mathbb{E}\big(\min(W, 2n)\big) = 2\mathbb{E}|S_{2n}| = 4n\mathbb{P}(S_{2n} = 0), \qquad n \geq 1.$$

---

[19] Alexandre-Théophile Vandermonde (1735–1796), French mathematician, musician, and chemist.

**Hint:** Consider (2.45), (2.35), Theorem 2.6 (b), and Exercise 2.1 (a).

**Exercise 2.11** Using (2.67) and (7.25), prove the first part of statement (b) of Theorem 2.12.

**Exercise 2.12** Show that in the situation of Theorem 2.12:

$$\mathbb{E}\left(M_{2n}^2\right) = 2n + \frac{1}{2} - \left(2n + \frac{1}{2}\right) \frac{\binom{2n}{n}}{2^{2n}}.$$

Hint: Start with (2.69).

**Exercise 2.13** Show that in the situation of Sect. 2.7:

$$\mathbb{P}(M_{2n-1} = k, S_{2n} = 0) = \mathbb{P}(S_{2n} = 2k) - \mathbb{P}(S_{2n} = 2k+2), \quad k \in \{0, 1, \ldots, n\}.$$

**Exercise 2.14** Let $N_{2n}$ and $M_{2n}$ be the number of returns to zero and the maximum of a simple symmetric random walk of length $2n$ on $\mathbb{Z}$, respectively. Show that

$$\left|\mathbb{E}(N_{2n}) - \mathbb{E}(M_{2n})\right| \leq \frac{1}{2}.$$

**Exercise 2.15** Let $a, b$ and $v$ be integers with $a \leq 0 \leq b$, $a < b$, $a \leq v \leq b$, $v \neq a$ and $v \neq b$. In addition, let $v$ be odd. Furthermore, let $X$ be a random variable with $\mathbb{P}(X = 1) = \mathbb{P}(X = -1) = \frac{1}{2}$. Show that

$$\mathbb{P}(a < \min(0, X) \leq \max(0, X) < b, X = v)$$
$$= \sum_{k=-\infty}^{\infty} \mathbb{P}\bigl(X = v + 2k(b-a)\bigr) - \sum_{k=-\infty}^{\infty} \mathbb{P}\bigl(X = 2b - v + 2k(b-a)\bigr).$$

(2.128)

**Exercise 2.16** Show using (2.77):

$$\mathbb{V}(Q_{2n}) = 2 - \frac{2(3n+2)}{n+2} \cdot u_{2(n+1)} - 4\, u_{2(n+1)}^2.$$

**Exercise 2.17** Show that the first-passage time $V_1$ satisfies

$$\mathbb{P}(V_1 = 2n+1) \sim \frac{C}{n^{3/2}} \quad \text{as } n \to \infty,$$

and determine $C$.

## 2.15 Outlook: Brown–Wiener Process and Law of the Iterated Logarithm

**Exercise 2.18** Show that the distribution of the $k$-th passage time $V_k$ satisfies the recursion formula

$$\mathbb{P}(V_k = n+2) = \frac{n(n+1)}{(n+2+k)(n+2-k)} \mathbb{P}(V_k = n), \qquad n \geq k,$$

and the initial condition $\mathbb{P}(V_k = n) = 2^{-k}$.

**Exercise 2.19** Using (2.90), show that for each $t$ with $|t| < 1$ and $t \neq 0$:

$$\frac{2\left(1 - \sqrt{1-t}\right) - t}{t} = \sum_{n=1}^{\infty} \frac{\binom{2n}{n}}{(n+1)\, 2^{2n}} t^n.$$

**Exercise 2.20** Show that for the number $C_{2n+1}$ of changes of sign in a simple symmetric random walk:

$$\mathbb{V}(C_{2n+1}) = \frac{n+1}{2} - \frac{1}{4} - \left((n+1)\frac{\binom{2n+1}{n}}{2^{2n+1}}\right)^2.$$

**Exercise 2.21** Let $n \geq 2$ and $k \in \{1, \ldots, n-1\}$. Show that in the case of the symmetric random walk on $\mathbb{Z}$:

$$\mathbb{P}\big((S_1 \neq S_{2n},\, S_2 \neq S_{2n},\, \ldots,\, S_{2k-1} \neq S_{2n},\, S_{2k} = S_{2n}\big)$$
$$= \mathbb{P}(S_{2k} = 0)\, \mathbb{P}(S_{2(n-k)} = 0).$$

**Hint:** Duality.

# 3 Bridges: The Tied-down Random Walk

This chapter addresses various questions related to simple symmetric random walks of even length $2n$ that start at the origin and end at the point $(2n, 0)$. In Chap. 2 we called such a random walk a $2n$-*bridge*. In what follows, we will sometimes also use the term *tied-down random walk*. Figure 3.1 shows such a tied-down random walk of length 500.

According to the main lemma, there are $\binom{2n}{n}$ tied-down random walks of length $2n$, and we will throughout assume that each of these $2n$-bridges has the same probability, i.e., we choose a uniform distribution on the set

$$W_{2n}^\circ := \{w := (a_1, \ldots, a_{2n}) \in \{-1, 1\}^{2n} : a_1 + \ldots + a_{2n} = 0\}$$

of all $2n$-bridges. As in Chap. 2, $a_j$ denotes the direction of the $j$-th step of the random walk, i.e., 1 for an upward step and $-1$ for a downward step. The superscript "$\circ$" in the notation $W_{2n}^\circ$ highlights that the random walk of length $2n$ returns to zero at the end. We use a similar notation for random variables defined on $W_{2n}^\circ$, writing, for example, $M_{2n}^\circ$ for the maximum of a $2n$-bridge. An exception is the random variable $X_j$ on $W_{2n}^\circ$, which, as in Chap. 2, denotes the $j$-th step of the random walk; hence $X_j(w) = a_j$ for each $w = (a_1, \ldots, a_{2n}) \in W_{2n}^\circ$. We proceed similarly with the partial sums $S_k := X_1 + \ldots + X_k$ for $k = 1, \ldots, 2n$, which indicate the heights of the random walk at times $1, 2, \ldots, 2n$. As before, let $S_0 := 0$.

Tied-down random walks arise naturally in the so-called *two-sample problem* of *nonparametric statistics*. In this problem, realizations of independent random variables $U_1, \ldots, U_n$ and $V_1, \ldots, V_n$ are given, where $U_1, \ldots, U_n$ share the same distribution function $F$, and $V_1, \ldots, V_n$ share the same distribution function $G$. Thus, $F(x) = \mathbb{P}(U_1 \le x)$ and $G(x) = \mathbb{P}(V_1 \le x)$ for $x \in \mathbb{R}$. The functions $F$ and $G$ are assumed to be continuous but otherwise unknown. The hypothesis to be tested is $H_0 : F = G$, which means the distributions of $U_1$ and $V_1$ are equal.

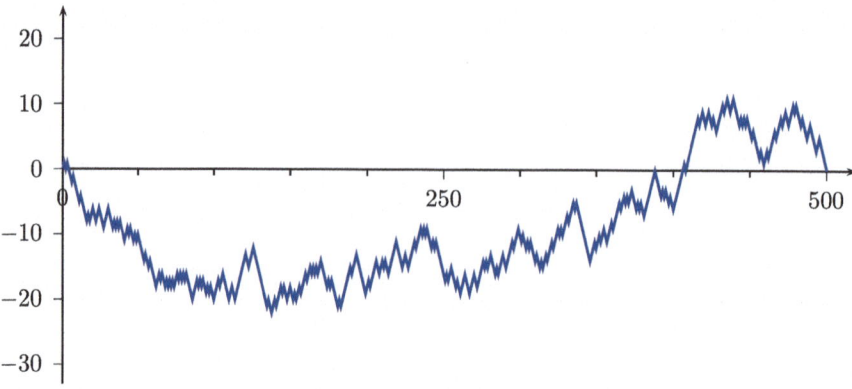

**Fig. 3.1** A tied-down random walk of length 500

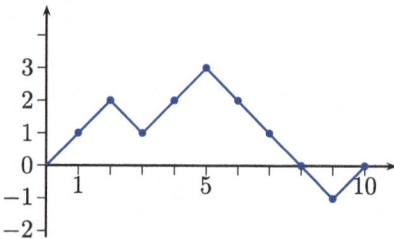

**Fig. 3.2** The bridge corresponding to the sequence $u\,u\,v\,u\,u\,v\,v\,v\,v\,u$

Due to the continuity of $F$ and $G$, the realizations of the random variables $U_1, \ldots, U_n$ and $V_1, \ldots, V_n$ are pairwise distinct with probability one. If one arranges these realizations in ascending order, and starts a random walk at 0 by taking an upward or downward step for each $j = 1, \ldots, 2n$, depending on whether the $j$-th smallest value is a realization of $U_1, \ldots, U_n$ or $V_1, \ldots, V_n$, a $2n$-bridge is created. Under the hypothesis $H_0$, all $\binom{2n}{n}$ such bridges are equally likely.

Figure 3.2 shows a specific bridge in the case $n = 5$, which corresponds to a realized sequence $u\,u\,v\,u\,u\,v\,v\,v\,v\,u$ of $U_1, \ldots, U_5, V_1, \ldots, V_5$. Thus, both the smallest and the second smallest of all 10 values are realizations of the $U_j$'s, the third smallest value is a realization of the $V_j$'s, the fourth smallest is a realization of the $U_j$'s, and so on.

Francis Galton[1] and Charles Darwin[2] proposed testing the hypothesis $H_0$ of the equality of the underlying distributions by counting how many $j \in \{1, \ldots, n\}$ have the $j$-th-smallest realization of $U_1, \ldots, U_n$ smaller than the $j$-th-smallest realization of $V_1, \ldots, V_n$ (see [DA], p. 16). In the situation shown in Fig. 3.2, the value for this test statistic, known as *Galton's rank order test* (see [HO]) is 4, because only the largest value among all the $u$'s is not smaller than the largest value among all the $v$'s. The value 4 is half of the total time the path shown in Fig. 3.2 spends above the horizontal axis. This statement is generally valid, because if $S_1, \ldots, S_{2n-1}, S_{2n}(= 0)$ are the heights of the path consisting of $n$ upward and downward steps of the ordered realizations of $U_1, \ldots, U_n, V_1, \ldots, V_n$, then the $j$-th-smallest value of $U_1, \ldots, U_n$ is smaller than the $j$-th-smallest value of $V_1, \ldots, V_n$ if and only if $S_{2j-1} > 0$. Since for each such $j$, $S_{2j} \geq 0$ also holds, the test statistic of Galton's rank order test is equal to half of the sojourn time of a $2n$-bridge above the $x$-axis, studied in Sect. 3.2. In Sect. 3.7, we will discuss the nonparametric two-sample problem in more detail.

The assumption of a uniform distribution over all paths $w = (a_1, \ldots, a_{2n})$ from $W_{2n}^\circ$ corresponds to the joint distribution of $X_1, \ldots, X_{2n}$, defined by

$$\mathbb{P}(X_1 = a_1, \ldots, X_{2n} = a_{2n}) := \binom{2n}{n}^{-1}, \quad \text{if } \sum_{j=1}^{2n} \mathbf{1}\{a_j = 1\} = n, \qquad (3.1)$$

and $\mathbb{P}(X_1 = a_1, \ldots, X_{2n} = a_{2n}) := 0$ otherwise. Note that $X_1, \ldots, X_{2n}$ are not independent, unlike in Chap. 2. It still holds that $\mathbb{P}(X_j = 1) = \mathbb{P}(X_j = -1) = \frac{1}{2}$ for each $j = 1, \ldots, 2n$, but, for example,

$$\mathbb{P}(X_1 = 1, X_2 = 1) = \mathbb{P}(X_1 = 1) \cdot \mathbb{P}(X_2 = 1 | X_1 = 1)$$

$$= \frac{\binom{2n-2}{n-2}\binom{n}{n}}{\binom{2n}{n}} = \frac{n}{2n} \cdot \frac{n-1}{2n-1}$$

$$\neq \frac{1}{4} = \mathbb{P}(X_1 = 1) \cdot \mathbb{P}(X_2 = 1).$$

---

[1] Francis Galton (1822–1911), a cousin of Charles Darwin, initially studied medicine at Cambridge. He then embarked on exploratory journeys, earning the Gold Medal of the Royal Geographical Society in 1853. Galton eventually turned his attention to meteorology, where he was the first to recognize the significance of what he termed anticyclones. From 1865 onward, he became deeply involved in inheritance theory and statistics, contributing to the development of regression and correlation, and inventing the "Galton board." Additionally, he pioneered the use of fingerprints for personal identification.

[2] Charles Robert Darwin (1809–1882), British naturalist. His nearly five-year-long second voyage with the HMS Beagle laid the foundation for his later work. Because of his contributions to the theory of evolution, Darwin is considered one of the most significant natural scientists.

Interpreting 1 as drawing a red ball and $-1$ as drawing a black ball from an urn containing $n$ red and $n$ black balls, where the drawing is completely random without replacement, the random vector $(X_1, \ldots, X_{2n})$ with the distribution given in (3.1) models the colors of the drawn balls in the order of drawing. All $2n$ balls are drawn sequentially. It is helpful to connect the temporal progression of a $2n$-bridge with this concept.

As in Chap. 2, we will derive some interesting limit theorems. For asymptotic investigations as $n$ tends to infinity, we would technically need to introduce a second index and write $X_{n,1}, X_{n,2}, \ldots, X_{n,2n}$ instead of $X_1, \ldots, X_{2n}$, thus indicating the underlying uniform distribution on the set $W_{2n}^\circ$ with the index $n$. However, we will use this cumbersome notation only in the last section. Since a tied-down random walk of length $2n$, in contrast to a "free" random walk of the same length as in Chap. 2, whose steps are chosen independently of each other, is ultimately "forced onto the $x$-axis," it is to be expected that such a path will be dampened in its volatility. Consequently, the random walk statistics considered in Chap. 2, such as the number of visits to zero, sojourn times, number of changes of sign, etc., may now exhibit less erratic behavior. We will see to what extent this assumption holds true.

## 3.1 The Number of Interior Zeros

In this section, we consider the number

$$N_{2n}^\circ := \sum_{j=1}^{n-1} \mathbf{1}\{S_{2j} = 0\} \tag{3.2}$$

of non-trivial visits to zero of a $2n$-bridge. We will term a visit to zero at one of the time points $2, 4, \ldots, 2n-2$ an *interior zero* of the $2n$-bridge. The random variable $N_{2n}^\circ$ thus counts the number of interior zeros, and it takes the values $0, 1, \ldots, n-1$. To determine the probability $\mathbb{P}(N_{2n}^\circ = k)$, we need the number of all $2n$-bridges that have exactly $k$ interior zeros. This task becomes somewhat simpler if we first ask for the number of all $2n$-bridges that have *at least* $k$ interior zeros, i.e., for which $N_{2n}^\circ \geq k$. Since each such bridge can be continued in two ways, both at the beginning and directly after each of the $k$ interior zeros, the number sought is equal to $2^{k+1}$ times the number of $2n$-bridges with at least $k$ interior zeros, which are non-negative up to the $k$-th interior zero and subsequently take an upward step (see Fig. 3.3). The latter set of $2n$-bridges is denoted by $M_1$.

We consider an arbitrarily chosen bridge from $M_1$. If we omit the first upward step and, for each $j$ from 1 to $k$, the upward step directly following the $k$-th interior zero, and then concatenate the remaining parts, we obtain a path from $(0, 0)$ to the point $(2n - (k+1), -(k+1))$. In Fig. 3.3, the upward steps to be omitted are marked with blue arrows, and Fig. 3.4 shows the resulting path leading from $(0, 0)$

## 3.1 The Number of Interior Zeros

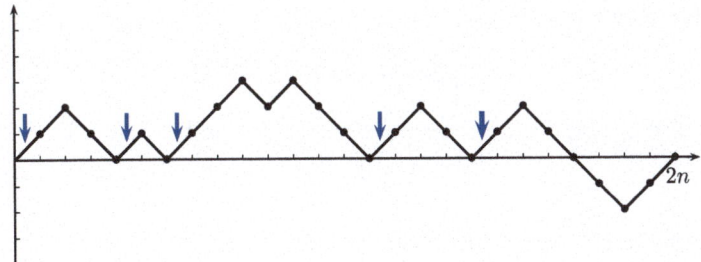

**Fig. 3.3** $2n$-bridge with at least 4 interior zeros, which is non-negative up to the fourth such zero and then takes an upward step

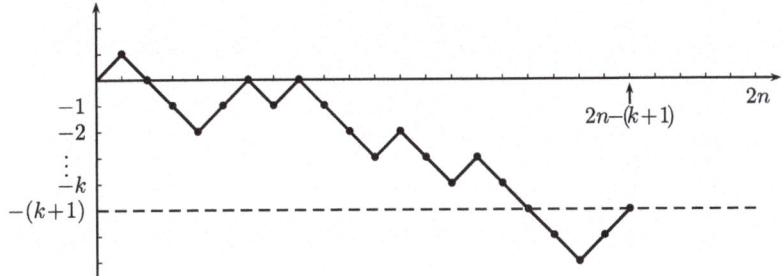

**Fig. 3.4** Path leading from $(0, 0)$ to $(2n - (k + 1), -(k + 1))$

to $(2n-(k+1), -(k+1))$. This mapping of paths, which we will term the "upward-steps-omission-mapping," is clearly an injective function on $M_1$.

Let $M_2$ denote the set of all paths leading from $(0, 0)$ to $(2n - (k + 1), -(k + 1))$. Note that, according to Lemma 2.1,

$$|M_2| = \binom{2n - (k + 1)}{n}.$$

We will now show that the "upward-steps-omission-mapping" from $M_1$ to $M_2$ is also surjective and thus bijective overall. To this end, consider an arbitrary path leading from $(0, 0)$ to $(2n - (k + 1), -(k + 1))$. We add exactly $k + 1$ upward steps to this path: one at the beginning and, for each $j$ from 1 to $k$, an upward step directly following the first time the path reaches the height $-j$. In this way, the path from Fig. 3.4 is transformed back into the path from Fig. 3.3. Since $|M_1| = |M_2|$, it follows that

$$|\{N^\circ_{2n} \geq k\}| = 2^{k+1} |M_1| = 2^{k+1} |M_2| = 2^{k+1} \binom{2n - (k + 1)}{n},$$

and thus (cf. [DW], p. 1046),

$$\mathbb{P}(N^\circ_{2n} \geq k) = 2^{k+1} \frac{\binom{2n-k-1}{n}}{\binom{2n}{n}}, \quad k = 0, 1, \ldots, n-1. \tag{3.3}$$

The following result gives the distribution of $N^\circ_{2n}$.

**Theorem 3.1 (Distribution of the Number of Interior Zeros of a Bridge)** *If $N^\circ_{2n}$ denotes the number of interior zeros of a $2n$-bridge, then:*

(a) $\mathbb{P}(N^\circ_{2n} = k) = \dfrac{2^{k+1}(k+1)}{n} \cdot \dfrac{\binom{2n-k-2}{n-1}}{\binom{2n}{n}}, \quad k = 0, \ldots, n-1,$

(b) $\mathbb{E}\left(N^\circ_{2n}\right) = \dfrac{2^{2n}}{\binom{2n}{n}} - 2,$

(c) $\lim\limits_{n \to \infty} \dfrac{\mathbb{E}\left(N^\circ_{2n}\right)}{\sqrt{2n}} = \sqrt{\dfrac{\pi}{2}} \approx 1.253,$

(d) $\mathbb{V}\left(N^\circ_{2n}\right) = 2(2n+1) - \dfrac{2^{2n}}{\binom{2n}{n}} \left(1 + \dfrac{2^{2n}}{\binom{2n}{n}}\right).$

*Proof*

(a) Using $\mathbb{P}(N^\circ_{2n} = k) = \mathbb{P}(N^\circ_{2n} \geq k) - \mathbb{P}(N^\circ_{2n} \geq k+1)$ for $k = 0, 1, \ldots, n-2$ and $\mathbb{P}(N^\circ_{2n} \geq n-1) = \mathbb{P}(N^\circ_{2n} = n-1)$, the assertion follows from (3.3).

(b) We provide two different proofs. The first uses (3.2), according to which

$$\mathbb{E}(N^\circ_{2n}) = \sum_{j=1}^{n-1} \mathbb{P}(S_{2j} = 0).$$

Among all $\binom{2n}{n}$ paths, the favorable paths for the event $\{S_{2j} = 0\}$ are characterized by the fact that among the first $2j$ steps, $j$ must be selected as upward steps, and among the last $2n - 2j$ steps, $n - j$ must be chosen as upward steps. The steps not selected are downward steps. Thus, the number of favorable

## 3.1 The Number of Interior Zeros

paths is $\binom{2j}{j}\binom{2(n-j)}{n-j}$, and it follows that

$$\mathbb{E}(N^\circ_{2n}) = \sum_{j=1}^{n-1} \frac{\binom{2j}{j}\binom{2(n-j)}{n-j}}{\binom{2n}{n}}.$$

Since the sum of the probabilities in Theorem 2.4 is equal to 1, we obtain

$$\mathbb{E}(N^\circ_{2n}) = \frac{2^{2n}}{\binom{2n}{n}}\left(\sum_{j=0}^{n}\frac{\binom{2j}{j}\binom{2(n-j)}{n-j}}{2^{2n}} - 2\frac{\binom{2n}{n}}{2^{2n}}\right)$$

$$= \frac{2^{2n}}{\binom{2n}{n}}\left(1 - 2\frac{\binom{2n}{n}}{2^{2n}}\right).$$

Thus, the assertion holds. Another proof can be provided using the relationship shown in Sect. 7.6: $\mathbb{E}(N^\circ_{2n}) = \sum_{k=1}^{n-1}\mathbb{P}(N^\circ_{2n} \geq k)$. The assertion then follows from (3.3) and the fact that the sum of the probabilities in (2.32) is equal to 1, which implies

$$\sum_{k=1}^{n-1} 2^{k+1}\binom{2n-k-1}{n} = 2^{2n} - \binom{2n}{n} - 2\binom{2n-1}{n}.$$

(c) Note that

$$\frac{\mathbb{E}(N^\circ_{2n})}{\sqrt{2n}} = \sqrt{\frac{\pi}{2}} \cdot \frac{1}{\sqrt{\pi n} \cdot u_{2n}} - \frac{2}{\sqrt{2n}},$$

with $u_{2n}$ as in (2.15). The assertion now follows from (2.29).

(d) Let $q_k := \mathbb{P}(N^\circ_{2n} \geq k)$ with $\mathbb{P}(N^\circ_{2n} \geq k)$ as in (3.3). A direct calculation yields

$$(2n-k-1)q_{k+1} = 2(n-k-1)q_k, \qquad k = 0, 1, \ldots, n-2. \qquad (3.4)$$

> **SAQ 1** What does this direct calculation look like?

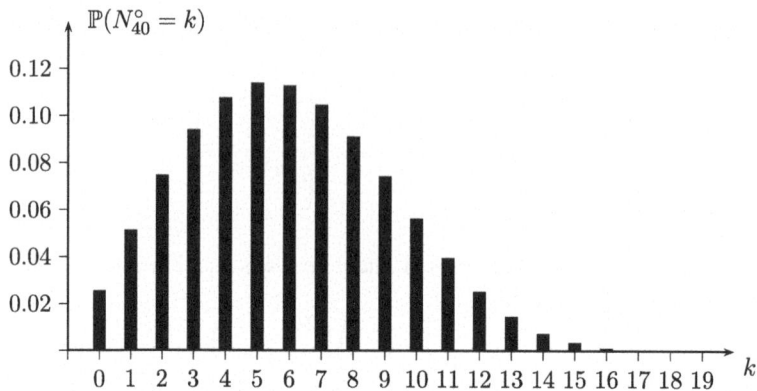

**Fig. 3.5** Bar chart of the distribution of $N_{40}^\circ$

Using (7.12) and (7.13), we have $\mathbb{E}(N_{2n}^\circ) = \sum_{k=1}^{n-1} q_k$ and $\mathbb{E}(N_{2n}^{\circ\,2}) = \sum_{k=1}^{n-1}(2k-1)q_k$. Summing the equations in (3.4) over $k$, it follows—after some patience—that

$$\mathbb{E}\left(N_{2n}^{\circ\,2}\right) = 4(n-1) - 5\mathbb{E}(N_{2n}^\circ), \tag{3.5}$$

proving the assertion (Exercise 3.3). ∎

Figure 3.5 shows a bar chart of the distribution of $N_{40}^\circ$. Comparing this figure with the bar chart of the distribution of the number $N_{2n}$ of visits to zero of a simple random walk, as seen in Chap. 2 (Fig. 2.18), reveals a striking qualitative difference: while the probabilities $\mathbb{P}(N_{40} = k)$ decrease monotonically with $k$, the probabilities $p_k := \mathbb{P}(N_{2n}^\circ = k)$ initially increase rapidly (where $p_{k+1} > p_k$ if and only if $(k+1)(k+2) < 2n$), and then decrease more slowly. This effect, known as *right-skewness*, is also observed for larger values of $n$. The most likely number of interior zeros is 5, with a probability of 0.114. A bridge of length 40 remains without a single interior zero with a probability of only 0.0256.

Comparing Theorem 3.1 (c) with Theorem 2.6 (c), it is noticeable that, on average, there are significantly fewer visits to zero in a long simple random walk (as discussed in Chap. 2) than there are interior zeros in a bridge of the same length. For large $n$, $\mathbb{E}(N_{2n}) \approx 0.798\sqrt{2n}$, whereas $\mathbb{E}(N_{2n}^\circ) \approx 1.253\sqrt{2n}$. This difference arises because a bridge is "tied down" on the $x$-axis at the end, creating a "stronger urge for balance," which manifests itself in a larger number of interior zeros.

The next result states that the limit distribution of $N_{2n}^\circ/\sqrt{2n}$ as $n \to \infty$ is a Weibull[3] distribution with a shape parameter of 2.

---

[3] Ernst Hjalmar Waloddi Weibull (1887–1979) was a Swedish engineer and mathematician. Since 1941, he served as a Professor for Technical Physics at the Royal Institute of Technology in Stockholm. His method of marine seismic exploration remains in use for oil exploration. His

## 3.1 The Number of Interior Zeros

**Theorem 3.2 (Limit Distribution of the Number of Interior Zeros)** *For the number $N_{2n}^\circ$ of interior zeros of a $2n$-bridge,*

$$\lim_{n\to\infty} \mathbb{P}\left(\frac{N_{2n}^\circ}{\sqrt{2n}} \leq x\right) = 1 - \exp\left(-\frac{x^2}{2}\right), \qquad x \geq 0. \tag{3.6}$$

*Proof* We can assume $x > 0$, as the statement obviously holds for $x = 0$.

**SAQ 2** Why does the statement hold for $x = 0$?

Let $k_n := k_n(x)$ denote the smallest integer that is greater than or equal to $x\sqrt{2n}$. Since $N_{2n}^\circ$ is integer-valued, (3.3) yields

$$\mathbb{P}\left(\frac{N_{2n}^\circ}{\sqrt{2n}} \geq x\right) = \mathbb{P}(N_{2n}^\circ \geq k_n) = 2^{k_n+1} \frac{\binom{2n-k_n-1}{n}}{\binom{2n}{n}}$$

$$= \prod_{j=1}^{k_n}\left(1 - \frac{j}{2n-j}\right).$$

If we take the logarithm and use (7.10) and (7.11), we get

$$\log \mathbb{P}(N_{2n}^\circ \geq k_n) \leq -\sum_{j=1}^{k_n}\frac{j}{2n-j} \leq -\sum_{j=1}^{k_n}\frac{j}{2n-1} \leq -\frac{k_n(k_n+1)}{2(2n-1)}, \tag{3.7}$$

$$\log \mathbb{P}(N_{2n}^\circ \geq k_n) \geq -\sum_{j=1}^{k_n}\frac{j}{2n-2j} \geq -\sum_{j=1}^{k_n}\frac{j}{2n-2k_n} \geq -\frac{k_n(k_n+1)}{4(n-k_n)}. \tag{3.8}$$

By the definition of $k_n$, the rightmost terms in (3.7) and (3.8) converge to $-x^2/2$ as $n \to \infty$, which yields the assertion. ∎

**Remark 3.3** As already mentioned prior to Theorem 3.2, the limit distribution appearing in (3.6) is a Weibull distribution, with a shape parameter of 2 and a scale parameter of $\frac{1}{2}$. In general, a random variable $Z$ follows a *Weibull distribution with*

---

primary areas of research included material fatigue, material strength, and the fracture behavior of solids.

*shape parameter $\alpha > 0$ and scale parameter $\lambda > 0$, if $Z$ has the distribution function*

$$G_{\alpha,\lambda}(x) := 1 - \exp\left(-\lambda x^{\alpha}\right), \qquad x \geq 0, \tag{3.9}$$

and $G_{\alpha,\lambda}(x) := 0$ for $x < 0$. We briefly write this as

$$Z \sim \text{Wei}(\alpha, \lambda).$$

A random variable $Z$ with the Weibull distribution $\text{Wei}(\alpha, \lambda)$ can be obtained from a random variable $Z_0$ with the exponential distribution $\text{Exp}(\lambda)$ through the power transformation $Z := Z_0^{1/\alpha}$.

**SAQ 3** Why does the last statement hold?

The density $g_{\alpha,\lambda}$ of the Weibull distribution $\text{Wei}(\alpha, \lambda)$ is given by

$$g_{\alpha,\lambda}(x) = \alpha \lambda x^{\alpha-1} \exp\left(-\lambda x^{\alpha}\right), \qquad x > 0,$$

and $g_{\alpha,\lambda}(x) := 0$ for $x \leq 0$.

Theorem 3.2 states that, as $n \to \infty$, the sequence of distribution functions of $N_{2n}^{\circ}/\sqrt{2n}$ converges to the distribution function of a $\text{Wei}(2, 0.5)$-distributed random variable $Z$. We write this briefly as

$$\frac{N_{2n}^{\circ}}{\sqrt{2n}} \xrightarrow{\mathcal{D}} Z, \qquad Z \sim \text{Wei}(2, 0.5), \qquad \text{as } n \to \infty.$$

Figure 3.6 shows the density (left) and the distribution function (right) of the Weibull distribution $\text{Wei}(2, 0.5)$. Table 3.1 provides some values of the distribution function $G_{2,0.5}$. Comparing these values with the corresponding ones of the distribution

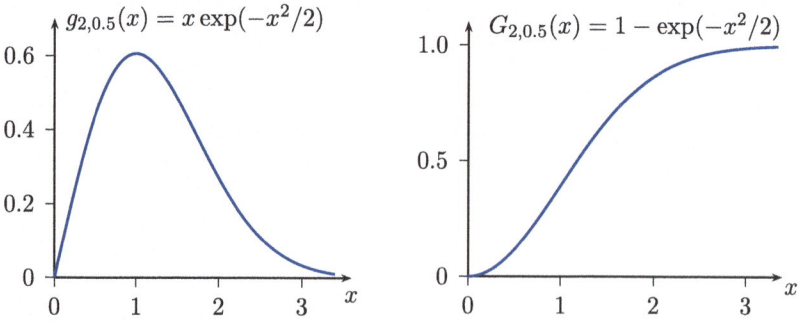

**Fig. 3.6** Density (left) and distribution function (right) of the Weibull distribution $\text{Wei}(2, 0.5)$

## 3.1 The Number of Interior Zeros

**Table 3.1** Values of the distribution function $G_{2,0.5}(x) = 1 - \exp(-x^2/2)$

| $x$ | $G_{2,0.5}(x)$ | $x$ | $G_{2,0.5}(x)$ | $x$ | $G_{2,0.5}(x)$ |
|---|---|---|---|---|---|
| 0.00 | 0.000 | 0.90 | 0.333 | 2.146 | 0.900 |
| 0.20 | 0.020 | 1.00 | 0.394 | 2.20 | 0.911 |
| 0.30 | 0.044 | 1.177 | 0.500 | 2.40 | 0.944 |
| 0.40 | 0.077 | 1.20 | 0.513 | 2.448 | 0.950 |
| 0.50 | 0.118 | 1.40 | 0.625 | 2.60 | 0.966 |
| 0.60 | 0.165 | 1.60 | 0.722 | 2.80 | 0.980 |
| 0.70 | 0.217 | 1.80 | 0.802 | 3.00 | 0.989 |
| 0.80 | 0.274 | 2.00 | 0.865 | 3.034 | 0.990 |

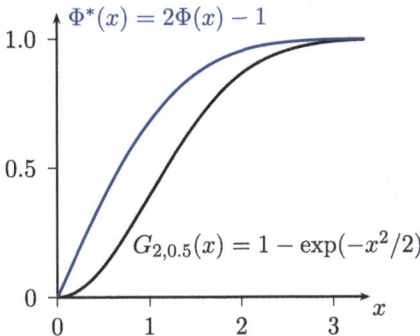

**Fig. 3.7** Limit distribution functions $G_{2,0.5}$ of $N^\circ_{2n}/\sqrt{2n}$ and $\Phi^*$ of $N_{2n}/\sqrt{2n}$

function $\Phi^*$ of the half-normal distribution from Table 2.2 confirms the previously observed phenomenon: when comparing the expectations of $N_{2n}$ and $N^\circ_{2n}$, we find significantly more interior zeros in a bridge compared to the number of visits to zero in a simple random walk of the same length. Because $\Phi^*(0.655) \approx 0.5$ and $G_{2,0.5}(1.177) \approx 0.5$, simple random walks of length 10,000 will, in about half of all cases, have at most 66 visits to zero, whereas in about every second bridge of the same length, more than 118 interior zeros occur.

Figure 3.7 illustrates this effect using the graphs of the limit distribution functions of $N^\circ_{2n}/\sqrt{2n}$ and $N_{2n}/\sqrt{2n}$: For each $x > 0$, the inequality $G_{2,0.5}(x) < \Phi^*(x)$ and thus $1 - \Phi^*(x) < 1 - G_{2,0.5}(x)$ holds (note: $\Phi^*(x) = 2\Phi(x) - 1$ for each $x > 0$). This follows from defining the function $h : \mathbb{R}_{\geq 0} \to \mathbb{R}$ by

$$h(x) := \Phi^*(x) - G_{2,0.5}(x) = 2\Phi(x) + \exp(-x^2/2) - 2, \quad x \geq 0,$$

subjecting it to a curve analysis (Exercise 3.5), and considering Exercise 3.4. Therefore, the limit distribution of $N^\circ_{2n}/\sqrt{2n}$ is stochastically larger than the limit distribution of $N_{2n}/\sqrt{2n}$.

### How Often Does a Father Beat His Kid at a Game?

The following problem was discussed in [LE] and [ZA]: A card game with $2n$ cards, consisting of $n$ red and $n$ black cards, is well shuffled, and the cards are revealed one by one. The color of the next card must be guessed. A child makes the prediction *red* or *black* based on a fair coin toss. The father, on the other hand, remembers how many red and black cards have already been drawn and only tosses a coin like his child when an equal number of red and black cards remain in the deck. Otherwise, he always bets on the color that is predominantly represented in the deck. The winner is the one who has guessed correctly more often. Obviously, the father has an advantage at this game, but by how much?

If we denote the number of correct predictions of the father as $V_n$ and those of the child as $K_n$, we will see in particular that $\lim_{n \to \infty} \big(\mathbb{E}(V_n) - \mathbb{E}(K_n)\big) = \infty$. For the win at this game, however, it is only decisive whether the event $\{V_n > K_n\}$ occurs. Here, the surprising result is:

$$\lim_{n \to \infty} \mathbb{P}(V_n > K_n) = \frac{1}{2} + \frac{1}{\sqrt{8}} \approx 0.854. \qquad (3.10)$$

Although the average number of correct predictions by the father compared to his child increases without limit as the deck size grows, his probability of winning converges to a value close to $\frac{6}{7}$. Thus, in the long run, the child should triumph over the father once a week, if one game is played daily.

Apparently, the situation is conceptually equivalent to observing a $2n$-bridge over time, where, without loss of generality, an upward step corresponds to drawing a red card and a downward step means drawing a black card. Given that a *a specific* one of all $\binom{2n}{n}$ possible paths is present, the number $K_n$ of the child's correct predictions is binomially distributed with parameters $2n$ and $\frac{1}{2}$. The specific path only determines before each coin toss which side of the coin is to be considered a hit and which a miss. According to the law of total probability (7.4), $\mathbb{P}(K_n = k) = \binom{2n}{k}(1/2)^{2n}$ for $k = 0, 1, \ldots, 2n$, which means that $K_n$ follows the binomial distribution $\text{Bin}\big(2n, \frac{1}{2}\big)$.

> **SAQ 4** How does the law of total probability come into play?

At the beginning and directly after each interior zero of the $2n$-bridge, the father tosses a coin independently of the child to make the prediction "up" or "down." If $T_j$ denotes the even time of the $j$-th interior zero ($j = 1, \ldots, N^\circ_{2n}$), the father, due to his strategy, definitely makes

$$\frac{T_1}{2} + \frac{T_2 - T_1}{2} + \ldots + \frac{T_{N^\circ_{2n}} - T_{N^\circ_{2n}-1}}{2} + \frac{2n - T_{N^\circ_{2n}}}{2} = n$$

correct predictions.

## 3.1 The Number of Interior Zeros

> **SAQ 5** Why is half of the father's predictions correct between any two equal counts of red and black cards in the deck?

Additionally, there is a random component based on the results of $1 + N_{2n}^\circ$ coin tosses. Therefore, we obtain the equality in distribution

$$V_n \sim n + \sum_{j=1}^{1+N_{2n}^\circ} U_j. \tag{3.11}$$

Here, $U_1, U_2, \ldots, U_n$ are i.i.d. random variables that follow the binomial distribution $\text{Bin}(1, \frac{1}{2})$, which are also independent of $N_{2n}^\circ$. Note that only a random number of the $U_j$'s is needed.

Up to an additive constant, the random variable $V_n$ is therefore a randomized sum, as we have already encountered in (2.97). Since $N_{2n}^\circ$ and $U_1, \ldots, U_n$ are independent, iterated expectation (see, e.g., [RO], Section 7.5) yields

$$\mathbb{E}(V_n) = n + \mathbb{E}(U_1) \cdot \left(1 + \mathbb{E}\left(N_{2n}^\circ\right)\right) = n + \frac{1}{2}\left(\frac{2^{2n}}{\binom{2n}{n}} - 1\right)$$

(see also (2.99) and Theorem 3.1 (b)). Since $\mathbb{E}(K_n) = n$, it follows that

$$\mathbb{E}(V_n) - \mathbb{E}(K_n) = \frac{1}{2}\left(\frac{2^{2n}}{\binom{2n}{n}} - 1\right).$$

Hence, using (2.15) and (2.29),

$$\lim_{n \to \infty} \frac{\mathbb{E}(V_n) - \mathbb{E}(K_n)}{\sqrt{n}} = \frac{\sqrt{\pi}}{2}.$$

In particular, $\lim_{n \to \infty} \left(\mathbb{E}(V_n) - \mathbb{E}(K_n)\right) = \infty$, as initially claimed.

The distribution of $V_n$ is obtained via the law of total probability (7.4) by conditioning on the realizations of the random variable $N_{2n}^\circ$ and noting that, under the condition $N_{2n}^\circ = \ell$, the randomized sum in (3.11) follows the binomial distribution $\text{Bin}(\ell+1, \frac{1}{2})$. Since $\ell$ must be at least equal to $k-1$ for the conditional probability $\mathbb{P}\left(\sum_{j=1}^{1+N_{2n}^\circ} U_j = k \mid N_{2n}^\circ = \ell\right)$ to be positive, Theorem 3.1 (a) implies that for every $k \in \{0, \ldots, n\}$:

$$\mathbb{P}(V_n = n+k) = \mathbb{P}\left(\sum_{j=1}^{1+N_{2n}^\circ} U_j = k\right)$$

$$= \sum_{\ell=k-1}^{n-1} \mathbb{P}\left(\sum_{j=1}^{1+N_{2n}^\circ} U_j = k \,\Big|\, N_{2n}^\circ = \ell\right) \cdot \mathbb{P}\left(N_{2n}^\circ = \ell\right)$$

$$= \sum_{\ell=k-1}^{n-1} \binom{\ell+1}{k} \left(\frac{1}{2}\right)^{\ell+1} \frac{2^{\ell+1}(\ell+1)}{n} \cdot \frac{\binom{2n-\ell-2}{n-1}}{\binom{2n}{n}}$$

$$= \frac{1}{n} \sum_{\ell=k-1}^{n-1} \binom{\ell+1}{k} (\ell+1) \frac{\binom{2n-\ell-2}{n-1}}{\binom{2n}{n}}.$$

Figure 3.8 shows bar charts of the distributions of $V_n$ and $K_n$ for the case $n = 40$. It is clearly evident that the father has an advantage.

We now prove (3.10) and briefly set $L_n := 1 + N_{2n}^\circ$. According to Theorem 3.2 and (7.8), $L_n/\sqrt{2n}$ converges in distribution to a random variable $Z$ that follows the Weibull distribution Wei(2, 0.5). Now, fix any $\varepsilon > 0$, and choose real numbers $a, b$ such that $\mathbb{P}(Z < a) = \frac{\varepsilon}{4}$ and $\mathbb{P}(Z > b) = \frac{\varepsilon}{4}$. Setting $A_n := \{a \leq L_n/\sqrt{2n} \leq b\}$

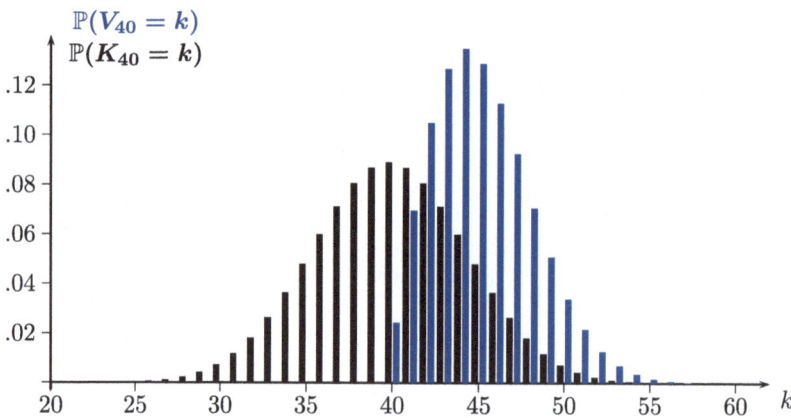

**Fig. 3.8** Bar charts of the distributions of $V_{40}$ (blue) and $K_{40}$ (black)

## 3.1 The Number of Interior Zeros

and using $L_n/\sqrt{2n} \xrightarrow{D} Z$ as $n \to \infty$, there is an integer $n_0$ with $\mathbb{P}(A_n) \geq 1 - \varepsilon$ for each $n \geq n_0$. For each such $n$,

$$\mathbb{P}(\{V_n > K_n\} \cap A_n) \leq \mathbb{P}(V_n > K_n)$$
$$= \mathbb{P}(\{V_n > K_n\} \cap A_n) + \mathbb{P}(\{V_n > K_n\} \cap A_n^c) \quad (3.12)$$
$$\leq \mathbb{P}(\{V_n > K_n\} \cap A_n) + \varepsilon.$$

This shows that we can examine the probability $\mathbb{P}(\{V_n > K_n\} \cap A_n)$ in what follows. Setting $I_n := \{k \in \mathbb{N} : a \leq k/\sqrt{2n} \leq b\}$, we have for each $n \geq n_0$:

$$\mathbb{P}(\{V_n > K_n\} \cap A_n) = \sum_{k \in I_n} \mathbb{P}(V_n > K_n, L_n = k)$$
$$= \sum_{k \in I_n} \mathbb{P}(L_n = k) \cdot \mathbb{P}(V_n > K_n | L_n = k).$$

Using elementary transformations for binomial coefficients and Theorem 3.1 (a), we get

$$\mathbb{P}(L_n = k) = \mathbb{P}(N_{2n}^\circ = k - 1) = \frac{k}{2n} \cdot \frac{\prod_{j=1}^{k-1}\left(1 - \frac{j}{n}\right)}{\prod_{j=1}^{k}\left(1 - \frac{j}{2n}\right)} \quad (3.13)$$

(see Exercise 3.6). Now, set

$$h(t) := t \cdot \exp\left(-\frac{t^2}{2}\right), \quad t \geq 0. \quad (3.14)$$

Using the inequalities (7.10) and (7.11), direct calculation yields the existence of a constant $C > 0$ (depending on $\varepsilon$ via $a$ and $b$), such that

$$\sup_{k \in I_n} \left| \frac{\mathbb{P}(L_n = k)}{\frac{1}{\sqrt{2n}} \cdot h\left(\frac{k}{\sqrt{2n}}\right)} - 1 \right| \leq \frac{C}{\sqrt{n}}$$

for each sufficiently large $n$. For such $n$, putting $x_{n,k} := k/\sqrt{2n}$, we thus have

$$\left| \mathbb{P}(L_n = k) - \frac{h(x_{n,k})}{\sqrt{2n}} \right| \leq \frac{C}{\sqrt{n}} \cdot \frac{h(x_{n,k})}{\sqrt{2n}}, \quad k \in I_n.$$

Since the probabilities $p_{n,k} := \mathbb{P}(V_n > K_n | L_n = k)$ are at most 1, it follows that

$$\sum_{k \in I_n} \left| \mathbb{P}(L_n = k) \cdot p_{n,k} - \frac{h(x_{n,k})}{\sqrt{2n}} \cdot p_{n,k} \right| \leq \frac{C}{\sqrt{n}} \sum_{k \in I_n} \frac{h(x_{n,k})}{\sqrt{2n}}.$$

Due to $x_{n,k+1} - x_{n,k} = 1/\sqrt{2n}$, the sum on the right-hand side is a Riemann sum that converges as $n \to \infty$ to the integral $\int_a^b h(t)dt$. Hence, the sums $\sum_{k \in I_n} \mathbb{P}(L_n = k) \cdot p_{n,k}$ and $\sum_{k \in I_n} (h(x_{n,k})/\sqrt{2n}) \cdot p_{n,k}$ have the same limit as $n \to \infty$. To derive the limit of the second sum, we need to know the asymptotic behavior of the probabilities $p_{n,k} = \mathbb{P}(V_n > K_n | L_n = k)$ as $n \to \infty$. Now,

$$\mathbb{P}(V_n > K_n | L_n = k) = \mathbb{P}(n + Z_{n,k} > K_n), \quad (3.15)$$

where $Z_{n,k}$ denotes a random variable following the binomial distribution $\text{Bin}(k, \frac{1}{2})$, which is independent of $K_n$. By double indexing with $n$ and $k$, we emphasize that there is also a dependency on $n$ because, due to the condition $k \in I_n$, $k$ is not fixed but increases (of the order of magnitude $\sqrt{n}$) with $n$. If we write $K_n^* = (K_n - n)/\sqrt{n/2}$ and $Z_{n,k}^* = (Z_{n,k} - k/2)/\sqrt{k/4}$ for the random variables resulting from $K_n$ and $Z_{n,k}$ through standardization, and if we set $c_{n,k} := \sqrt{k/(2n)}$, then a direct calculation (Exercise 3.8) yields

$$\mathbb{P}(V_n > K_n | L_n = k) = \mathbb{P}\left(Z_{n,k}^* \cdot c_{n,k} + x_{n,k} > K_n^*\right). \quad (3.16)$$

The random variable $\widetilde{Z}_{n,k} := Z_{n,k}^* \cdot c_{n,k}$ has expectation 0 and variance $c_{n,k}^2$. Since $c_{n,k} \to 0$ as $n \to \infty$, it follows from Chebyshev's inequality that $\mathbb{P}(|\widetilde{Z}_{n,k}| > \varepsilon) \to 0$ as $n \to \infty$. For sufficiently large $n$ (we may have to increase $n_0$ here), $\mathbb{P}(|\widetilde{Z}_{n,k}| > \varepsilon) \leq \varepsilon$. If one decomposes the event $\{\widetilde{Z}_{n,k} + x_{n,k} > K_n^*\}$ according to whether $\{|\widetilde{Z}_{n,k}| > \varepsilon\}$ or $\{|\widetilde{Z}_{n,k}| \leq \varepsilon\}$, then for such $n$:

$$\mathbb{P}\left(Z_{n,k}^* \cdot c_{n,k} + x_{n,k} > K_n^*\right) \leq \mathbb{P}(K_n^* \leq x_{n,k} + \varepsilon) + \varepsilon.$$

Similarly, $\mathbb{P}\left(Z_{n,k}^* \cdot c_{n,k} + x_{n,k} > K_n^*\right) \geq \mathbb{P}(K_n^* \leq x_{n,k} - \varepsilon) - \varepsilon$. According to the De Moivre–Laplace central limit theorem, $K_n^*$ converges in distribution to the standard normal distribution $N(0, 1)$. Since the distribution function $\Phi$ of the distribution $N(0, 1)$ is continuous, the convergence of the respective distribution functions is uniform on the whole real line (see (7.7)). It follows that, for sufficiently large $n$ (it may be necessary to increase $n_0$ once again):

$$\mathbb{P}(K_n^* \leq x_{n,k} + \varepsilon) \leq \Phi(x_{n,k} + \varepsilon) + \varepsilon \leq \Phi(x_{n,k}) + 2\varepsilon,$$
$$\mathbb{P}(K_n^* \leq x_{n,k} - \varepsilon) \geq \Phi(x_{n,k} - \varepsilon) - \varepsilon \geq \Phi(x_{n,k}) - 2\varepsilon.$$

In each case, the second inequality arises from the fact that the derivative of $\Phi$ is at most equal to $1/\sqrt{2\pi}$ and thus less than 1. Overall, for sufficiently large $n$:

$$\mathbb{P}(V_n > K_n | L_n = k) \leq \Phi(x_{n,k}) + 3\varepsilon, \qquad \mathbb{P}(V_n > K_n | L_n = k) \geq \Phi(x_{n,k}) - 3\varepsilon.$$

## 3.1 The Number of Interior Zeros

Thus,

$$\sum_{k \in I_n} \frac{h(x_{n,k})}{\sqrt{2n}} \cdot \mathbb{P}(V_n > K_n | L_n = k) \le \sum_{k \in I_n} \frac{h(x_{n,k})}{\sqrt{2n}} \cdot \Phi(x_{n,k}) + 3\varepsilon \sum_{k \in I_n} \frac{h(x_{n,k})}{\sqrt{2n}}.$$

Here, the first sum on the right-hand side converges to $\int_a^b h(t)\Phi(t)dt$. The second sum converges to $\int_a^b h(t)dt$, which, due to $\int_0^\infty h(t)dt = 1$, yields

$$\limsup_{n \to \infty} \sum_{k \in I_n} \frac{h(x_{n,k})}{\sqrt{2n}} \cdot \mathbb{P}(V_n > K_n | L_n = k) \le \int_a^b h(t)\Phi(t)dt + 3\varepsilon.$$

Similarly, it follows that

$$\liminf_{n \to \infty} \sum_{k \in I_n} \frac{h(x_{n,k})}{\sqrt{2n}} \cdot \mathbb{P}(V_n > K_n | L_n = k) \ge \int_a^b h(t)\Phi(t)dt - 3\varepsilon.$$

Therefore, (3.12) implies

$$\int_a^b h(t)\Phi(t)dt - 4\varepsilon \le \liminf_{n \to \infty} \mathbb{P}(V_n > K_n)$$

$$\le \limsup_{n \to \infty} \mathbb{P}(V_n > K_n) \le \int_a^b h(t)\Phi(t)dt + 4\varepsilon.$$

If we now let $\varepsilon$ tend to 0, then $a$ converges to 0 and $b$ to $\infty$, and it follows that as claimed

$$\lim_{n \to \infty} \mathbb{P}(V_n > K_n) = \int_0^\infty h(t)\Phi(t)dt$$

$$= \int_0^\infty h(t) \int_{-\infty}^t \frac{1}{\sqrt{2\pi}} \exp\left(-\frac{x^2}{2}\right) dx\, dt$$

$$= \int_0^\infty h(t) \left(\frac{1}{2} + \int_0^t \frac{1}{\sqrt{2\pi}} \exp\left(-\frac{x^2}{2}\right) dx\right) dt$$

$$= \frac{1}{2} + \frac{1}{\sqrt{2\pi}} \int_0^\infty \left(\int_x^\infty t \exp\left(-\frac{t^2}{2}\right) dt\right) \exp\left(-\frac{x^2}{2}\right) dx$$

$$= \frac{1}{2} + \frac{1}{\sqrt{2\pi}} \int_0^\infty \exp\left(-x^2\right) dx$$

$$= \frac{1}{2} + \frac{1}{\sqrt{2\pi}} \cdot \frac{\sqrt{\pi}}{2} = \frac{1}{2} + \frac{1}{\sqrt{8}}.$$

## 3.2 Sojourn Times

According to the results of Sect. 2.6, simple symmetric random walks tend to stay either long or short periods above the $x$-axis (see Fig. 2.25): the distribution of the sojourn time $O_{2n} = \sum_{j=1}^{2n} \mathbf{1}\{S_j \geq 0, S_{j-1} \geq 0\}$ has a U-shaped bar chart (Fig. 2.12). Moreover, in the limit as $n \to \infty$, the proportion of time $O_{2n}/(2n)$ that the random walk spends above the $x$-axis converges in distribution to the arcsine distribution having the density shown in Fig. 2.14. We now examine the time span, denoted by

$$O_{2n}^\circ := \sum_{j=1}^{2n} \mathbf{1}\{S_j \geq 0, \ S_{j-1} \geq 0\},$$

which a purely random $2n$-bridge spends above the $x$-axis, as motivated at the beginning of this chapter in connection with the Galton rank order test. The following result, going back to Chung and Feller ([CF]), and apparently independently discovered by Gnedenko[4] and Mihalevič[5] [GM], stands in stark contrast to Theorem 2.10.

**Theorem 3.4 (Distribution of the Sojourn Time of a Bridge)** *For the sojourn time $O_{2n}^\circ$ of a $2n$-bridge,*

$$\mathbb{P}(O_{2n}^\circ = 2k) = \frac{1}{n+1}, \qquad k = 0, \ldots, n. \tag{3.17}$$

***Proof*** We first consider the cases $k = n$ and $k = 0$. For symmetry reasons, we can restrict ourselves to the case where the bridge runs entirely above the $x$-axis, that is, $k = n$. Among all $\binom{2n}{n}$ possible $2n$-bridges, the paths favorable to the occurrence of the event $\{O_{2n}^\circ = 2n\}$ are those that, after an initial upward step, run from the point $(1, 1)$ to the point $(2n, 0)$ and do not hit the height $y = -1$. According to (2.11), there are $\binom{2n-1}{n}$ paths from $(1, 1)$ to $(2n, 0)$, and by the reflection principle (see Fig. 2.6), each such path that hits the height $y = -1$ corresponds uniquely to a path from $(1, -3)$ to $(2n, 0)$. Since (again according to (2.11)) exactly $\binom{2n-1}{n+1}$ paths lead from $(1, -3)$ to $(2n, 0)$,

$$|\{O_{2n}^\circ = 2n\}| = \binom{2n-1}{n} - \binom{2n-1}{n+1} = \frac{\binom{2n}{n}}{n+1},$$

---

[4] Boris Vladimirovich Gnedenko (1912–1997) was one of the leading members of the Russian school for probability theory and statistics.

[5] V.S. Mihalevič.

## 3.2 Sojourn Times

and we obtain (3.17) for the cases $k = n$ and $k = 0$.

We now prove (3.17) by induction on $n$. Note that the base case $n = 1$ has already been shown. In the following, we assume $n \geq 2$ and that (3.17) is correct for $2s$-bridges with $s \leq n - 1$ (induction hypothesis). Setting

$$c(s) := \binom{2s}{s} \cdot \frac{1}{s+1}, \quad s = 0, 1, 2, \ldots,$$

the induction hypothesis states:

$$|\{O_{2s}^\circ = 2\ell\}| = c(s), \quad s \in \{1, \ldots, n-1\}, \quad \ell \in \{0, 1, \ldots, s\}. \tag{3.18}$$

Since according to the above considerations, (3.17) always holds (i.e., for every $n \geq 1$) for the extreme cases $k = 0$ and $k = n$, we assume $1 \leq k \leq n - 1$ in what follows. The paths favorable to the occurrence of $\{O_{2n}^\circ = 2k\}$ thus have at least one and therefore also a *first* interior zero, whose random time is denoted by $T^\circ$. We now decompose the set of all $2n$-bridges with the property $O_{2n}^\circ = 2k$ based on whether $S_1 = 1$ or $S_1 = -1$, and on the possible values of $T^\circ$ in each case. If the first step moves upward, then $T^\circ$ can only take the values $2r$ with $r \in \{1, 2, \ldots, k\}$; otherwise $O_{2n}^\circ > 2k$ would hold. For each such $r$, the $2n$-bridge decomposes into two sub-paths. The first path leads from $(0, 0)$ to $(1, 1)$, then proceeds to the point $(2r-1, 1)$ without hitting the $x$-axis, and finally moves from $(2r-1, 1)$ to the point $(2r, 0)$, which is the first interior zero of the entire path. The second sub-path is a $(2n - 2r)$-bridge that leads from $(2r, 0)$ to $(2n, 0)$, spending the remaining $2k - 2r$ time steps above the $x$-axis (see Fig. 3.9).

In the remaining case $S_1 = -1$, $T^\circ$ can only take the values $2r$ with $r \in \{1, 2, \ldots, n-k\}$.

> **SAQ 6** Why does the last statement hold?

For each such $r$, the $2n$-bridge decomposes into two sub-paths as described above. The first path leads from $(0, 0)$ to $(1, -1)$, then proceeds to the point $(2r-1, -1)$ without hitting the $x$-axis, and then moves from $(2r-1, -1)$ to the point $(2r, 0)$. The second sub-path is a $(2n - 2r)$-bridge that leads from $(2r, 0)$ to $(2n, 0)$, spending $2k$ time steps above the $x$-axis (Fig. 3.10).

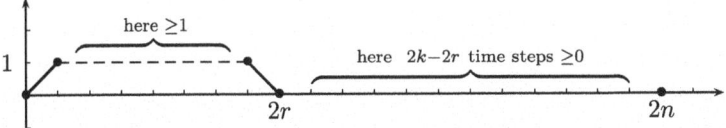

**Fig. 3.9** Decomposition of a $2n$-bridge with $O_{2n}^\circ = 2k$ and $S_1 = 1$ based on the first interior zero

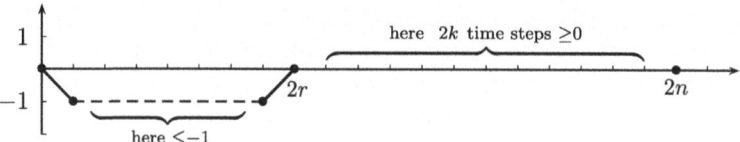

**Fig. 3.10** Decomposition of a $2n$-bridge with $O_{2n}^\circ = 2k$ and $S_1 = -1$ based on the first interior zero

Since each of the resulting sub-paths has a maximum length of $2n - 2$, by the induction hypothesis (3.18) and the multiplication rule of combinatorics:

$$|\{O_{2n}^\circ = 2k\}| = \sum_{r=1}^{k} c(r-1) \cdot c(n-r) + \sum_{r=1}^{n-k} c(r-1) \cdot c(n-r).$$

Using the index shift $t := n + 1 - r$, the second sum becomes

$$\sum_{t=k+1}^{n} c(t-1) \cdot c(n-t).$$

Hence, setting $r := t$ again, we obtain

$$|\{O_{2n}^\circ = 2k\}| = \sum_{r=1}^{n} c(r-1) \cdot c(n-r).$$

Note that the right-hand side does not depend on $k \in \{1, 2, \ldots, n-1\}$. Therefore,

$$\binom{2n}{n} = |W_{2n}^\circ| = \sum_{k=0}^{n} |\{O_{2n}^\circ = 2k\}|$$

$$= 2 \cdot |\{O_{2n}^\circ = 2n\}| + (n-1) \cdot |\{O_{2n}^\circ = 2\}|$$

$$= 2 \cdot \binom{2n}{n} \cdot \frac{1}{n+1} + (n-1) \cdot |\{O_{2n}^\circ = 2\}|,$$

and we get $|\{O_{2n}^\circ = 2k\}| = \binom{2n}{n}/(n+1)$ for each $k = 0, 1, \ldots, n$, which was to be shown. ∎

Theorem 3.4 states that the sojourn time $O_{2n}^\circ$ above the $x$-axis of a $2n$-bridge is *uniformly distributed* over the possible values. Naturally, one might have expected that, due to the condition $S_{2n} = 0$, which dampens the volatility compared to a simple symmetric random walk, a very short or very long sojourn time above the $x$-axis would be unlikely for a $2n$-bridge. However, the fact that this results in a uniform distribution is surprising. If we consider the proportion of time $O_{2n}^\circ/(2n)$,

## 3.3 Last Visit to Zero and First Return Time

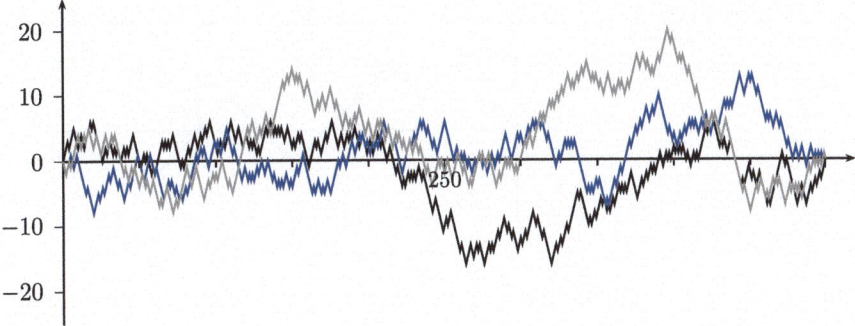

**Fig. 3.11** Sojourn times of bridges are uniformly distributed

the discrete uniform distribution on the values $0, 2, 4, \ldots, 2n$ is transformed into a uniform distribution on the values $0, \frac{1}{n}, \frac{2}{n}, \ldots, 1$. In the limit as $n \to \infty$, this directly leads to the following result.

**Theorem 3.5 (Limit Distribution of the Sojourn Time of a Bridge)** *For the sojourn time $O_{2n}^\circ$ of a $2n$-bridge above the x-axis,*

$$\lim_{n \to \infty} \mathbb{P}\left(\frac{O_{2n}^\circ}{2n} \leq x\right) = x, \qquad 0 \leq x \leq 1.$$

Thus, as $n \to \infty$, the proportion of time that a purely random $2n$-bridge spends above the x-axis has a continuous uniform distribution $U(0, 1)$ on the unit interval, i.e.,

$$\frac{O_{2n}^\circ}{2n} \xrightarrow{\mathcal{D}} Z, \qquad Z \sim U(0, 1), \qquad \text{as } n \to \infty.$$

Figure 3.11 shows the qualitatively different behavior of the sojourn time above the x-axis for bridges compared to simple symmetric random walks (cf. Fig. 2.25), as demonstrated by three bridges of length 500.

### 3.3 Last Visit to Zero and First Return Time

In Sect. 2.5, we studied the time of the first return to zero of a simple symmetric random walk. A $2n$-bridge, by definition, returns to zero at the latest after $2n$ steps, but it can also return earlier. In this section, we first examine the number of steps, denoted by

$$T_{2n}^\circ := \min\{2j : 1 \leq j \leq n \text{ and } S_{2j} = 0\},$$

**Fig. 3.12** Times of the first return and the last visit to zero of a 2n-bridge

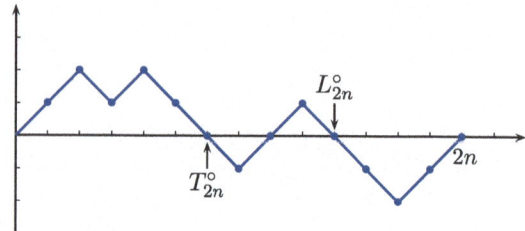

until a 2n-bridge returns to zero for the first time. As in the case of a simple symmetric random walk, we also refer to $T_{2n}^\circ$ as a *first return time*. Note that $T_{2n}^\circ$ can take each of the values $2, 4, \ldots, 2n$.

Closely related to $T_{2n}^\circ$ is the time of the last return to zero, denoted by

$$L_{2n}^\circ := \max\{2j : 0 \leq j \leq n-1 \text{ and } S_{2j} = 0\}.$$

This occurs at the latest by the time $2n - 2$, i.e., before the trivial return to zero at the end. These concepts are illustrated in Fig. 3.12.

Note that the origin $(0, 0)$ is counted as a visit to zero, so $L_{2n}^\circ$ takes the possible values $0, 2, \ldots, 2n - 2$. Regarding the distribution of $T_{2n}^\circ$, there is the following result:

**Theorem 3.6 (Distribution of the First Return Time of a Bridge)** *For the first return time $T_{2n}^\circ$ of a 2n-bridge, the following hold:*

(a) $\mathbb{P}(T_{2n}^\circ = 2k) = \dfrac{2}{k} \cdot \dfrac{\binom{2(k-1)}{k-1}\binom{2(n-k)}{n-k}}{\binom{2n}{n}}, \quad k = 1, \ldots, n,$

(b) $\mathbb{E}(T_{2n}^\circ) = \dfrac{2^{2n}}{\binom{2n}{n}} = \dfrac{1}{u_{2n}},$

(c) $\lim\limits_{n \to \infty} \mathbb{P}(T_{2n}^\circ = 2k) = \dfrac{1}{2k} \cdot \dfrac{\binom{2(k-1)}{k-1}}{2^{2(k-1)}}, \quad k = 1, 2, \ldots$

(d) $\mathbb{V}(T_{2n}^\circ) = \dfrac{2^{2n}}{\binom{2n}{n}}\left(n + 1 - \dfrac{2^{2n}}{\binom{2n}{n}}\right).$

## 3.3 Last Visit to Zero and First Return Time

### Proof

(a) Among all $\binom{2n}{n}$ possible $2n$-bridges, the favorable cases for the event $\{T_{2n}^\circ = 2k\}$ each split into two segments, namely a $2k$-bridge leading from $(0,0)$ to $(2k, 0)$ and a bridge from $(2k, 0)$ to $(2n, 0)$, with no restrictions on the latter. The bridge from $(0,0)$ to $(2k, 0)$ does not hit zero in the interval $[1, 2k-1]$. Thus, after the first step, it either stays entirely above the height $1$ or entirely below the height $-1$ until time $2k-1$. Then it descends or ascends to the point $(2k, 0)$ of the first return to zero. According to the considerations in Sect. 3.2,

$$|\{T_{2n}^\circ = 2k\}| = 2 \cdot \binom{2(k-1)}{k-1} \cdot \frac{1}{k} \cdot \binom{2(n-k)}{n-k},$$

from which the assertion follows.

(b) Using $u_{2m}$ as in (2.15), we get

$$\mathbb{E}(T_{2n}^\circ) = \sum_{k=1}^{n} 2k\, \mathbb{P}(T_{2n}^\circ = 2k) = 4 \cdot \frac{2^{2(n-1)}}{\binom{2n}{n}} \sum_{k=1}^{n} u_{2(k-1)}\, u_{2(n-k)}.$$

Since the last sum, after the index shift $j = k-1$, is equal to the sum of the probabilities $\mathbb{P}(L_{2(n-1)} = 2j)$ over $j = 0, 1, \ldots, n-1$, which equals $1$ (see Theorem 2.4), the claim follows.

(c) This assertion follows from (a) and

$$\lim_{n \to \infty} \frac{\binom{2(n-k)}{n-k}}{\binom{2n}{n}} = \lim_{n \to \infty} \left( \prod_{j=0}^{k-1} \frac{n-j}{2n-j} \prod_{j=0}^{k-1} \frac{n-j}{2n-k-j} \right) = \frac{1}{2^{2k}}.$$

(d) Using $u_{2m}$ as in (2.15), we obtain

$$\mathbb{E}\left(T_{2n}^{\circ\,2}\right) = \sum_{k=1}^{n}(2k)^2 \mathbb{P}(T_{2n}^\circ = 2k) = \frac{2^{2n+1}}{\binom{2n}{n}} \sum_{k=1}^{n} k\, u_{2(k-1)}\, u_{2(n-k)}.$$

Since $u_{2(k-1)} u_{2(n-k)} = \mathbb{P}(L_{2(n-1)} = 2(k-1))$ (see Theorem 2.4), the last sum equals

$$\frac{1}{2} \sum_{k=1}^{n}[2(k-1)+2]\cdot \mathbb{P}(L_{2(n-1)} = 2(k-1)) = \frac{1}{2}\left(\mathbb{E}(L_{2(n-1)}) + 2\right) = \frac{n+1}{2}.$$

Thus, the claim follows with $\mathbb{V}(T_{2n}^\circ) = \mathbb{E}\left(T_{2n}^{\circ\,2}\right) - (\mathbb{E}(T_{2n}^\circ))^2$ and part (b). ∎

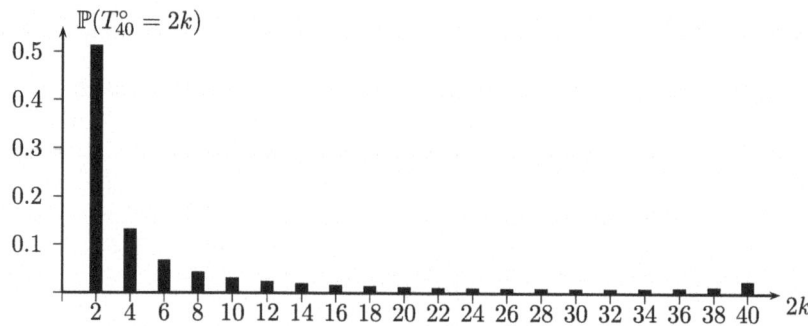

**Fig. 3.13** Bar chart of the distribution of the first return time $T_{40}^\circ$

By Theorem 2.9, the limit in statement (c) equals the probability that the time of the first return to zero in a simple symmetric random walk is $2k$. This result is plausible, as it reflects the intuitive fact that the stochastic behavior of a $2n$-bridge near its start increasingly resembles that of a simple symmetric random walk as $n$ grows. When interpreting the temporal progression of a $2n$-bridge as successive random draws from an urn containing $n$ red and $n$ black balls, the distinction between drawing with or without replacement becomes increasingly insignificant as $n$ increases. The first case represents the simple symmetric random walk, while the second describes a $2n$-bridge.

Figure 3.13 shows the bar chart of the distribution of $T_{2n}^\circ$ for the case $n = 20$. We now turn to the distribution of $L_{2n}^\circ$. One might be tempted to derive this analogously to that of $T_{2n}^\circ$. However, it can be done much more easily through a symmetry argument. We have already seen that the temporal development of a $2n$-bridge can be described by an urn model. In this model, all balls are randomly drawn one after the other without replacement from an urn containing $n$ red and $n$ black balls, where drawing a red or black ball is noted as an upward or downward step of a random walk, respectively. Imagine all balls are made up of two halves and look the same externally; one could mentally place all $2n$ balls in the imagined drawing order in a row and then open each ball to see the color inside. For symmetry reasons, one could also reverse the temporal order, starting with the last ball, then continuing with the second last ball, and so on. This is equivalent to starting the $2n$-bridge at the point $(2n, 0)$ and then "developing it to the left" until it ends at the origin. This possibility arises directly from the fact that the joint distribution of $X_1, \ldots, X_{2n}$ given in (3.1) is symmetric in $X_1, \ldots, X_{2n}$. In particular, there is the equality in distribution

$$(X_1, X_2, \ldots, X_{2n-1}, X_{2n}) \sim (X_{2n}, X_{2n-1}, \ldots, X_2, X_1).$$

The $2n$-bridge traversed in reversed time thus has the same stochastic behavior as the original $2n$-bridge. Now, the number of steps until the last visit to zero up to time $2n - 2$ of a $2n$-bridge is the same as the number of steps until the first non-trivial

## 3.4 Maximum and Minimum

visit to zero of the bridge traversed in reverse. The random time $L_{2n}^\circ$ of this last visit to zero thus has the same distribution as $2n - T_{2n}^\circ$; for this reason,

$$\mathbb{P}(L_{2n}^\circ = 2k) = \mathbb{P}(2n - T_{2n}^\circ = 2k) = \mathbb{P}(T_{2n}^\circ = 2n - 2k),$$
$$\mathbb{E}(L_{2n}^\circ) = \mathbb{E}(2n - T_{2n}^\circ) = 2n - \mathbb{E}(T_{2n}^\circ),$$
$$\mathbb{V}(L_{2n}^\circ) = \mathbb{V}(T_{2n}^\circ).$$

Thus, Theorem 3.6 immediately provides the following result:

**Theorem 3.7 (Distribution of the Last Visit to Zero of a Bridge)** *For the time $L_{2n}^\circ$ of the last visit to zero of a $2n$-bridge, the following hold:*

(a) $\mathbb{P}(L_{2n}^\circ = 2k) = \dfrac{2}{n-k} \cdot \dfrac{\binom{2k}{k}\binom{2(n-k-1)}{n-k-1}}{\binom{2n}{n}}$, $\quad k = 0, \ldots, n-1$,

(b) $\mathbb{E}\left(L_{2n}^\circ\right) = 2n - \dfrac{2^{2n}}{\binom{2n}{n}}$,

(c) $\lim\limits_{n \to \infty} \mathbb{P}(L_{2n}^\circ = 2n - 2k) = \dfrac{1}{2k} \cdot \dfrac{\binom{2(k-1)}{k-1}}{2^{2(k-1)}}$, $\quad k = 1, 2, \ldots$

(d) $\mathbb{V}(L_{2n}^\circ) = \dfrac{2^{2n}}{\binom{2n}{n}} \left( n + 1 - \dfrac{2^{2n}}{\binom{2n}{n}} \right)$.

### 3.4 Maximum and Minimum

The distribution of the maximum

$$M_{2n}^\circ := \max(S_0, S_1, \ldots, S_{2n})$$

of a $2n$-bridge follows directly from the reflection principle presented in Sect. 2.1: If we reflect a $2n$-bridge with the property $M_{2n}^\circ \geq k$, where $k \in \{1, \ldots, n\}$, from the time it first reaches the height $k$ across the line $y = k$ and leave the path segment before it unchanged, we obtain a path leading from $(0, 0)$ to $(2n, 2k)$. Conversely, every path leading from $(0, 0)$ to $(2n, 2k)$ reaches the height $k$ for the first time. By reflecting the path from this point across the line $y = k$, while keeping the first segment unchanged, we obtain a $2n$-bridge (Fig. 3.14). Thus, there is a bijective mapping between the set of all $2n$-bridges with $M_{2n}^\circ \geq k$ and the set of all paths from $(0, 0)$ to $(2n, 2k)$.

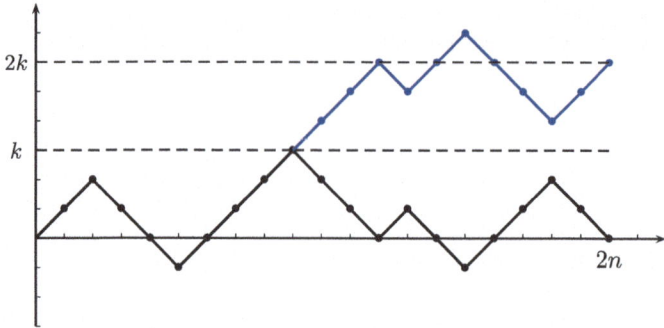

**Fig. 3.14** Bijection between $2n$-bridges with $M_{2n}^\circ \geq k$ and paths to $(2n, 2k)$

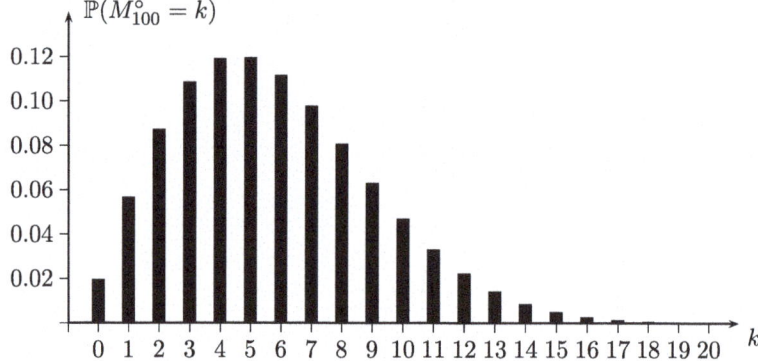

**Fig. 3.15** Bar chart of the distribution of $M_{100}^\circ$

According to Lemma 2.1,

$$\mathbb{P}(M_{2n}^\circ \geq k) = \frac{\binom{2n}{n+k}}{\binom{2n}{n}}, \qquad k = 0, \ldots, n, \tag{3.19}$$

and we obtain the following result, for part (a), based on B.V. Gnedenko and V.S. Korolyuk [GK] (see also [REN], p. 496 ff.) (Fig. 3.15):

**Theorem 3.8 (Distribution of the Maximum of a Bridge)** *For the maximum $M_{2n}^\circ$ of a $2n$-bridge, the following hold:*

## 3.4 Maximum and Minimum

(a) $\mathbb{P}(M_{2n}^\circ = k) = \dfrac{2k+1}{n+k+1} \cdot \dfrac{\binom{2n}{n+k}}{\binom{2n}{n}}, \quad k = 0, \ldots, n,$

(b) $\mathbb{E}(M_{2n}^\circ) = \dfrac{2^{2n-1}}{\binom{2n}{n}} - \dfrac{1}{2},$

(c) $\lim\limits_{n\to\infty} \dfrac{\mathbb{E}(M_{2n}^\circ)}{\sqrt{2n}} = \dfrac{\sqrt{\pi}}{2\sqrt{2}},$

(d) $\mathbb{V}(M_{2n}^\circ) = n + \dfrac{1}{4} - \dfrac{2^{4n-2}}{\binom{2n}{n}^2}.$

**Proof** (a) follows from (3.19) and $\mathbb{P}(M_{2n}^\circ = k) = \mathbb{P}(M_{2n}^\circ \geq k) - \mathbb{P}(M_{2n}^\circ \geq k+1)$. Because of (7.12), $\binom{2n}{n+k} = \binom{2n}{n-k}$ for $k = 0, \ldots, n$. Moreover, $\sum_{j=0}^{2n} \binom{2n}{j} = 2^{2n}$. Hence,

$$\mathbb{E}(M_{2n}^\circ) = \sum_{k=1}^n \mathbb{P}(M_{2n}^\circ \geq k) = \binom{2n}{n}^{-1} \sum_{k=1}^n \binom{2n}{n+k}$$

$$= \binom{2n}{n}^{-1} \cdot \frac{1}{2} \cdot \left(2^{2n} - \binom{2n}{n}\right),$$

which proves (b). Due to $\mathbb{E}(M_{2n}^\circ) = 1/(2u_{2n}) - \frac{1}{2}$ with $u_{2n}$ as in (2.15), (c) follows from (2.29). According to (7.13),

$$\mathbb{E}\left[(M_{2n}^\circ)^2\right] = \sum_{k=1}^n (2k-1)\mathbb{P}(M_{2n}^\circ \geq k)$$

$$= 2\sum_{k=1}^n \frac{k\binom{2n}{n+k}}{\binom{2n}{n}} - \mathbb{E}(M_{2n}^\circ). \qquad (3.20)$$

Using (7.25), part (d) then follows from direct calculation (Exercise 3.10). ∎

The next result, proved in a more general form by N.W. Smirnov[6] [SM], describes the asymptotic behavior of $M_{2n}^\circ$ as $n \to \infty$ (see also [REN], p. 496 ff.):

---

[6] Nikolai Vasilyevich Smirnov (1900–1966), was a leading Russian mathematical statistician. From 1938 he worked at the Steklov Institute, where in his last year of life he succeeded A.N. Kolmogorov as head of the department of mathematical statistics.

**Theorem 3.9 (Limit Distribution of the Maximum of a Bridge)** *For the maximum $M_{2n}^\circ$ of a $2n$-bridge,*

$$\lim_{n\to\infty} \mathbb{P}\left(\frac{M_{2n}^\circ}{\sqrt{2n}} \leq x\right) = 1 - \exp\left(-2x^2\right), \qquad x \geq 0.$$

*Proof* We can assume $x > 0$, as the statement obviously holds for $x = 0$. If we write $k_n$ for the smallest integer greater than or equal to $x\sqrt{2n}$, then, due to the fact that $M_{2n}^\circ$ is integer-valued and Eq. (3.19),

$$\mathbb{P}\left(\frac{M_{2n}^\circ}{\sqrt{2n}} \geq x\right) = \mathbb{P}(M_{2n}^\circ \geq k_n) = \frac{\binom{2n}{n+k_n}}{\binom{2n}{n}} = \prod_{j=0}^{k_n-1}\left(1 - \frac{k_n}{n-j+k_n}\right).$$

Switching to the logarithm and using the inequalities (7.10) and (7.11), we get

$$\log \mathbb{P}\left(\frac{M_{2n}^\circ}{\sqrt{2n}} \geq x\right) \leq -k_n \sum_{j=0}^{k_n-1} \frac{1}{n-j+k_n} \leq -\frac{k_n^2}{n+k_n}, \qquad (3.21)$$

$$\log \mathbb{P}\left(\frac{M_{2n}^\circ}{\sqrt{2n}} \geq x\right) \geq -k_n \sum_{j=0}^{k_n-1} \frac{1}{n-j} \geq -\frac{k_n^2}{n-k_n+1}. \qquad (3.22)$$

By the definition of $k_n$, the rightmost terms in (3.21) and (3.22) converge to $-2x^2$ as $n \to \infty$, so the assertion follows. ∎

Figure 3.16 shows the distribution function $G_{2,2}$ (right) and the density $g_{2,2}$ (left) of the distribution Wei(2, 2) (cf. (3.9)). Table 3.2 contains some values of $G_{2,2}$.

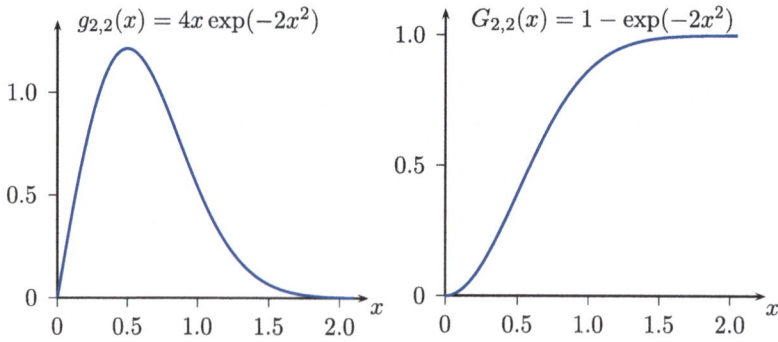

**Fig. 3.16** Density and distribution function of the Weibull distribution Wei(2, 2)

## 3.4 Maximum and Minimum

**Table 3.2** Values of the distribution function $G_{2,2}(x) := 1 - \exp(-2x^2)$

| $x$ | $G_{2,2}(x)$ | $x$ | $G_{2,2}(x)$ | $x$ | $G_{2,2}(x)$ |
|---|---|---|---|---|---|
| 0.00 | 0.000 | 0.589 | 0.500 | 1.20 | 0.944 |
| 0.05 | 0.005 | 0.60 | 0.513 | 1.224 | 0.950 |
| 0.10 | 0.020 | 0.70 | 0.625 | 1.30 | 0.966 |
| 0.15 | 0.044 | 0.80 | 0.722 | 1.40 | 0.980 |
| 0.20 | 0.077 | 0.90 | 0.802 | 1.50 | 0.989 |
| 0.30 | 0.165 | 1.00 | 0.865 | 1.52 | 0.990 |
| 0.40 | 0.274 | 1.073 | 0.900 | 1.60 | 0.994 |
| 0.50 | 0.394 | 1.10 | 0.911 | 1.70 | 0.997 |

It is interesting to compare the asymptotic behavior of the maximum $M_{2n}$, studied in Sect. 2.7 for a simple random walk, with that of a bridge of the same length. We suspected that a bridge, compared to a simple random walk, is restricted in its volatility, which should be noticeable in the stochastic behavior of the "maximum upward deflection." According to Theorems 2.13 and 3.9, for large $n$ and $x > 0$,

$$\mathbb{P}\left(M_{2n} \leq x\sqrt{2n}\right) \approx \Phi^*(x), \qquad \mathbb{P}\left(M_{2n}^\circ \leq x\sqrt{2n}\right) \approx W_{2,2}(x),$$

where $\Phi^*(x)$ and $W_{2,2}(x)$ can be compared using Tables 3.2 and 2.2 for $x = 0.1$ or $x = 1$:

$$\mathbb{P}\left(M_{10,000} \leq 10\right) \approx 0.08, \qquad \mathbb{P}\left(M_{10,000}^\circ \leq 10\right) \approx 0.02,$$

$$\mathbb{P}\left(M_{10,000} \leq 100\right) \approx 0.683, \qquad \mathbb{P}\left(M_{10,000}^\circ \leq 100\right) \approx 0.865.$$

Simple symmetric random walks therefore have a greater chance of a small maximum, compared to bridges. On the other hand, bridges have a smaller chance of a large maximum compared to equally long simple random walks. A free random walk can, after the start, "drift into the area of negative integers without being affected by any reset mechanism" and might never reach the $x$-axis again, which happens with a substantial probability. For example, according to Theorems 2.12 (c) and 3.8,

$$\mathbb{P}(M_{2n} = 0) = \binom{2n}{n} 2^{-2n} \approx \frac{1}{\sqrt{\pi n}}, \qquad \mathbb{P}(M_{2n}^\circ = 0) = \frac{1}{n+1}.$$

On the other hand, a symmetric random walk of length $2n$ can significantly exceed the maximum possible height $n$ for $2n$-bridges. Figure 3.17 shows that for small $x$, the inequality $\Phi^*(x) > W_{2,2}(x)$ holds, whereas for large $x$, the reverse inequality applies. The numerically determined value $x_0$, for which $\Phi^*(x_0) = W_{2,2}(x_0)$, is $x_0 = 0.4791$.

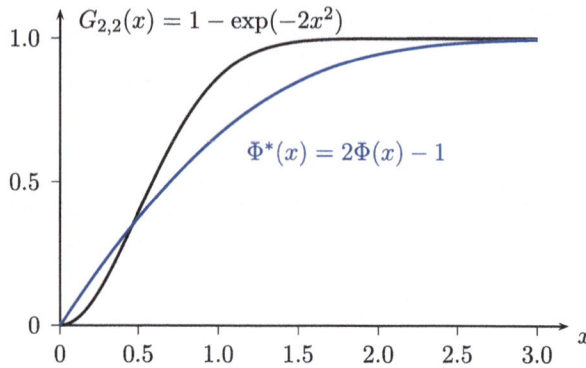

**Fig. 3.17** Graphs of the functions $\Phi^*$ and $W_{2,2}$

Due to

$$m_{2n}^\circ := \min(S_0, S_1, \ldots, S_{2n}) = -\max(-S_0, -S_1, \ldots, -S_{2n}),$$

the distribution of the minimum $m_{2n}^\circ$ directly follows from that of the maximum, since $M_{2n}^\circ$ and $\max(-S_0, \ldots, -S_{2n})$ have the same distribution. This is because the latter holds for the random vectors $(X_1, \ldots, X_{2n})$ and $(-X_1, \ldots, -X_{2n})$.

We conclude this section by answering a pertinent question: At which point $k \in \{0, 1, \ldots, 2n-1\}$ is the maximum of a $2n$-bridge attained? For the path in Fig. 3.14, this occurs at exactly one point, namely at time 9. However, in general, there can be more than one such maximizer.

Let

$$A := \left\{ k \in \{0, 1, \ldots, 2n-1\} : S_k = \max_{0 \le j \le 2n-1} S_j \right\} \tag{3.23}$$

be the random set of time points at which the maximum of a $2n$-bridge is attained. Note that the set $A$ can have at most $n$ elements. This maximum number is achieved, for example, by the path that alternates between down and then up by one step. We agree to select one element from the set $A$ purely at random. Denoting the result of this selection by $T$, the random variable $T$, referred to as the *random maximizer* of the $2n$-bridge, takes on one of the values $0, 1, \ldots, 2n-1$. The following surprising result states that $T$ has a uniform distribution (cf. [FE1], p. 97).

**Theorem 3.10 (Distribution of the Random Maximizer of a Bridge)** *The random maximizer $T$ of a $2n$-bridge is uniformly distributed, i.e.,*

$$\mathbb{P}(T = k) = \frac{1}{2n}, \quad k = 0, 1, \ldots, 2n-1.$$

***Proof*** On the set

$$\Omega := \{(a_1, \ldots, a_{2n}) \in \{-1, 1\}^{2n} : a_1 + \ldots + a_{2n} = 0\}$$

of all $2n$-bridges, we call two paths $(a_1, \ldots, a_{2n})$ and $(b_1, \ldots, b_{2n})$ from $\Omega$ *equivalent*, if they can be transformed into one another by a cyclic permutation, i.e., if there exists a $j \in \{0, 1, \ldots, 2n-1\}$ such that $b_k = a_{k+j}$ for $k = 1, \ldots, 2n-j$ and $b_k = a_{j-2n+k}$ for $k = 2n-j+1, \ldots, 2n$.

Geometrically, the paths resulting from cyclic permutation are formed by extending the bridge path once more at the point $(2n, 0)$ and then viewing a time segment of length $2n$ of this "doubled path" for each $j = 0, 1, \ldots, 2n-1$ from the point $(j, S_j)$ as the new origin. The cyclic permutation defines an equivalence relation on $\Omega$, where each equivalence class contains exactly $2n$ paths.

Any two equivalent paths possess the same maximum, and the corresponding sets $A$ of maximizers as in (3.23) are transformed into one another by the same cyclic permutation as the two paths. Each of the maximizers takes on each of the values $0, 1, \ldots, 2n-1$. It follows that $T$ is uniformly distributed on each of the equivalence classes; thus, $\mathbb{P}(T = k | B_i) = \frac{1}{2n}$ for $k = 0, 1, \ldots, 2n-1$, for each equivalence class $B_i \subset \Omega$. This establishes the claim using the law of total probability (7.4). ∎

## 3.5 Changes of Sign

In this section, we examine the distribution of

$$C_{2n}^\circ := \sum_{j=1}^{n-1} \mathbf{1}\{S_{2j+1} \cdot S_{2j-1} = -1\},$$

which counts the changes of sign in a $2n$-bridge. Clearly, this random variable can take the values $0, 1, \ldots, n-1$. We begin by determining $\mathbb{P}(C_{2n}^\circ \geq k)$, and consequently, the size of the set $\{C_{2n}^\circ \geq k\}$, representing all $2n$-bridges with *at least* $k$ changes of sign. Due to symmetry,

$$|\{C_{2n}^\circ \geq k\}| = 2 \cdot |\{C_{2n}^\circ \geq k, S_1 = 1\}|. \qquad (3.24)$$

Thus, we assume that the bridge starts with an upward step.

Let $M_{1,k}$ denote the set appearing on the right-hand side of (3.24), i.e., the set of all $2n$-bridges with $C_{2n}^\circ \geq k$ and $S_1 = 1$. Let $M_{2,k}$ be the set of all paths leading from $(1, 1)$ to $(2n, -2k)$. We will show that, for each $k \geq 0$, the set $M_{1,k}$ can be mapped one-to-one onto the set $M_{2,k}$. Since, by (2.11), the set $M_{2,k}$ contains $\binom{2n-1}{n+k}$ elements, we can, in conjunction with (3.24), conclude that

$$\mathbb{P}(C_{2n}^\circ \geq k) = 2 \cdot \frac{\binom{2n-1}{n+k}}{\binom{2n}{n}}, \qquad k = 0, 1, \ldots, n-1, \qquad (3.25)$$

(cf. [DW], p. 1049). To construct a bijection between $M_{1,k}$ and $M_{2,k}$, we assume, without loss of generality, the case $k \geq 1$.

**SAQ 7** Why can we assume w.l.o.g. $k \geq 1$?

For the case $n = 10$, Fig. 3.18 shows a $2n$-bridge that, for each $k \in \{1, 2, 3, 4\}$, belongs to $M_{1,k}$ (note that $M_{1,1} \supset M_{1,2} \supset M_{1,3} \supset \ldots$).

We first show that there is a bijection between the sets $M_{1,1}$ and $M_{2,1}$. To this end, consider any bridge from $M_{1,1}$. After the first change of sign, the bridge reaches the height $-1$. By reflecting the subsequent segment of the path across the line $y = 1$, and combining it with the segment from $(1, 1)$ to the point of the first change of sign, we obtain a path that leads from $(1, 1)$ to $(2n, -2)$, which belongs to $M_{2,1}$. This assignment of paths from $M_{1,1}$ to paths in $M_{2,1}$ is clearly injective.

Conversely, any path leading from $(1, 1)$ to $(2n, -2)$ eventually reaches the height $-1$ for the first time. Reflecting the subsequent path segment across the line $y = -1$ and adding an upward step at the beginning yields a $2n$-bridge from $M_{1,1}$ (see Fig. 3.19). This demonstrates that the mapping of paths from $M_{1,1}$ to paths in $M_{2,1}$, as described above, is surjective.

We now consider the case $k = 2$, where there are at least two changes of sign. A path from $M_{1,2}$ reaches the height $-1$ after the first change of sign and the height $1$ after the second change of sign. Reflecting the path segment between the first and second change of sign across the line $y = -1$, the reflected segment reaches the height $-3$. If we then attach the segment following the second change of sign (by

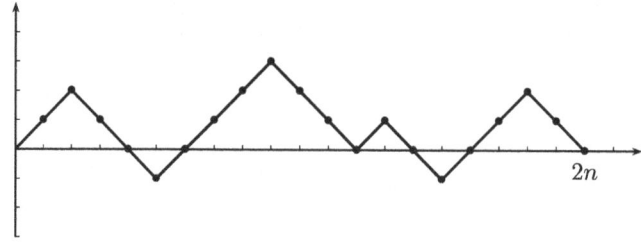

**Fig. 3.18** $2n$-bridge with at least $k$ changes of sign ($k = 1, 2, 3, 4$)

## 3.5 Changes of Sign

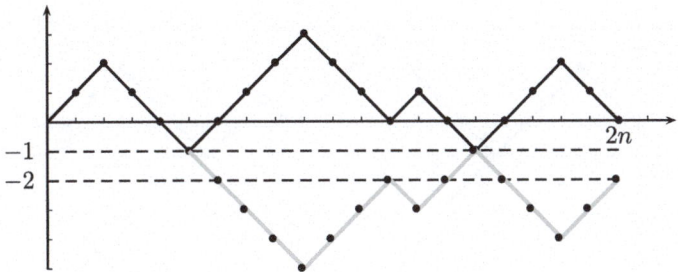

**Fig. 3.19** Illustrating the bijection between $M_{1,1}$ and $M_{2,1}$

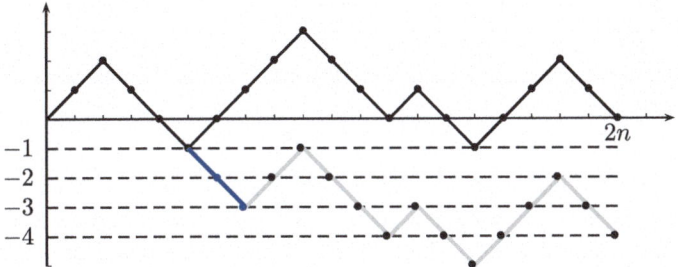

**Fig. 3.20** Bijection between $M_{1,2}$ and $M_{2,2}$ by reflection (blue path segment) and shifting (gray path segment)

shifting it downward by 4 units), we obtain a path leading from $(1, 1)$ to $(2n, -4)$, which is a path from $M_{2,2}$ (see Fig. 3.20). This assigment of paths from $M_{1,2}$ to paths in $M_{2,2}$ is clearly injective.

Conversely, starting with any path leading from $(1, 1)$ to $(2n, -4)$, we can uniquely construct a $2n$-bridge with at least two changes of sign. A path from $M_{2,2}$ reaches the height $-1$ after the first change of sign, and then hits the height $-3$. Reflecting the sub-path between these two heights across the line $y = -1$ introduces another change of sign. The subsequent sub-path is then shifted upward by 4 units and attached to the reflected segment (see Fig. 3.20). Adding an upward step at the beginning, we obtain a path from $M_{1,2}$. Thus, the mapping of paths from $M_{1,2}$ to paths in $M_{2,2}$, as described above, is surjective. Combined with the previously established injectivity, this proves that the mapping is bijective.

We now consider the case $k \geq 3$. A path from $W_{1,k}$ with $k \geq 3$ is mapped to a path leading from $(1, 1)$ to $(2n, -2k)$ by alternating between reflection and shifting as follows: The first upward step is removed, and the segment of the path up to the first change of sign is left unchanged. The next segment, up to the second change of sign, is reflected across the line $y = -1$, so that the reflected segment ends at height $-3$. The subsequent segment, up to the third change of sign, remains unchanged and is attached, resulting in a path ending at height $-5$.

If $k = 3$, the remaining segment is simply attached, ending at the point $(2n, -2k)$. Otherwise, for $k > 3$, the segment up to the next change of sign is

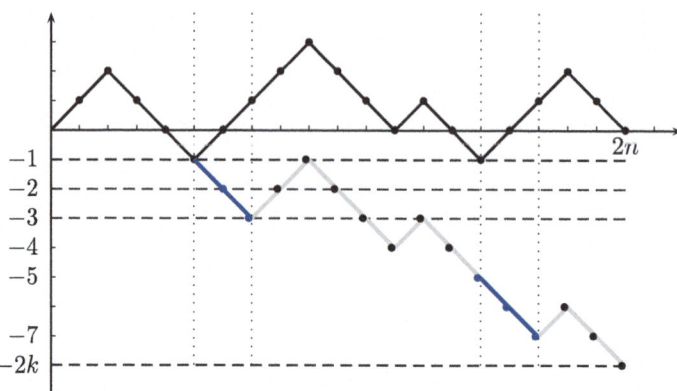

**Fig. 3.21** Bijection from $M_{1,4}$ to $M_{2,4}$ through alternating reflection (blue sub-paths) and shifting (gray sub-paths)

reflected across the line $y = -5$, ending at height $-7$. For $k = 4$, the remaining segment is attached, reaching the point $(2n, -2k)$, and the process continues similarly for larger values of $k$. Figure 3.21 illustrates the procedure for the case $k = 4$.

The alternating reflections and shifts ensure that the mapping from paths in $M_{1,k}$ to paths in $M_{2,k}$ is injective. Surjectivity follows as in the previously discussed case $k = 2$, by dividing the path from $(1, 1)$ to $(2n, -2k)$ into $k + 1$ segments. The first segment runs up to the first time it reaches the height $-1$, while the subsequent segments run from the first time they reach the height $-2j + 1$ to the first time they reach the height $-2j - 1$ for $j = 1, \ldots, k - 1$. The final segment then reaches the endpoint $(2n, -2k)$. The construction proceeds as follows: the first segment is preceded by an upward step, the next segment is reflected across the line $y = -1$ and attached to the first segment, the third segment is attached to the end of the reflected segment, the fourth segment is reflected across the line $y = -3$ and attached, and so on. In this manner, we obtain a $2n$-bridge from $M_{1,k}$.

Equation (3.25) is the key for subsequent results about the stochastic behavior of $C_{2n}^\circ$ (see [CV]).

**Theorem 3.11 (Distribution of the Number of Changes of Sign in Bridges)** *For the number $C_{2n}^\circ$ of changes of sign in a $2n$-bridge, the following hold:*

(a) $\mathbb{P}(C_{2n}^\circ = k) = \dfrac{2(k+1)}{n} \cdot \dfrac{\binom{2n}{n+k+1}}{\binom{2n}{n}}, \quad k = 0, \ldots, n-1,$

## 3.5 Changes of Sign

(b) $\mathbb{E}(C_{2n}^\circ) = \dfrac{2^{2n-1}}{\binom{2n}{n}} - 1,$

(c) $\lim\limits_{n\to\infty} \dfrac{\mathbb{E}(C_{2n}^\circ)}{\sqrt{2n}} = \dfrac{\sqrt{\pi}}{2\sqrt{2}},$

(d) $\mathbb{V}(C_{2n}^\circ) = n - \left(\dfrac{2^{2n-1}}{\binom{2n}{n}}\right)^2.$

**Proof** Using (3.25), part (a) follows from $\mathbb{P}(C_{2n}^\circ = k) = \mathbb{P}(C_{2n}^\circ \geq k) - \mathbb{P}(C_{2n}^\circ \geq k+1)$. To prove (b), we use (3.25) and

$$\mathbb{E}(C_{2n}^\circ) = \sum_{k=1}^{n-1} \mathbb{P}(C_{2n}^\circ \geq k) = 2\binom{2n-1}{n+k}\binom{2n}{n}^{-1}.$$

Since $\sum_{j=0}^{2n-1} \binom{2n-1}{j} = 2^{2n-1}$ and $\binom{2n-1}{j} = \binom{2n-1}{2n-1-j}$, we get

$$2\sum_{k=1}^{n-1}\binom{2n-1}{n+k} = 2^{2n-1} - 2\binom{2n-1}{n},$$

from which (b) follows. Since $2^{2n-1}/\binom{2n}{n} = 2/u_{2n}$ with $u_{2n}$ as in (2.15) and

$$\dfrac{\mathbb{E}(C_{2n}^\circ)}{\sqrt{2n}} = \dfrac{\sqrt{\pi}}{2\sqrt{2}} \cdot \dfrac{1}{\sqrt{\pi n}\cdot u_{2n}} - \dfrac{1}{\sqrt{2n}},$$

part (c) is obtained using (2.29). To show part (d), we employ (3.25) and (7.13), according to which

$$\mathbb{E}\left[(C_{2n}^\circ)^2\right] = \dfrac{1}{\binom{2n}{n}}\sum_{k=1}^{n-1}(2k-1)\cdot 2 \cdot \binom{2n-1}{n+k}. \qquad (3.26)$$

Using (7.26) and

$$\sum_{k=1}^{n-1}\binom{2n-1}{n+k} = 2^{2n-2} - \binom{2n-1}{n-1}, \qquad (3.27)$$

together with part b), the variance of $C_{2n}^\circ$ can then be obtained by direct calculation (Exercise 3.12). ■

**SAQ 8**  Why does Eq. (3.27) hold?

Using (2.29), one also obtains the asymptotic behavior of the variance:

$$\lim_{n \to \infty} \mathbb{V}\left(\frac{C_{2n}^\circ}{\sqrt{2n}}\right) = \frac{1}{2} - \frac{\pi}{8}.$$

**SAQ 9**  Can you justify this equation?

**Theorem 3.12 (Limit Distribution of $C_{2n}^\circ$, [CV])**  *For the number $C_{2n}^\circ$ of changes of sign in a $2n$-bridge,*

$$\lim_{n \to \infty} \mathbb{P}\left(\frac{C_{2n}^\circ}{\sqrt{2n}} \leq x\right) = 1 - \exp\left(-2x^2\right), \qquad x \geq 0.$$

***Proof***  Since the case $x = 0$ is immediately apparent, we assume $x > 0$.

**SAQ 10**  Do you see that the case $x = 0$ is immediately apparent?

Let $k_n$ denote the smallest integer greater than or equal to $x\sqrt{2n}$. The fact that $C_{2n}^\circ$ is integer-valued and (3.25) yield

$$\mathbb{P}\left(\frac{C_{2n}^\circ}{\sqrt{2n}} \geq x\right) = 2 \cdot \frac{\binom{2n-1}{n+k_n}}{\binom{2n}{n}} = \prod_{j=0}^{k_n}\left(1 - \frac{k_n}{n + k_n - j}\right).$$

Taking logarithms and using (7.10) and (7.11), it follows that

$$\log \mathbb{P}\left(\frac{C_{2n}^\circ}{\sqrt{2n}} \geq x\right) \leq -k_n \sum_{j=0}^{k_n} \frac{1}{n + k_n - j} \leq -\frac{k_n(k_n + 1)}{n + k_n}, \qquad (3.28)$$

$$\log \mathbb{P}\left(\frac{C_{2n}^\circ}{\sqrt{2n}} \geq x\right) \geq -k_n \sum_{j=0}^{k_n} \frac{1}{n - j} \geq -\frac{k_n(k_n + 1)}{n - k_n}. \qquad (3.29)$$

By definition of $k_n$, the rightmost terms of (3.28) and (3.29) converge to the same limit $-2x^2$ as $n \to \infty$, from which the assertion follows. ■

## 3.5 Changes of Sign

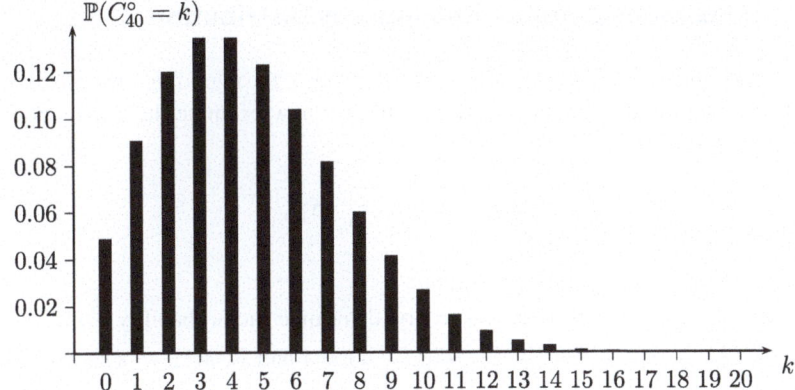

**Fig. 3.22** Bar chart of the distribution of $C^\circ_{40}$

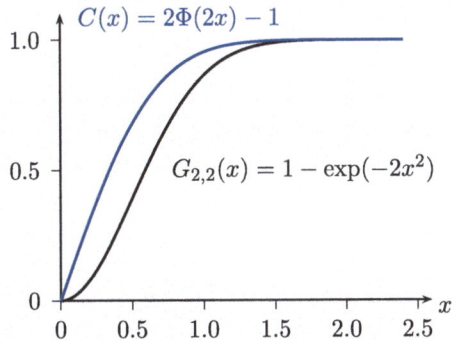

**Fig. 3.23** Graphs of the functions $C$ and $W_{2,2}$

The bar chart of the distribution of $C^\circ_{40}$, shown in Fig. 3.22, has a different shape compared to the bar chart in Fig. 2.44. This difference is reflected in the distinct limit distributions for the number of changes of sign in "free" random walks (Theorem 2.34) versus bridges (Theorem 3.12). Notably, the Weibull limit distributions of $C^\circ_{2n}/\sqrt{2n}$ and $M^\circ_{2n}/\sqrt{2n}$ are identical.

Figure 3.23 displays the graphs of the two distribution functions $C(x) = 2\Phi(2x) - 1$ from Theorem 2.34 and $G_{2,2}(x) = 1 - \exp(-2x^2)$ from Theorem 3.12. This figure is identical to Fig. 3.7, except for the scale change $x \mapsto 2x$.

The effect already observed in the number of visits to zero is evident here as well: a bridge tends to have significantly more changes of sign compared to a simple symmetric random walk. The limit distribution of the number of changes of sign in bridges is stochastically larger than the corresponding limit distribution in "free" random walks.

## 3.6 Maximum Modulus, Kolmogorov Distribution

Quite analogously to the considerations in Sect. 2.12 regarding the maximum modulus of a simple symmetric random walk, we now examine the distribution of the random variable

$$|M^\circ|_{2n} = \max_{j=1,\ldots,2n} |S_j|,$$

i.e., the *maximum modulus* of a $2n$-bridge.

Here, too, it proves advantageous to first determine the probability $\mathbb{P}(|M^\circ|_{2n} \geq k)$ for $k = 2, \ldots, n-1$. Clearly, $\mathbb{P}(|M^\circ|_{2n} \geq 1) = 1$ and $\mathbb{P}(|M^\circ|_n \geq n) = 2\binom{2n}{n}^{-1}$.

**SAQ 11** Why is $\mathbb{P}(|M^\circ|_n \geq n) = 2\binom{2n}{n}^{-1}$?

Thus, we now assume $2 \leq k \leq n-1$. The following reasoning mirrors that of Sect. 2.12, where, as before, we suppress the dependence on $k$ in the notation. For the number $b_n$ of all $2n$-bridges with the property $|M^\circ|_{2n} \geq k$, the inclusion-exclusion principle applies (the argument from Sect. 2.12 can be repeated verbatim here):

$$b_n = 2\sum_{s \geq 1} (-1)^{s-1} b_{n,s}^+. \tag{3.30}$$

Here, $b_{n,s}^+$ denotes the number of all $2n$-bridges that reach the height $k$ and thereafter fluctuate at least $s-1$ times between the respective opposite height, i.e., first $-k$, then $k$, and so on. Note that the above alternating sum terminates because a path of length $2n$ can only fluctuate between the heights $k$ and $-k$ a finite number of times.

**SAQ 12** How often can a $2n$-bridge fluctuate between $\pm k$?

From (3.19), we obtain

$$b_{n,1}^+ = \binom{2n}{n+k},$$

as illustrated in Fig. 3.14. To determine the number $b_{n,2}^+$, representing all $2n$-bridges that reach the height $k$ and subsequently the height $-k$, we construct a bijective mapping between the set of all $2n$-bridges with the aforementioned property and the set of all paths leading from $(0, 0)$ to $(2n, 4k)$. By Lemma 2.1, the number of paths

## 3.6 Maximum Modulus, Kolmogorov Distribution

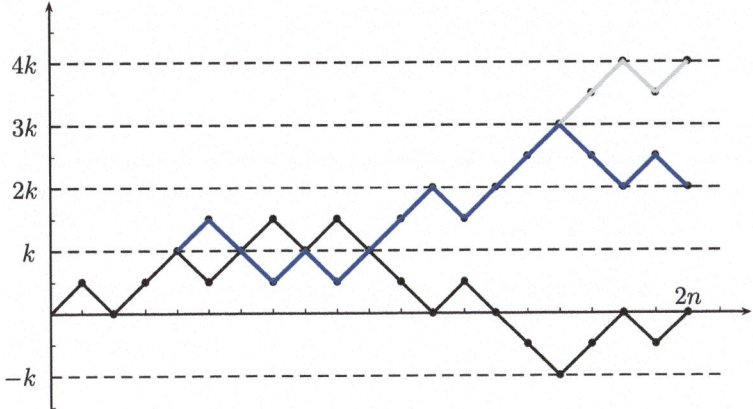

**Fig. 3.24** Double reflection of a path that first reaches the height $k$ and then the height $-k$ results in a path leading from $(0, 0)$ to $(2n, 4k)$

with this property is given by $\binom{2n}{n+2k}$. Thus, we have shown that

$$b_{n,2}^+ = \binom{2n}{n+2k}.$$

To construct the bijection, we reflect a $2n$-bridge that reaches the height $k$ and subsequently the height $-k$ at the first time it reaches the height $k$ across the line $y = k$, thereby obtaining a path leading from $(0, 0)$ to $(2n, 2k)$ (see Fig. 3.24). Since the original path also reaches the height $-k$, the path leading from $(0, 0)$ to $(2n, 2k)$ must reach the height $3k$. Reflecting this path at the *first* time it reaches the height $3k$ across the line $y = 3k$, we obtain a path leading from $(0, 0)$ to $(2n, 4k)$ (see Fig. 3.24).

It is evident that this mapping of paths is injective. It is also surjective because every path from $(0, 0)$ to $(2n, 4k)$ eventually reaches the height $3k$ for the first time. Reflecting the path from this point across the line $y = 3k$ creates a path from $(0, 0)$ to $(2n, 2k)$, which, since $2k > k$, must eventually reach the height $k$ for the first time. Reflecting the path from this point across the line $y = k$ results in a $2n$-bridge that reaches the height $k$ and subsequently the height $-k$ (see Fig. 3.24).

If a $2n$-bridge first reaches the height $k$, then $-k$, and subsequently reaches the height $k$ once more, the path resulting from two reflections as shown in Fig. 3.24, which leads from $(0, 0)$ to $(2n, 4k)$, will eventually reach the height $5k$. Reflecting this path at the first time it reaches this height across the line $y = 5k$ produces a path from $(0, 0)$ to $(2n, 6k)$. This mapping of $2n$-bridges, which first reach the height $k$, then $-k$, and later again $k$, to all paths leading from $(0, 0)$ to $(2n, 6k)$ is bijective, and the number of such paths is given by the binomial coefficient $\binom{2n}{n+3k}$. Further reflections across the lines $y = 7k$, $y = 9k$, and so on are required if the $2n$-bridge continues to oscillate between the heights $k$ and $-k$ more frequently. Thus, in general, we obtain

$$b_{n,s}^+ = \binom{2n}{n+sk}.$$

Together with (3.30), we derive the following distribution of the maximum modulus of a $2n$-bridge (see [GK] and [GRV] for a more general result on the joint distribution of $M_{2n}^\circ$ and $m_{2n}^\circ$).

**Theorem 3.13 (Distribution of the Maximum Modulus of a 2$n$-Bridge)** *For the maximum modulus $|M^\circ|_{2n}$ of a $2n$-bridge,*

$$\mathbb{P}(|M^\circ|_{2n} \geq k) = 2 \cdot \sum_{s=1}^{\lfloor \frac{n}{k} \rfloor} (-1)^{s-1} \frac{\binom{2n}{n+sk}}{\binom{2n}{n}}, \quad k = 1, \ldots, n. \quad (3.31)$$

By taking the difference

$$\mathbb{P}(|M^\circ|_{2n} = k) = \mathbb{P}(|M^\circ|_{2n} \geq k) - \mathbb{P}(|M^\circ|_{2n} \geq k+1),$$

one obtains the probability that the maximum modulus of a bridge takes a specific value. Figure 3.25 shows the bar chart of the distribution of $|M^\circ|_{100}$. Although any value from 1 to 50 is possible for $|M^\circ|_{100}$, we have $\mathbb{P}(3 \leq |M^\circ|_{100} \leq 19) > 0.999$, indicating that the distribution is essentially concentrated in the range $3 \leq k \leq 19$.

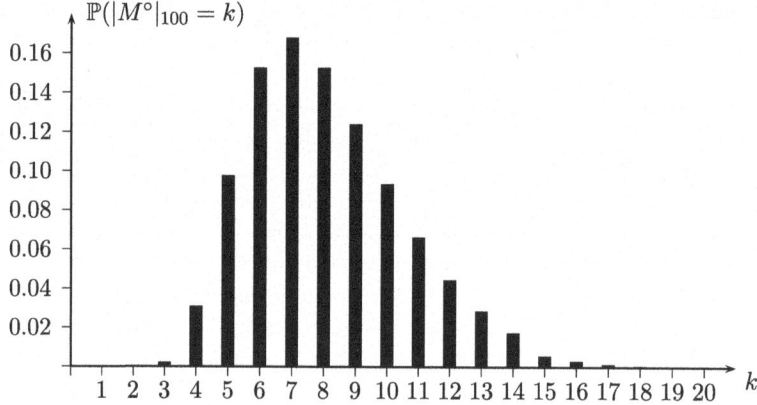

**Fig. 3.25** Bar chart of the distribution of $|M^\circ|_{100}$

## 3.6 Maximum Modulus, Kolmogorov Distribution

According to (7.12), the expectation of $|M^\circ|_{2n}$ is obtained by summing the probabilities in (3.31) over $k$. Due to (7.13), the expectation $\mathbb{E}(|M^\circ|_{2n}^2)$ is given by

$$\mathbb{E}(|M^\circ|_{2n}^2) = \sum_{k=1}^{n}(2k-1)\mathbb{P}(|M^\circ|_{2n} \geq k),$$

from which one can also derive a (not very convenient) expression for the variance of $|M^\circ|_{2n}$.

Katzenbeisser and Panny [KP] obtained the following expressions for $\mathbb{E}(|M^\circ|_{2n})$ and $\mathbb{V}(|M^\circ|_{2n})$:

$$\mathbb{E}(|M^\circ|_{2n}) = \log 2\sqrt{\pi n} - \frac{1}{2} + \frac{\log 2\sqrt{\pi}}{8} \cdot \frac{1}{\sqrt{n}} + o\left(\frac{1}{\sqrt{n}}\right),$$

$$\mathbb{V}(|M^\circ|_{2n}) = \left(\frac{\pi^2}{6} - \pi (\log 2)^2\right) n + \frac{1}{12} - \frac{\pi}{4}\left((\log 2)^2 - \frac{\pi}{9}\right) + o(1).$$

Here, $o(a_n)$ denotes a term that converges to zero when divided by $a_n$ as $n \to \infty$.

Regarding the asymptotic behavior of $|M^\circ|_{2n}$ as $n \to \infty$, the following limit theorem, proved in a more general form by N.W. Smirnov [SM], holds.

**Theorem 3.14 (Limit Distribution of $|M^\circ|_{2n}$)** Let $|M^\circ|_{2n}$ be the maximum modulus of a $2n$-bridge. Then, for every $x > 0$,

$$\lim_{n \to \infty} \mathbb{P}\left(\frac{|M^\circ|_{2n}}{\sqrt{2n}} \leq x\right) = K(x),$$

where

$$K(x) := 1 - 2\sum_{s=1}^{\infty}(-1)^{s-1}\exp\left(-2s^2 x^2\right). \tag{3.32}$$

**Proof** Fix any $x > 0$, and let $k_n = k_n(x) := \lceil x\sqrt{2n} \rceil$, which is the smallest integer greater than or equal to $x\sqrt{2n}$. Choose $n$ sufficiently large so that $1 \leq k_n \leq n$. Since $|M^\circ|_{2n}$ is integer-valued, (3.31) yields

$$\mathbb{P}\left(|M^\circ|_{2n} \geq x\sqrt{2n}\right) = \mathbb{P}(|M^\circ|_{2n} \geq k_n)$$

$$= 2\sum_{s=1}^{s_n}(-1)^{s-1}\frac{\binom{2n}{n+sk_n}}{\binom{2n}{n}}, \tag{3.33}$$

where we set $s_n = s_n(x) = \lfloor n/k_n(x) \rfloor$. We first claim that

$$\lim_{n\to\infty} \frac{\binom{2n}{n+sk_n}}{\binom{2n}{n}} = e^{-2s^2x^2}. \qquad (3.34)$$

To prove this, let us denote the quotient in (3.34) by $q_n$:

$$q_n = \frac{n!}{(n-sk_n)!} \cdot \frac{n!}{(n+sk_n)!}$$
$$= \prod_{j=0}^{sk_n-1} \left(\frac{n-j}{n+sk_n-j}\right) = \prod_{j=0}^{sk_n-1} \left(1 - \frac{sk_n}{n+sk_n-j}\right).$$

Using the inequality $\log t \leq t - 1$, we obtain

$$\log q_n \leq -\sum_{j=0}^{sk_n-1} \frac{sk_n}{n+sk_n-j} \leq -\frac{(sk_n)^2}{n+sk_n}.$$

Since $x\sqrt{2n} \leq k_n \leq 1 + x\sqrt{2n}$, the rightmost term converges to $-2s^2x^2$ as $n \to \infty$, leading to $\limsup_{n\to\infty} q_n \leq \exp(-2s^2x^2)$. On the other hand, using the inequality $\log t \geq 1 - \frac{1}{t}$ with $t = 1 - (sk_n)/(n+sk_n-j)$, we get

$$\log q_n \geq -\sum_{j=0}^{sk_n-1} \frac{sk_n}{n-j} \geq -\frac{(sk_n)^2}{n-sk_n+1}.$$

By definition of $k_n$, the rightmost term also converges to $-2s^2x^2$, implying that $\liminf_{n\to\infty} q_n \geq \exp(-2s^2x^2)$. Thus, combining the upper and lower bounds for $q_n$, we conclude that $\lim_{n\to\infty} q_n = \exp(-2s^2x^2)$, which proves (3.34).

To conclude the proof, we note that by applying the inclusion-exclusion principle for counting the favorable paths, the partial sums resulting from truncating the sum in (3.33) after an odd or even number of terms are alternately too large and too small. Thus, analogous to the proof of Theorem 2.38, we have:

$$\mathbb{P}\left(|M^\circ|_{2n} \geq x\sqrt{2n}\right) \leq 2 \sum_{s=1}^{2r+1} (-1)^{s-1} \frac{\binom{2n}{n+sk_n}}{\binom{2n}{n}},$$

## 3.6 Maximum Modulus, Kolmogorov Distribution

and

$$\mathbb{P}\left(|M^\circ|_{2n} \geq x\sqrt{2n}\right) \geq 2 \sum_{s=1}^{2r}(-1)^{s-1}\frac{\binom{2n}{n+sk_n}}{\binom{2n}{n}},$$

for each fixed $r$, provided that the upper summation limits are between 1 and $s_n$. Together with (3.34), it follows that

$$\limsup_{n\to\infty} \mathbb{P}\left(|M^\circ|_{2n} \geq x\sqrt{2n}\right) \leq 2 \sum_{s=1}^{2r+1}(-1)^{s-1} e^{-2s^2 x^2},$$

and

$$\liminf_{n\to\infty} \mathbb{P}\left(|M^\circ|_{2n} \geq x\sqrt{2n}\right) \geq 2 \sum_{s=1}^{2r}(-1)^{s-1} e^{-2s^2 x^2},$$

for any fixed $r$. Therefore, as $r \to \infty$, we obtain

$$\lim_{n\to\infty} \mathbb{P}\left(|M^\circ|_{2n} \geq x\sqrt{2n}\right) = 2 \sum_{s=1}^{\infty}(-1)^{s-1} e^{-2s^2 x^2}.$$

Taking the complement and using (7.9), the result follows. ∎

The continuous function $K : \mathbb{R} \to [0, 1]$, defined by (3.32) for $x > 0$ and extended to the entire real line by $K(x) := 0$ for $x \leq 0$, is called the *distribution function of the Kolmogorov distribution*, named after the Russian mathematician A. N. Kolmogorov.[7] Since the series in (3.32) converges uniformly on the interval $[a, \infty)$ for every $a > 0$, the function $K(x)$ is differentiable term-by-term for $x > 0$. The derivative,

$$k(x) := 8x \sum_{s=1}^{\infty}(-1)^{s-1} s^2 e^{-2s^2 x^2}, \qquad x > 0,$$

extended by the definition $k(x) := 0$ for $x \leq 0$, is called the *density of the Kolmogorov distribution*.

---

[7] Andrei Nikolaevich Kolmogorov (1903–1987), a professor in Moscow from 1930, was one of the most significant mathematicians of the twentieth century. He made fundamental contributions to several fields, including probability theory, mathematical statistics, mathematical logic, topology, measure and integration theory, functional analysis, and information and algorithm theory.

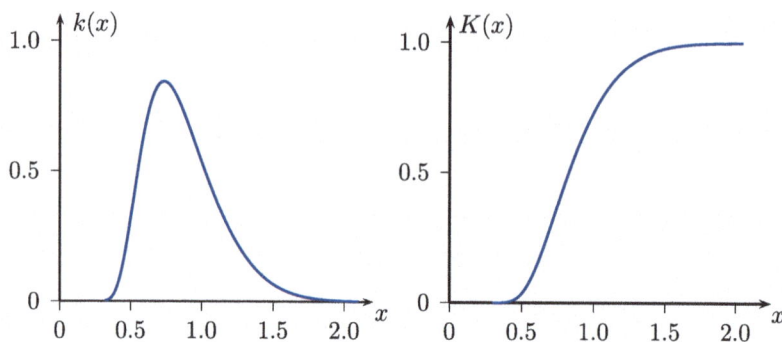

**Fig. 3.26** Density (left) and distribution function (right) of the Kolmogorov distribution

**Table 3.3** Values of the distribution function $K(x)$ from (3.32)

| $x$ | $K(x)$ | $x$ | $K(x)$ | $x$ | $K(x)$ |
|---|---|---|---|---|---|
| 0.40 | 0.0028 | 0.80 | 0.4559 | 1.15 | 0.8580 |
| 0.45 | 0.0126 | 0.83 | 0.5038 | 1.20 | 0.8877 |
| 0.50 | 0.0361 | 0.85 | 0.5347 | 1.23 | 0.9030 |
| 0.55 | 0.0772 | 0.90 | 0.6073 | 1.30 | 0.9319 |
| 0.60 | 0.1357 | 0.95 | 0.6725 | 1.40 | 0.9603 |
| 0.65 | 0.2080 | 1.00 | 0.7300 | 1.48 | 0.9750 |
| 0.70 | 0.2888 | 1.05 | 0.7798 | 1.63 | 0.9902 |
| 0.75 | 0.3728 | 1.10 | 0.8223 | 1.73 | 0.9950 |

Figure 3.26 shows the density $k(x)$ (left) and the distribution function $K(x)$ (right) of the Kolmogorov distribution. The density is right-skewed, and its shape resembles that of the bar chart of the distribution of $|M^\circ|_{100}$ from Fig. 3.25.

The median of the Kolmogorov distribution is 0.828, and the 95%-quantile equals 1.36 (Table 3.3). Thus, Theorem 3.14 implies, among other things, that a purely random bridge of length 10,000 has only a 0.05 probability of deviating more than 136 units from the $x$-axis at any point along its path, and that approximately every second bridge of this length has a maximum modulus of at least 83.

## 3.7 The Kolmogorov–Smirnov Test

The results on the distribution of the maximum modulus of a bridge have direct applications in the context of the *two-sample problem* in *nonparametric statistics*. In this scenario, the goal is to test the hypothesis

$$H_0 : F(x) = G(x) \quad \text{for each} \quad x \in \mathbb{R},$$

which asserts that two unknown distribution functions, $F$ and $G$, are equal. The testing of $H_0$ is based on data consisting of realizations of independent random variables

## 3.7 The Kolmogorov–Smirnov Test

$U_1, \ldots, U_n$ and $V_1, \ldots, V_n$. Each $U_i$ follows the *continuous* distribution function $F$, and each $V_j$ follows the *continuous* distribution function $G$. Continuity ensures that $U_1, \ldots, U_n$ and $V_1, \ldots, V_n$ take pairwise distinct values with probability one.

To test $H_0$, it is natural to estimate the unknown distribution functions $F$ and $G$ using the so-called *empirical distribution functions*

$$\widehat{F}_n(x) := \frac{1}{n}\sum_{j=1}^n \mathbf{1}\{U_j \leq x\}, \qquad \widehat{G}_n(x) := \frac{1}{n}\sum_{j=1}^n \mathbf{1}\{V_j \leq x\}, \qquad x \in \mathbb{R},$$

of $U_1, \ldots, U_n$ and $V_1, \ldots, V_n$, respectively. As arithmetic means of independent and identically distributed random variables with expectations $\mathbb{E}(\mathbf{1}\{U_j \leq x\}) = F(x)$ and $\mathbb{E}(\mathbf{1}\{V_j \leq x\}) = G(x)$, the random variables $\widehat{F}_n(x)$ and $\widehat{G}_n(x)$ converge almost surely to $F(x)$ or $G(x)$, respectively, as $n \to \infty$, according to the law of large numbers. Moreover, by the Glivenko–Cantelli theorem (see, e.g., [DU], p. 76), it holds with probability one that

$$\lim_{n\to\infty} \sup_{x\in\mathbb{R}} \left|\widehat{F}_n(x) - F(x)\right| = 0, \qquad \lim_{n\to\infty} \sup_{x\in\mathbb{R}} \left|\widehat{G}_n(x) - G(x)\right| = 0. \qquad (3.35)$$

Thus, the empirical distribution functions $\widehat{F}_n$ and $\widehat{G}_n$ converge uniformly on the real line to the underlying distribution functions $F$ and $G$ with probability one as $n \to \infty$. Note that all random variables $U_1, \ldots, U_n, V_1, \ldots, V_n$ (and for asymptotic considerations, additional random variables $U_j, V_j$ for $j \geq n+1$) are defined on a common probability space $(\Omega, \mathcal{A}, \mathbb{P})$, whose existence is ensured by general theorems from measure theory. For a fixed $\omega \in \Omega$, the functions $\widehat{F}_n^\omega : \mathbb{R} \to [0,1]$ and $\widehat{G}_n^\omega : \mathbb{R} \to [0,1]$ are then defined by

$$\widehat{F}_n^\omega(x) := \frac{1}{n}\sum_{j=1}^n \mathbf{1}\{U_j(\omega) \leq x\}, \qquad \widehat{G}_n^\omega(x) := \frac{1}{n}\sum_{j=1}^n \mathbf{1}\{V_j(\omega) \leq x\}, \qquad \omega \in \Omega.$$

These represent the realizations of the empirical distribution functions $\widehat{F}_n$ and $\widehat{G}_n$ for a given $\omega \in \Omega$. If all values $u_j := U_j(\omega)$, $1 \leq j \leq n$, are pairwise distinct, then $\widehat{F}_n^\omega$ is a step function that jumps by $\frac{1}{n}$ at the points $u_j$ and remains constant between any two distinct $u_j$ (see Fig. 3.27). The same applies to the function $\widehat{G}_n^\omega$. Note that the function $\widehat{F}_n^\omega$ is monotonically increasing, right-continuous (as indicated in Fig. 3.27 by marking the function values at the jump points), with $\lim_{x\to-\infty} \widehat{F}_n^\omega(x) = 0$ and $\lim_{x\to\infty} \widehat{F}_n^\omega(x) = 1$. This shows that $\widehat{F}_n^\omega$ and $\widehat{G}_n^\omega$ possess all the properties required of a distribution function.

It is now appropriate to test the hypothesis $H_0$ using the test statistic named after A.N. Kolmogorov and N.W. Smirnov:

$$KS_n := \sup_{x\in\mathbb{R}} \left|\widehat{F}_n(x) - \widehat{G}_n(x)\right| \qquad (3.36)$$

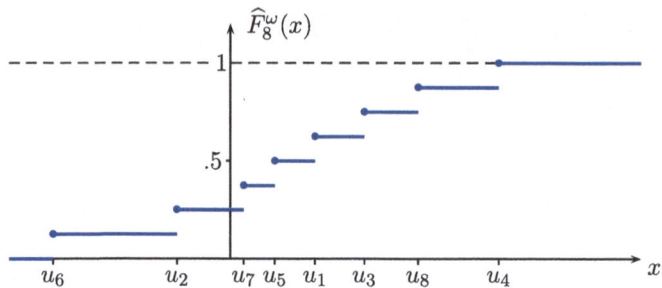

**Fig. 3.27** Realization of an empirical distribution function for data $u_j = U_j(\omega)$, $j = 1, \ldots, 8$

In this context, $H_0$ is rejected for large values of $KS_n$. What constitutes a "large" value depends on the predetermined significance level $\alpha$, which represents the maximum allowable probability of a Type I error—rejecting the hypothesis when it is actually true.

To determine the critical value for rejection, we need to understand the distribution of the test statistic $KS_n$ under the assumption that $H_0$ holds. Since under $H_0$, all the random variables $U_i$ and $V_j$ share the unknown distribution function $F$, it might initially seem that the distribution of $KS_n$ under $H_0$ would also depend on $F$. However, similar to the considerations in Sect. 2.13, this is fortunately not the case.

Due to the continuity of $F$, the random variables $U_1, \ldots, U_n$ and $V_1, \ldots, V_n$ take on pairwise distinct values with probability one. The function

$$\Delta_n(x) := n \cdot \left(\widehat{F}_n(x) - \widehat{G}_n(x)\right) = \sum_{j=1}^{n} \left(\mathbf{1}\{U_j \leq x\} - \mathbf{1}\{V_j \leq x\}\right), \qquad x \in \mathbb{R},$$

has the properties that $\Delta_n(x) = 0$ for $x < \min(U_1, \ldots, U_n, V_1, \ldots, V_n)$ and for $x \geq \max(U_1, \ldots, U_n, V_1, \ldots, V_n)$. Moreover, with probability one, it jumps by one unit up or down at the points $U_i$ and $V_j$. Since all random variables $U_i$, $V_j$ follow the same distribution, all $\binom{2n}{n}$ sequences of $n$ "+1-jumps" and $n$ "−1-jumps" are equally probable. Figure 3.28 shows a realization of $\Delta_n$ for the case $n = 5$.

Since $KS_n = \frac{1}{n} \sup_{x \in \mathbb{R}} |\Delta_n(x)|$ only depends on the maximum absolute value of $\Delta_n$, which is attained at one of its jump points, it is clear that the specific form of the unknown continuous distribution function $F$ has no effect on the distribution of $KS_n$. All random variables $U_i$ and $V_j$ can undergo the *probability integral transform* $U_i \mapsto F(U_i)$, $V_j \mapsto F(V_j)$, resulting in a uniform distribution over the interval $[0, 1]$, without changing the value of $KS_n$.

Crucially, under the hypothesis $H_0$, all sequences of $n$ "+1-jumps" and $n$ "−1-jumps" are equally probable. Thus, we can consider these jumps to occur at fixed time points $1, 2, \ldots, 2n$. Furthermore, for the supremum in (3.36), it does not matter whether one linearly interpolates between consecutive jump heights, or whether the first point, i.e., $(1, 1)$ or $(1, -1)$, is connected to the origin. Therefore, under $H_0$, the

## 3.7 The Kolmogorov–Smirnov Test

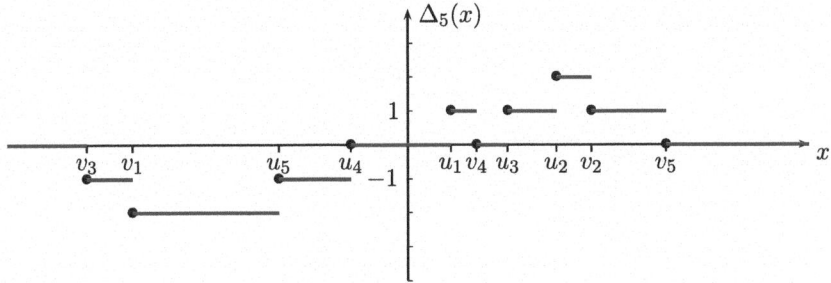

**Fig. 3.28** Graph of $\Delta_5$ for realizations $u_j = U_j(\omega)$ and $v_j = V_j(\omega)$, $1 \leq j \leq 5$

random variable $nKS_n$ has the same distribution as the absolute modulus $|M|_{2n}^\circ$ of a $2n$-bridge.

According to Theorem 3.13, for any $k \in \{1, \ldots, n\}$,

$$\mathbb{P}_{H_0}\left(KS_n \geq \frac{k}{n}\right) = 2 \sum_{s=1}^{\lfloor \frac{n}{k} \rfloor} (-1)^{s-1} \frac{\binom{2n}{n+sk}}{\binom{2n}{n}}. \tag{3.37}$$

Here, $\mathbb{P}_{H_0}$ emphasizes that the probability is calculated under the hypothesis $H_0$. If $k_0$ is the smallest number $k$ such that the above probability is less than or equal to $\alpha$, then $k_0$ is a critical value for the test statistic $KS_n$. We reject the hypothesis $H_0$ if $KS_n \geq k_0/n$; otherwise, we do not reject $H_0$.

Alternatively, one can calculate the probability in (3.37) as the "$p$-value $p^*$ for the observed value $\frac{k}{n}$" of the test statistic $KS_n$. The hypothesis $H_0$ is then rejected at a significance level $\alpha$ if $p^* \leq \alpha$.

As an application example of the Kolmogorov–Smirnov test, we consider the problem of comparing the effectiveness of two penicillin samples. An experiment was conducted using 14 agar plates for each of samples A and B under identical, independent conditions. The diameters of the inhibition zones were recorded (measured in millimeters). These diameters, considered as random variables $U_1, \ldots, U_n$ and $V_1, \ldots, V_n$ before the experiment, are assumed to be independent with unknown continuous distribution functions $F$ for sample A and $G$ for sample B.

After the experiment, the following values, sorted by size within each sample, were obtained (Table 3.4, source unkown):

**Table 3.4** Diameters of inhibition zones (in mm) for 14 values each from two penicillin samples

| Sample A: | 20.2 | 20.6 | 20.8 | 20.9 | 21.5 | 21.8 | 22.0 |
|---|---|---|---|---|---|---|---|
| | 22.7 | 23.8 | 24.1 | 24.3 | 25.0 | 25.5 | 25.6 |
| Sample B: | 21.0 | 21.3 | 22.4 | 23.2 | 23.3 | 23.5 | 24.4 |
| | 24.7 | 25.3 | 26.1 | 26.7 | 26.9 | 27.1 | 28.3 |

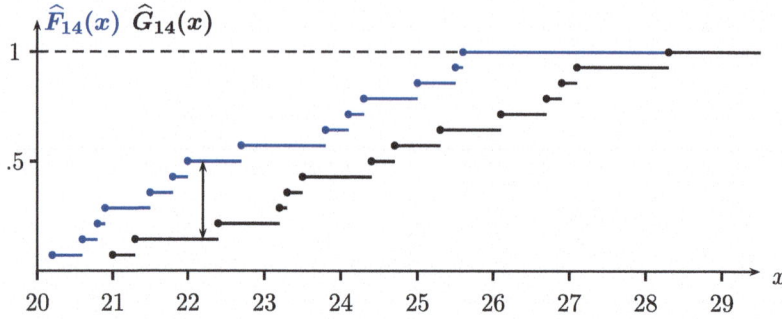

**Fig. 3.29** Graphs of the empirical distribution functions of the data from sample A (blue) and sample B (black)

Figure 3.29 shows the empirical distribution functions of both samples. The length of the double arrow indicates the maximum distance, $\frac{5}{14}$, which is attained at another position as well. This represents the realized value of the Kolmogorov–Smirnov test statistic $KS_{14}$. Under the hypothesis $H_0 : F = G$, it follows from (3.37) that

$$\mathbb{P}_{H_0}\left(KS_{14} \geq \frac{5}{14}\right) = 2\sum_{s=1}^{2}(-1)^{s-1}\frac{\binom{28}{14+s\cdot 5}}{\binom{28}{14}} = 0.343\ldots$$

The probability of observing a maximum deviation of at least $\frac{5}{14}$ under $H_0$ is sufficiently large, meaning that $H_0$ cannot be rejected at a 5% significance level (i.e., with $\alpha = 0.05$, the maximum allowable probability of making a Type I error). For large sample size $n$, we can utilize the equality in distribution of $nKS_n$ and $|M|^\circ_{2n}$, as well as Theorem 3.14. It follows that

$$\lim_{n\to\infty} \mathbb{P}_{H_0}\left(\sqrt{\frac{n}{2}}KS_n > x\right) = 1 - K(x), \qquad x > 0,$$

where $K(x)$ is the Kolmogorov distribution function defined in (3.32). Since $K(1.36) \approx 0.95$, for large $n$, the hypothesis $H_0 : F = G$ would be rejected at a 5% significance level if the inequality $\sqrt{n/2}KS_n > 1.36$ is satisfied.

If the hypothesis $H_0$ does not hold, then there exists at least one $x_0 \in \mathbb{R}$ such that $|F(x_0) - G(x_0)| > 0$. Given that $KS_n \geq |\widehat{F}_n(x_0) - \widehat{G}_n(x_0)|$, (3.35) implies

$$\lim_{n\to\infty} \mathbb{P}\left(\sqrt{\frac{n}{2}}KS_n > c\right) = 1$$

for any $c > 0$. Thus, if $F \neq G$, the probability of rejecting $H_0$ converges to one as $n \to \infty$. In this sense, the Kolmogorov–Smirnov test is consistent against any alternative to $H_0$.

In conclusion, we note that this testing problem also naturally arises when two samples of potentially different sizes are considered, that is, realizations of independent random variables $U_1, \ldots, U_m$ and $V_1, \ldots, V_n$ with $m \neq n$. Here, the hypothesis $H_0$ of equality between $F$ and $G$ can be tested using the Kolmogorov–Smirnov test statistic

$$KS_{m,n} := \sup_{x \in \mathbb{R}} |\widehat{F}_m(x) - \widehat{G}_n(x)|.$$

With the same reasoning as above, it can be shown that, under $H_0 : F = G$, the distribution of $KS_{m,n}$ does not depend on $F$. The combinatorial considerations for determining the distribution of $KS_{m,n}$ are more complex because the empirical distribution functions $\widehat{F}_m$ and $\widehat{G}_n$ have different jump heights, $\frac{1}{m}$ and $\frac{1}{n}$, respectively. These differences result in random walks that move from the point $(0, 0)$ to the point $(m + n, m - n)$.

## 3.8 Outlook: The Brownian Bridge

In Sect. 2.15, we saw that a symmetric random walk of length $n$ defines a random continuous function $W_n$ on $[0, 1]$ through the setting

$$W_n(t) := \frac{S_{\lfloor nt \rfloor}}{\sqrt{n}} + (nt - \lfloor nt \rfloor) \frac{X_{\lfloor nt \rfloor + 1}}{\sqrt{n}}, \qquad 0 \leq t \leq 1, \qquad (3.38)$$

and that the sequence $(W_n)$ of these so-called partial sum processes converges in distribution to the Brown–Wiener process $W = (W(t))_{0 \leq t \leq 1}$ as $n \to \infty$. For a purely random $2n$-bridge, an analogous situation holds, which we will now present.

Such a $2n$-bridge consists of $n$ upward and $n$ downward steps, where all $\binom{2n}{n}$ possible sequences of these steps are equally likely. This can be modeled by a vector $(X_{n,1}, \ldots, X_{n,2n})$ of random variables $X_{n,1}, \ldots, X_{n,2n}$ that takes values in the Cartesian product $\{-1, 1\}^{2n}$, and for which

$$\mathbb{P}(X_{n,1} = a_1, \ldots, X_{n,2n} = a_{2n}) := \frac{1}{\binom{2n}{n}} \quad \text{if} \quad \sum_{j=1}^{2n} \mathbf{1}\{a_j = 1\} = n, \qquad (3.39)$$

and $\mathbb{P}(X_{n,1} = a_1, \ldots, X_{n,2n} = a_{2n}) := 0$ otherwise. With $S_{2n,0} := 0$ and $S_{2n,k} := X_{n,1} + \ldots + X_{n,k}$ for $1 \leq k \leq 2n$, we now define, as in (3.38)—with the difference

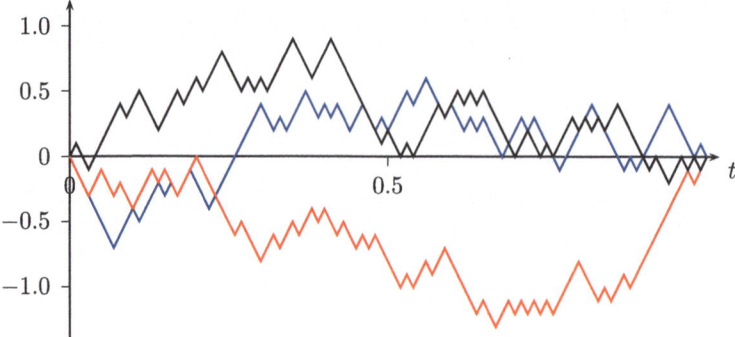

**Fig. 3.30** Three realizations of the process $W_{50}^{\circ}$

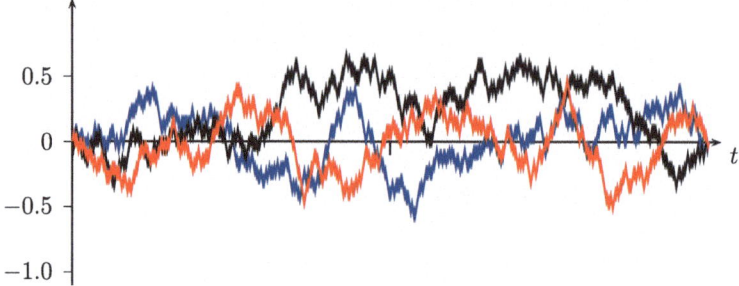

**Fig. 3.31** Three realizations of the process $W_{500}^{\circ}$

that the right-hand side now involves $2n$ instead of $n$—a stochastic process in the form of a random continuous function on $[0, 1]$ by

$$W_n^{\circ}(t) := \frac{S_{2n, \lfloor 2nt \rfloor}}{\sqrt{2n}} + (2nt - \lfloor 2nt \rfloor) \frac{X_{n, \lfloor 2nt \rfloor + 1}}{\sqrt{2n}}, \quad 0 \leq t \leq 1. \quad (3.40)$$

Here, we set $X_{n, 2n+1} := 0$.

Figure 3.30 shows three realizations of this process obtained via simulations for $n = 50$. Figure 3.31 illustrates the behavior of $W_n^{\circ}$ for $n = 500$. Similar to the partial sum process $W_n$, a stochastic stabilization appears to occur as $n \to \infty$. To better understand this behavior, we show that

$$\mathbb{E}(W_n^{\circ}(t)) = 0 \quad \text{and} \quad \lim_{n \to \infty} \mathbb{V}(W_n^{\circ}(t)) = t(1-t) \quad \text{for} \quad 0 \leq t \leq 1. \quad (3.41)$$

The first equation follows from $\mathbb{E}(X_{n,j}) = 0$, since $\mathbb{P}(X_{n,j} = 1) = \mathbb{P}(X_{n,j} = -1) = \frac{1}{2}$. For the second equation, we assume without loss of generality that $0 < t < 1$, since $W_n^{\circ}(0) = 0$ and $W_n^{\circ}(1) = 0$ imply that the random variables $W_n^{\circ}(0)$ and $W_n^{\circ}(1)$ are constant. To determine the variance of $W_n^{\circ}(t)$, we note that

## 3.8 Outlook: The Brownian Bridge

the joint distribution of the $\{0, 1\}$-valued random variables $Y_{n,j} := (X_{n,j} + 1)/2$, for $j = 1, \ldots, 2n$, is purely random according to (3.39). This distribution places $n$ ones and $n$ zeros into the tuple $(a_1, \ldots, a_{2n}) \in \{0, 1\}^{2n}$, which can be interpreted random sampling without replacement from a urn containing $n$ red and $n$ black balls. Viewing a 1 as drawing a red ball and a 0 as drawing a black ball, the sum $\sum_{j=1}^{\lfloor 2nt \rfloor} Y_{n,j}$ represents the number of red balls drawn after $\lfloor 2nt \rfloor$ draws, following a hypergeometric distribution (see, e.g., [FE1], p. 43 ff.). In particular,

$$\mathbb{V}\left(\sum_{j=1}^{\lfloor 2nt \rfloor} Y_{n,j}\right) = \lfloor 2nt \rfloor \cdot \frac{1}{2} \cdot \frac{1}{2} \cdot \left(1 - \frac{\lfloor 2nt \rfloor - 1}{2n - 1}\right)$$

(see, e.g., [FE1], p. 233). Since $S_{2n, \lfloor 2nt \rfloor} = 2 \sum_{j=1}^{\lfloor 2nt \rfloor} Y_{n,j} - \lfloor 2nt \rfloor$, it follows from the general rule $\mathbb{V}(aU + b) = a^2 \mathbb{V}(U)$ that

$$\mathbb{V}\left(\frac{S_{2n, \lfloor 2nt \rfloor}}{\sqrt{2n}}\right) = \frac{\lfloor 2nt \rfloor}{2n}\left(1 - \frac{\lfloor 2nt \rfloor - 1}{2n - 1}\right).$$

This leads to the second equation in (3.41), since the second term in (3.40) has a magnitude of at most $1/\sqrt{2n}$ and thus asymptotically does not affect the variance of $W_n^\circ(t)$ as $n \to \infty$. It can further be shown that a sequence of hypergeometric distributions whose variances tend to infinity converges in distribution to a standard normal distribution when standardized (see, for example, [FE1], p. 194). As a result, we have $W_n^\circ(t) \xrightarrow{\mathcal{D}} N(0, t(1-t))$, $0 \leq t \leq 1$. Here, the distribution $N(0, 0)$ is interpreted as a degenerate distribution at the point 0.

More generally, it can be shown that not only does $W_n^\circ(t)$ converge in distribution for fixed $t$, but also—similar to the partial sum process $W_n$—there is a suitably defined convergence in distribution of processes $W_n^\circ = (W_n^\circ(t))_{0 \leq t \leq 1}$ to a stochastic process $(W^\circ(t))_{0 \leq t \leq 1}$ as $n \to \infty$. Specifically,

$$\lim_{n \to \infty} \mathbb{P}\big(W_n^\circ(t_1) \leq x_1, \ldots, W_n^\circ(t_k) \leq x_k\big) = \mathbb{P}\big(W^\circ(t_1) \leq x_1, \ldots, W^\circ(t_k) \leq x_k\big)$$

for any choice of $k \geq 1$, $t_1, \ldots, t_k \in [0, 1]$ with $0 \leq t_1 < t_2 < \ldots < t_k \leq 1$ and $x_1, \ldots, x_k \in \mathbb{R}$.

The stochastic process $(W^\circ(t))_{0 \leq t \leq 1}$, whose realizations are continuous functions on $[0, 1]$, is called the *Brownian bridge*. It is related to the Brown–Wiener process $W(t)$ by the relation

$$W^\circ(t) = W(t) - t \cdot W(1).$$

This equation can be interpreted as assigning to each realization of $W$, viewed as a continuous function $f$ on $[0, 1]$, the continuous function $f^\circ$ defined by $f^\circ(t) := f(t) - t \cdot f(1)$, thus obtaining a realization of $W^\circ$; see, for example, [BI], p. 101 ff.

Analogous to the considerations for $W_n$ in Sect. 2.15, the convergence in distribution of $W_n$ to $W^\circ$ implies the convergence in distribution of certain real-valued functionals $h(W_n^\circ)$ to the distribution of $h(W^\circ)$. In particular, the following holds (see, for example, [BI], p. 103 ff.):

$$\lambda^1(\{t \in [0,1] : W_n^\circ(t) > 0\}) \xrightarrow{D} \lambda^1(\{t \in [0,1] : W^\circ(t) > 0\}), \quad (3.42)$$

$$\max_{0 \le t \le 1} W_n^\circ(t) \xrightarrow{D} \max_{0 \le t \le 1} W^\circ(t), \quad (3.43)$$

$$\max_{0 \le t \le 1} |W_n^\circ(t)| \xrightarrow{D} \max_{0 \le t \le 1} |W^\circ(t)|. \quad (3.44)$$

Since

$$\frac{O_{2n}^\circ}{2n} = \lambda^1(\{t \in [0,1] : W_n^\circ(t) > 0\}),$$

it follows from (3.42) and Theorem 3.5 that

$$\mathbb{P}\left(\lambda^1(\{t \in [0,1] : W^\circ(t) > 0\} \le x\right) = x, \quad 0 \le x \le 1.$$

Thus, the time that the Brownian bridge spends above the $x$-axis is uniformly distributed over the interval $[0, 1]$. Furthermore, the equations

$$\max_{0 \le t \le 1} W_n^\circ(t) = \frac{M_{2n}^\circ}{\sqrt{2n}}, \qquad \max_{0 \le t \le 1} |W_n^\circ(t)| = \frac{|M|_{2n}^\circ}{\sqrt{2n}}$$

together with (3.43), (3.44), as well as Theorems 3.9 and 3.14, yield the distributions of the maximum and the maximum modulus of the Brownian bridge, respectively. Specifically, we have:

$$\mathbb{P}\left(\max_{0 \le t \le 1} W^\circ(t) \le x\right) = 1 - \exp\left(-2x^2\right), \quad x \ge 0,$$

$$\mathbb{P}\left(\max_{0 \le t \le 1} |W^\circ(t)| \le x\right) = K(x), \quad x > 0.$$

**Answers to the Self-Assessment Questions**

**Answer 1** When forming the quotient $q_{k+1}/q_k$, the binomial coefficient $\binom{2n}{n}$ cancels out. Additionally, after cancelling the powers of two, only the factor 2 remains. Hence,

$$\frac{q_{k+1}}{q_k} = 2 \frac{\binom{2n-k-2}{n}}{\binom{2n-k-1}{n}} = 2 \cdot \frac{(2n-k-2)!\, n!\, (n-k-1)!}{(2n-k-1)!\, n!\, (n-k-2)!} = \frac{2(n-k-1)}{2n-k-1},$$

proving the assertion.

## 3.8 Outlook: The Brownian Bridge

**Answer 2** If we set $x = 0$, the left-hand side of (3.6) becomes $\lim_{n \to \infty} \mathbb{P}(N_{2n}^\circ = 0)$. According to part (a) of Theorem 3.1, $\mathbb{P}(N_{2n}^\circ = 0) = \frac{1}{2n-1}$, and so the assertion follows.

**Answer 3** A random variable $Z_0$ with the exponential distribution $\text{Exp}(\lambda)$ has the distribution function $F_0(t) = \mathbb{P}(Z_0 \leq t) = 1 - \exp(-\lambda t)$, for $t \geq 0$. If $F$ denotes the distribution function of $Z := Z_0^{1/\alpha}$, then

$$F(x) = \mathbb{P}(Z \leq x) = \mathbb{P}(Z_0^{1/\alpha} \leq x) = \mathbb{P}(Z_0 \leq x^\alpha) = 1 - \exp(-\lambda x^\alpha), \qquad x > 0.$$

**Answer 4** Set $m := \binom{2n}{n}$, and number the total $m$ of $2n$-tuples $(a_1, \ldots, a_{2n}) \in \{-1, 1\}^{2n}$ with $a_1 + \ldots + a_{2n} = 0$ in any order from 1 to $m$. For each $\ell \in \{1, \ldots, m\}$, let $A_\ell$ be the event that, after the $2n$ cards are well shuffled, the $\ell$-th tuple results when successively revealing all the cards. Here, a 1 or a $-1$ in the $j$-th component of each tuple means that the $j$-th revealed card is red or black, respectively, for $j = 1, \ldots, 2n$. The events $A_1, \ldots, A_m$ are pairwise disjoint, and each has a probability of $\frac{1}{m}$.

For the number $K_n$ of correct predictions made by the child, $\mathbb{P}(K_n = k | A_\ell) = \binom{n}{k}(1/2)^{2n}$, where $k \in \{0, \ldots, 2n\}$. This is because, no matter which color the child defines as "heads" or "tails" before the next throw, the probability of a hit is $\frac{1}{2}$, and this is independent of the results of earlier coin tosses. According to the law of total probability (7.4), it follows that

$$\mathbb{P}(K_n = k) = \sum_{\ell=1}^m \mathbb{P}(A_\ell) \mathbb{P}(K_n = k | A_\ell) = \frac{1}{m} \cdot m \cdot \binom{n}{k} \left(\frac{1}{2}\right)^{2n} = \binom{n}{k} \left(\frac{1}{2}\right)^{2n}.$$

**Answer 5** The father is always correct when the surplus of the more frequently represented cards in the stack is reduced. If $2k$ cards are drawn between any two equalizations of the two colors, the surplus builds up $k$ times and is reduced $k$ times.

**Answer 6** Otherwise, $O_{2n}^\circ < 2k$ would hold.

**Answer 7** If $k = 0$, both sides of (3.25) are equal to 1, and there is nothing to prove.

**Answer 8** Note that

$$2^{2n-1} = \sum_{j=0}^{2n-1} \binom{2n-1}{j} = \sum_{j=0}^{n-2} \binom{2n-1}{j} + \sum_{j=n+1}^{2n-1} \binom{2n-1}{j} + \binom{2n-1}{n-1} + \binom{2n-1}{n}.$$

Due to the symmetry property $\binom{m}{\ell} = \binom{m}{m-\ell}$ of binomial coefficients, both the first two sums and the last two binomial coefficients on the right-hand side are equal. Thus, the assertion follows.

**Answer 9** According to Theorem 3.11 (d), $\mathbb{V}(C_{2n}^\circ) = n - 1/(4u_{2n}^2)$, and thus

$$\mathbb{V}\left(\frac{C_{2n}^\circ}{\sqrt{2n}}\right) = \frac{\mathbb{V}(C_{2n}^\circ)}{2n} = \frac{1}{2} - \frac{1}{8} \cdot \frac{1}{nu_{2n}^2}.$$

In view of (7.17), $nu_{2n}^2 \to \frac{1}{\pi}$, and so the assertion follows.

**Answer 10** In the case where $x = 0$, the right-hand side equals 1, and the probability on the left-hand side is $\mathbb{P}(C_{2n}^\circ = 0)$, which, according to part (a) of Theorem 3.11, is equal to $\frac{2}{n+1}$. The claim follows from this.

**Answer 11** There are exactly two of all $\binom{2n}{n}$ bridges of length $2n$ for which $|M^\circ|_{2n}$ takes the maximum possible value $n$, namely the path that steadily ascends to the height $n$ and then returns to zero, as well as its reflected image across the $x$-axis.

**Answer 12** The walk must take at least $k$ steps to reach the height $k$ or $-k$, and it requires at least $2\ell k$ steps to subsequently oscillate $\ell$ times between the heights $k$ and $-k$ (or vice versa). Then, at least $k$ steps are needed to reach the final level of zero. It follows that $2k + 2k\ell \leq 2n$, which implies $\ell \leq \lfloor \frac{n}{k} \rfloor - 1$.

**Exercises**

**Exercise 3.1** Let $X_1, \ldots, X_{2n}$ be independent random variables, where $\mathbb{P}(X_j = 1) = p = 1 - \mathbb{P}(X_j = -1)$ for $j = 1, \ldots, 2n$. Here, $p$ is any number satisfying $0 < p < 1$. Show that

$$\mathbb{P}(X_1 = a_1, \ldots, X_{2n} = a_{2n} | X_1 + \ldots + X_{2n} = 0) = \frac{1}{\binom{2n}{n}}$$

for each $(2n)$-tuple $(a_1, \ldots, a_{2n}) \in \{-1, 1\}^{2n}$ with $a_1 + \ldots + a_{2n} = 0$.

The joint distribution of the random step directions of a $2n$-bridge is thus the conditional distribution of the step directions of a simple random walk of length $2n$ (with not necessarily equally likely step directions), under the condition that this random walk returns to the starting point at time $2n$.

**Exercise 3.2** In the context of Exercise 3.1, show that

$$\mathbb{P}(X_1 = a_1, \ldots, X_{2n} = a_{2n} | X_1 + \ldots + X_{2n} = 2k) = \frac{1}{\binom{2n}{n+k}}$$

## 3.8 Outlook: The Brownian Bridge

for each $k \in \{0, 1, \ldots, n\}$ and any choice of $(a_1, \ldots, a_{2n}) \in \{-1, 1\}^{2n}$ with $a_1 + \ldots + a_{2n} = 2k$.

**Exercise 3.3** Derive Eq. (3.5), and use it to obtain the formula for the variance of $N_{2n}^\circ$ in part (d) of Theorem 3.1.

**Exercise 3.4** Let $\varphi(x) := (2\pi)^{-1/2} \exp(-x^2/2)$, $x \in \mathbb{R}$, and let $\Phi : \mathbb{R} \to \mathbb{R}$ be the distribution function of the standard normal distribution, defined by

$$\Phi(x) := \int_{-\infty}^{x} \varphi(t)\,dt, \quad x \in \mathbb{R}.$$

Prove the inequality

$$1 - \Phi(x) < \frac{\varphi(x)}{x}, \quad x > 0.$$

**Note:** $\frac{t}{x} > 1$ if $t > x > 0$.

**Exercise 3.5** Let the function $h : [0, \infty) \to \mathbb{R}$ be defined by $h(x) := 2\Phi(x) + \exp(-x^2/2) - 2$, $x \geq 0$: Show that $h(x) > 0$ for every $x > 0$.
**Hint:** The result of Exercise 3.4 might be useful for part of the domain of definition of $h$.

**Exercise 3.6** Let $N_{2n}^\circ$ be the number of interior zeros of a purely random $2n$-bridge. Show that

$$\mathbb{P}(N_{2n}^\circ = k - 1) = \frac{k}{2n} \cdot \frac{\prod_{j=1}^{k-1}\left(1 - \frac{j}{n}\right)}{\prod_{j=1}^{k}\left(1 - \frac{j}{2n}\right)}, \quad k \in \{1, \ldots, n\}.$$

**Exercise 3.7** As in Chap. 2, let $X_1, \ldots, X_{2n}$ be independent random variables with $\mathbb{P}(X_j = \pm 1) = \frac{1}{2}$ for $j = 1, \ldots, 2n$. Furthermore, let $S_0 := 0$ and $S_j := X_1 + \ldots + X_j$ for $j = 1, \ldots, 2n$. Define $N_{2\ell} := \sum_{j=1}^{\ell} \mathbf{1}\{S_{2j} = 0\}$, where $\ell \in \{1, \ldots, n\}$. Show that

$$\mathbb{P}(N_{2n} = k + 1) = \mathbb{P}(S_{2n} = 0, N_{2n-2} \geq k), \quad k \in \{0, \ldots, n-1\}.$$

**Hint:** According to Exercise 3.1, a $2n$-bridge is present under the condition $S_{2n} = 0$.

**Exercise 3.8** Derive Eq. (3.16).

**Exercise 3.9** An urn contains 20 red and 30 black balls. All balls are drawn randomly, one after the other, without replacement. For each $j \in \{1, \ldots, 50\}$, let $A_j$ be the event that the $j$-th ball drawn is red. Then the random variable $X := \sum_{j=1}^{10} \mathbf{1}\{A_j\}$, which counts the number of red balls among the first 10 selected, follows the *hypergeometric distribution* Hyp(10, 20, 30). Specifically,

$$\mathbb{P}(X = k) = \frac{\binom{20}{k}\binom{30}{10-k}}{\binom{50}{10}}, \qquad k = 0, \ldots, 10.$$

What is the distribution of the random variable $Y := \sum_{j=41}^{50} \mathbf{1}\{A_j\}$, which represents the number of red balls drawn in the last 10 draws?

**Exercise 3.10** Use Eq. (3.20) to derive part (d) of Theorem 3.8.

**Exercise 3.11** Let $n$ be a positive integer and $a$ a non-negative integer with the same parity as $n$, where $0 \leq a \leq n$.

(a) Show: There are $\binom{n}{\frac{n+a}{2}}$ $n$-step polygonal paths leading from $(0, 0)$ to $(n, a)$.
(b) What is the maximum possible height of such a path?
(c) Let $k$ be an integer such that $a \leq k \leq \frac{n+a}{2}$. Show that the probability that the maximum of a random polygonal path from $(0, 0)$ to $(n, a)$ (where all such paths are equally likely) is greater that or equal to $k$ is

$$\frac{\binom{n}{\frac{n-a+2k}{2}}}{\binom{n}{\frac{n+a}{2}}}.$$

**Exercise 3.12** Derive the formula for the variance of the number of changes of sign in a $2n$-bridge (part (d) of Theorem 3.11).
**Hint:** Start with Eq. (3.26).

# Asymmetric Random Walks on $\mathbb{Z}$ and Related Topics

**4**

In this chapter, we move beyond the assumption made in Chap. 2 that the directions of a Bernoulli random walk on $\mathbb{Z}$ are equally likely, thus relaxing the symmetry assumption. Specifically, let $X_1, X_2, \ldots$ be a sequence of independent and identically distributed random variables, where $\mathbb{P}(X_1 = 1) := p$ and $\mathbb{P}(X_1 = -1) = q := 1 - p$, assuming $0 < p < 1$. The random walk is then defined by $S_0 := 0$ and $S_n := X_1 + \ldots + X_n$ for $n \geq 1$, as in Chap. 2. However, we now allow the probabilities $p$ and $q$ for upward or downward steps to differ. In this case, the random walk is referred to as *asymmetric* or *biased* (*Bernoulli*) *random walk*.

A generalization of (2.7) gives

$$\mathbb{P}(S_n = j) = \binom{n}{\frac{n+j}{2}} p^{\frac{n+j}{2}} q^{\frac{n-j}{2}}, \qquad n \in \mathbb{N}, \ j \in \mathbb{Z} \text{ with } |j| \leq n. \tag{4.1}$$

Here, $n$ and $j$ must have the same parity.

**SAQ 1** Why does Eq. (4.1) hold?

Since $\mathbb{E}(X_j) = p - q = 2p - 1$, it follows that

$$\mathbb{E}(S_n) = \sum_{j=1}^{n} \mathbb{E}(X_j) = (2p - 1) n.$$

Asymmetric random walks therefore exhibit a trend in the expected heights over time. The points $(n, \mathbb{E}(S_n))$ with $n \geq 0$ lie on the line $x \mapsto (2p - 1)x$, which describes this trend. When $p > \frac{1}{2}$, this line has a positive slope; when $p < \frac{1}{2}$, the slope is negative. Figure 4.1 illustrates two random walks of length 2500. One,

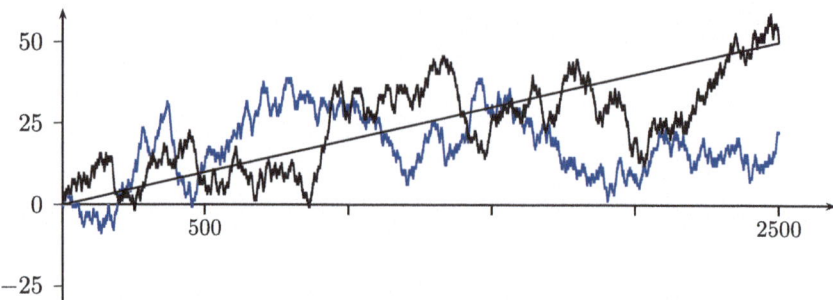

**Fig. 4.1** Symmetric (blue) and asymmetric (black) random walk with $p = 0.51$ and trend line

shown in blue, is symmetric, while the other, in black, is asymmetric with $p = 0.51$. Additionally, the trend line is depicted.

## 4.1 First-passage Times

In this section, we explore the question of when an asymmetric random walk, starting at the origin, first reaches a height of 1. Interpreting the random walk as a game between two players, where the game is unfair when $p \neq \frac{1}{2}$, we examine when one of the players first takes the lead. As in Sect. 2.9, the first-passage time to height 1 is denoted by

$$V_1 := \inf\{n \geq 1 : S_n = 1\}.$$

We previously noted that, in the case of a symmetric random walk, $\mathbb{P}(V_1 < \infty) = 1$ and $\mathbb{E}(V_1) = \infty$. This means that the height of 1 (or any other height) will certainly be reached in finite time, but the expected time required is infinite.

For an asymmetric random walk with a negative trend, i.e., $p < \frac{1}{2}$, the likelihood of reaching a height of 1 diminishes significantly. In fact, when $p < \frac{1}{2}$, the random walk may remain strictly below the line $y = 1$ with positive probability, never attaining the height of 1. On the other hand, for $p > \frac{1}{2}$, it is expected that $\mathbb{P}(V_1 < \infty) = 1$, and perhaps $V_1$ even has a finite expectation. The following result confirms these assumptions.

**Theorem 4.1 (Distribution of the First-passage Time $V_1$)** *For the first-passage time $V_1$, the following hold:*

(a) $\mathbb{P}(V_1 = 2\ell - 1) = \dfrac{1}{q\ell} \dbinom{2\ell - 2}{\ell - 1}(pq)^\ell, \qquad \ell \geq 1,$

(b) $\mathbb{P}(V_1 = \infty) = \begin{cases} 0, & \text{if } p \geq \frac{1}{2}, \\ 1 - \frac{p}{q}, & \text{if } p < \frac{1}{2}, \end{cases}$

## 4.1 First-passage Times

(c) $\mathbb{E}(V_1) = \begin{cases} \frac{1}{p-q}, & \text{if } p > \frac{1}{2}, \\ \infty, & \text{if } p \leq \frac{1}{2}. \end{cases}$

**Proof**

(a) Let $a_n := \mathbb{P}(V_1 = n)$ for $n \geq 0$, and let $g$ be the generating function of $V_1$, defined by

$$g(t) := \sum_{n=0}^{\infty} a_n t^n, \qquad |t| \leq 1. \tag{4.2}$$

We will derive a closed expression for $g$, from which all the above assertions follow. Clearly, $a_{2j} = 0$ for every $j \geq 0$, because $V_1$ (aside from possibly taking the value $\infty$) can only take odd values. Since $\mathbb{P}(V_1 = 1) = \mathbb{P}(X_1 = 1)$, we have $a_1 = p$. If for some $n \geq 3$ the event $\{V_1 = n\}$ occurs, it implies that $X_1 = -1$. Therefore, after the first step, the random walk must reach a height of 1 for the first time at time $n$, starting from the height $-1$ and having hit 0 at least once before. We decompose the event $\{V_1 = n\}$ based on the *first* time point, denoted by $r$, where $S_r = 0$, by setting

$$\{V_1 = n\} = \biguplus_{r=2}^{n-1} A \cap B_r \cap C_r, \tag{4.3}$$

with

$$A := \{S_1 = -1\},$$
$$B_r := \{S_1 < 0, \ldots, S_{r-1} < 0, S_r = 0\},$$
$$C_r := \{S_{r+1} \leq 0, \ldots, S_{n-1} \leq 0, S_n = 1\}$$

for $r = 2, \ldots, n-1$. The sets on the right-hand side of (4.3) are pairwise disjoint for different $r$. By the additivity of $\mathbb{P}$ and the multiplication rule for the probability of intersections of events, it follows that

$$\mathbb{P}(V_1 = n) = \sum_{r=2}^{n-1} \mathbb{P}(A) \cdot \mathbb{P}(B_r | A) \cdot \mathbb{P}(C_r | A \cap B_r).$$

Here, $\mathbb{P}(A) = q$. Under the condition $A$, a random walk starts at the point $(1, -1)$. The conditional probability that this random walk first reaches the height 0 at time $r$ is equal to the probability that a random walk starting at the origin first reaches the height 1 at time $r - 1$; hence $\mathbb{P}(B_r | A) = \mathbb{P}(V_1 = r - 1) = a_{r-1}$.

Given the occurrence of $A \cap B_r$, the random walk hits zero at time $r$. Therefore, given the event $C_r$, starting from $(r, 0)$, the random walk must reach height 1 at time $n$. By the temporal homogeneity of the random walk, this is equivalent to a random walk starting at the origin reaching the height 1 at time $n - r$; thus $\mathbb{P}(C_r | A \cap B_r) = \mathbb{P}(V_1 = n - r) = a_{n-r}$. Since $\mathbb{P}(V_1 = n) = a_n$, we obtain the recurrence relation

$$a_n = q \sum_{r=2}^{n-1} a_{r-1} a_{n-r}, \qquad n \geq 2. \tag{4.4}$$

Note that this equation also holds for $n = 2$, since an empty sum is defined to be zero. By multiplying both sides of this equation by $t^n$ and summing over $n \geq 2$, while considering $a_0 = 0$ and $a_1 = p$, the generating function $g$ of $V_1$ satisfies the quadratic equation

$$g(t) - pt = qt\,g^2(t), \qquad |t| \leq 1,$$

as shown in Exercise 4.2. One of the two solutions of this equation is unbounded as $t \to 0$, which is not possible for a generating function. Therefore, the uniquely determined solution of the above equation in closed form is

$$g(t) = \frac{1 - \sqrt{1 - 4pqt^2}}{2qt}, \qquad |t| \leq 1,\ t \neq 0, \tag{4.5}$$

with the continuous extension $g(0) = 0 = \lim_{t \to 0} g(t)$. Using the binomial series $(1 + x)^\alpha = \sum_{n=0}^{\infty} \binom{\alpha}{n} x^n$, $|x| < 1$ (see Sect. 7.11), the right-hand side of (4.5), after direct calculation, takes the form $\sum_{n=0}^{\infty} b_n t^n$, where

$$b_{2\ell-1} = \frac{(-1)^{\ell-1}}{2q} \binom{\frac{1}{2}}{\ell} (4pq)^\ell = \frac{1}{\ell q} \binom{2\ell - 2}{\ell - 1} (pq)^\ell, \qquad \ell \geq 1,$$

and $b_{2\ell} = 0$ for $\ell \geq 0$ (Exercise 4.3). By the definition of $g$, part (a) now follows due to the uniqueness of the coefficients in the power series expansion.

(b) According to (4.2), $g(1) = \sum_{n=0}^{\infty} \mathbb{P}(V_1 = n) = \mathbb{P}(V_1 < \infty)$. Using equation (4.5) and the fact that $1 - 4pq = (p - q)^2$, we obtain

$$g(1) = \frac{1 - |p - q|}{2q}.$$

Thus, $g(1) = 1$ when $p \geq \frac{1}{2}$, and $g(1) = \frac{p}{q}$ if $p < \frac{1}{2}$, which completes the proof.

(c) It was shown in Sect. 2.9 that $\mathbb{E}(V_1) = \infty$ when $p = \frac{1}{2}$. According to part (b), if $p < \frac{1}{2}$, then $\mathbb{P}(V_1 = \infty) > 0$, implying that $\mathbb{E}(V_1) = \infty$ in this case as well. For $p > \frac{1}{2}$, a direct calculation gives

$$g'(t) = \frac{2p}{\sqrt{1-4pqt^2}} - \frac{1}{2qt^2} + \frac{\sqrt{1-4pqt^2}}{2qt^2}, \quad t \neq 0, \ |t| \leq 1. \quad (4.6)$$

Thus, since $\sqrt{1-4pq} = p - q$, we have

$$\mathbb{E}(V_1) = g'(1) = \frac{2p}{p-q} - \frac{1}{2q} + \frac{p-q}{2q} = \frac{1}{p-q}. \quad (4.7)$$

The first equality holds according to Sect. 7.9.

∎

**SAQ 2** Why does $\sqrt{1-4pq} = p - q$ hold?

In the context of a game repeatedly played between two players, A and B, where A and B have success probabilities $p$ and $q = 1 - p$, respectively, Theorem 4.1 (b) implies that if the game is disadvantageous for player A, there is a positive probability $1 - \frac{p}{q}$ that A never gains the lead. Figure 4.2 (left) shows the probability $\mathbb{P}_p(V_1 = \infty)$, which depends on $p$ (as indicated by the index $p$). For instance, if $p = \frac{1}{3}$, there is a 50% chance that player A will never take the lead. Figure 4.2 (right) illustrates the expectation of $V_1$ as a function of $p$ for $p > \frac{1}{2}$. In the case where $p = 0.55$, player A has to wait, on average, 10 games before taking the lead for the first time.

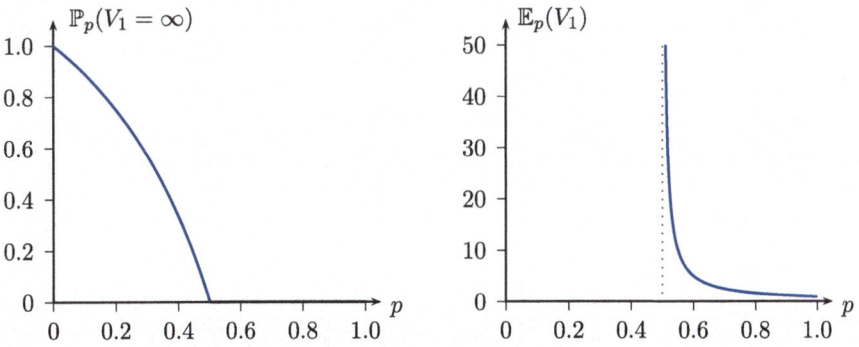

**Fig. 4.2** Probability of $\{V_1 = \infty\}$ (left) and expectation of $V_1$ (right) as a function of $p$

When $p > \frac{1}{2}$, the first-passage time $V_1$ is a random variable that takes on the values $1, 3, 5, \ldots$ and has a finite expectation. From (4.6), the second derivative of the generating function $g$ of $V_1$, after some calculation, is given by

$$g''(t) = \frac{8p^2qt}{(1-4pqt^2)^{3/2}} + \frac{1-\sqrt{1-4pqt^2}}{qt^3} - \frac{2p}{t\sqrt{1-4pqt^2}}, \quad |t| < 1, \, t \neq 0. \quad (4.8)$$

Since $g''(t)$ remains bounded as $t \uparrow 1$, $\mathbb{E}(V_1^2)$ exists. Furthermore, because $\mathbb{E}(V_1^2) = g''(1) + g'(1)$ (cf. Sect. 7.9), the variance of $V_1$ can be directly calculated as

$$\mathbb{V}(V_1) = \frac{4pq}{(p-q)^3} \quad (4.9)$$

(see Exercise 4.4). Due to the memoryless property of the random walk (discussed in Sect. 2.9), the first-passage time $V_k := \inf\{n \geq 1 : S_n = k\}$ to the height $k$ is distributed as the sum $V_{1,1} + V_{1,2} + \ldots + V_{1,k}$ of independent random variables $V_{1,j}$ ($j = 1, \ldots, k$), each distributed like $V_1$. This implies that when $p \geq \frac{1}{2}$, every value $\ell \geq 1$ is reached with probability one in finite time, and when $p \leq \frac{1}{2}$, the random walk hits any given height $-\ell$ (with $\ell \geq 1$) with probability one in finite time. In the case where $p < \frac{1}{2}$,

$$\mathbb{P}(V_k < \infty) = \left(\frac{p}{q}\right)^k, \quad k \geq 1$$

(see Exercise 4.6). Moreover, the above insight allows us to determine the distribution of $V_k$ for general $k$ from the generating function $g$ of $V_1$, because the generating function of $V_k$ is, according to Sect. 7.9, the $k$-th power $g^k$ of $g$. For example,

$$\mathbb{P}(V_2 = 2\ell) = \frac{1}{\ell+1}\binom{2\ell}{\ell} p^{\ell+1} q^{\ell-1}, \quad \ell \geq 1 \quad (4.10)$$

(see Exercise 4.5). However, there is a convenient alternative way to obtain an explicit expression for the distribution of $V_k$ for general $k$, namely by using the reflection principle (see, e.g., [GRS], pp. 78–79). To this end, let $M_n := \max\{S_j : j = 0, 1, \ldots, n\}$ be the maximum height of the random walk up to time $n$. The following lemma provides the key to obtaining a simple expression for $\mathbb{P}(V_k = n)$.

**Lemma 4.2** *Let $r, n \in \mathbb{N}$ and $\ell \in \mathbb{N}_0$, where $n$ and $\ell$ have the same parity. Then*

$$\mathbb{P}(M_n \geq r, S_n = \ell) = \begin{cases} \mathbb{P}(S_n = \ell), & \text{if } \ell \geq r, \\ \left(\frac{q}{p}\right)^{r-\ell} \mathbb{P}(S_n = 2r - \ell), & \text{if } \ell < r. \end{cases}$$

## 4.1 First-passage Times

**Proof** Since the statement is obvious when $\ell \geq r$, we will assume without loss of generality that $\ell < r$.

**SAQ 3** Why is the case $\ell \geq r$ obvious?

If we reflect every path from $(0, 0)$ to $(n, \ell)$ whose maximum is at least $r$ across the line $y = r$ at the first time the height $r$ is reached, we create a path from $(0, 0)$ to $(n, 2r - \ell)$. This transformation assigns each path with $S_n = \ell$ and $M_n \geq r$ to a path with $S_n = 2r - \ell$ in a bijective manner. The inverse mapping is achieved by reflecting a path from $(0, 0)$ to $(n, 2r - \ell)$ across the line $y = r$ at the first time it reaches the height $r$. The number of paths leading from $(0, 0)$ to $(n, 2r - \ell)$ is given by the binomial coefficient

$$\binom{n}{\frac{n+2r-\ell}{2}}.$$

Since every path from $(0, 0)$ to $(n, \ell)$ has a probability of $p^{(n+\ell)/2} q^{(n-\ell)/2}$, we obtain

$$\mathbb{P}(M_n \geq r, S_n = \ell) = \binom{n}{\frac{n+2r-\ell}{2}} p^{\frac{n+\ell}{2}} q^{\frac{n-\ell}{2}} = \left(\frac{q}{p}\right)^{r-\ell} \mathbb{P}(S_n = 2r - \ell).$$

The second equality follows from (4.1) with $j = 2r - \ell$. ∎

Why does Lemma 4.2 help to obtain the probability $\mathbb{P}(V_k = n)$ for general $k$? To answer this, note that the event $\{V_k = n\}$ occurs if and only if $M_{n-1} = S_{n-1} = k-1$ and $S_n = k$. Specifically, this implies that $X_n = 1$, and therefore

$$\mathbb{P}(V_k = n) = \mathbb{P}(M_{n-1} = S_{n-1} = k - 1, S_n = k)$$
$$= p \left( \mathbb{P}(M_{n-1} \geq k-1, S_{n-1} = k-1) - \mathbb{P}(M_{n-1} \geq k, S_{n-1} = k-1) \right)$$
$$= p \left( \mathbb{P}(S_{n-1} = k - 1) - \frac{q}{p} \mathbb{P}(S_{n-1} = k + 1) \right). \tag{4.11}$$

In this case, Lemma 4.2 was used at the third equality.

**SAQ 4** Why does the second equality hold?

With the help of (4.1) and direct calculation, it follows that the expression in (4.11) simplifies to $\frac{k}{n}\mathbb{P}(S_n = k)$. Surprisingly, the distribution of the first-passage time $V_k$ is thus linked to that of $S_n$ through the equation

$$\mathbb{P}(V_k = n) = \frac{k}{n}\mathbb{P}(S_n = k).$$

By applying (4.1), we arrive at the following result:

**Theorem 4.3 (Distribution of the First-passage Time $V_k$)** *If $k$ and $n$ are positive integers of the same parity, then*

$$\mathbb{P}(V_k = n) = \frac{k}{n}\binom{n}{\frac{n+k}{2}}p^{\frac{n+k}{2}}q^{\frac{n-k}{2}}, \qquad n = k, k+1, \ldots \qquad (4.12)$$

This result generalizes both the result obtained for the symmetric random walk from Theorem 2.24, as well as part (a) of Theorem 4.1 in the special case $k = 1$, and (4.10) for the case $k = 2$. For symmetry reasons, it also holds for the first-passage times $V_{-k} := \inf\{j \in \mathbb{N} : S_j = -k\}$ with $k \geq 1$, if on the right-hand side of the equality in (4.12), $k$ is replaced by $|k|$, and $p$ and $q$ are swapped.

Due to the equality in distribution $V_k \sim V_{1,1} + \ldots + V_{1,k}$, where $V_{1,1}, \ldots, V_{1,k}$ are independent random variables, each distributed like $V_1$, the following asymptotic behavior of $V_k$ as $k \to \infty$ holds for the case $p > \frac{1}{2}$:

**Theorem 4.4 (Limit Distribution of the First-passage Time $V_k$)** *Let $V_k$ be the first-passage time to height $k$. If $p > \frac{1}{2}$, then*

$$\lim_{k \to \infty} \mathbb{P}\left(\frac{V_k - \frac{k}{p-q}}{\sqrt{\frac{4pqk}{(p-q)^3}}} \leq x\right) = \Phi(x), \qquad x \in \mathbb{R}. \qquad (4.13)$$

*As before, $\Phi$ denotes the distribution function of the standard normal distribution. With (4.7) and (4.9), this implies convergence in distribution as $k \to \infty$:*

$$\frac{V_k - \mathbb{E}(V_k)}{\sqrt{\mathbb{V}(V_k)}} \xrightarrow{\mathcal{D}} Z, \qquad \text{where } Z \sim N(0, 1).$$

**Proof** The assertion follows from Theorem 4.1 (c) and (4.9), due to the Lindeberg–Lévy central limit theorem in Sect. 7.4. ∎

Statement (4.13) allows for a justified prediction regarding the time a random walk with a positive trend needs to reach a large height, as

$$\mathbb{P}\left(V_k \leq \frac{k}{p-q} + x\sqrt{k\left(\frac{1}{p-q} + \frac{8pq - 1}{(p-q)^3}\right)}\right) \approx \Phi(x)$$

for large $k$. In particular, setting $x = 0$, we find that

$$\mathbb{P}\left(V_k \leq \frac{k}{p-q}\right) \approx \frac{1}{2}$$

for large $k$.

This result stands in stark contrast to Theorem 2.26, which states that in a *symmetric* random walk, the first-passage time to height $k$ grows approximately proportional to $k^2$. Even a very slight positive trend, in the sense that $p > \frac{1}{2}$, is sufficient to change the growth of $V_k$ as $k \to \infty$ by an entire order of magnitude!

For example, if we specifically choose $x = \Phi^{-1}(0.95) = 1.645$ and $p = 0.6$, with $\mathbb{E}(V_1) = 5$ and $\mathbb{V}(V_1) = 120$, we obtain

$$\mathbb{P}\left(V_{100} \leq 500 + 1.645\sqrt{100 \cdot 120}\right) \approx 0.95,$$

which implies $\mathbb{P}(V_{100} \leq 681) \approx 0.95$. Thus, only about one in twenty random walks with $p = 0.6$ will require more than 681 steps to reach a height of 100. By comparison, for a symmetric random walk, approximately *every second walk* needs around 22,000 steps to reach the same height, as discussed following Fig. 2.36.

## 4.2 Visits to Zero

In this section, we first demonstrate that an asymmetric random walk has a positive probability of never returning to the starting point 0. Next, we examine the distribution of the time until the first return to zero, as well as the expectation and distribution of the number of visits to zero.

In Sect. 2.5, we saw that a *symmetric* random walk returns to the starting point with probability one: according to Theorem 2.9 (b), $\mathbb{P}(W < \infty) = 1$, where

$$W := \inf\{2k : k \in \mathbb{N} \text{ and } S_{2k} = 0\} \tag{4.14}$$

denotes the first return time. However, as the following theorem shows, this *recurrence property* is lost in the case where $p \neq \frac{1}{2}$.

**Theorem 4.5 (Transience of the Asymmetric Random Walk on $\mathbb{Z}$)** *The asymmetric random walk on $\mathbb{Z}$ is transient, i.e., if $p \neq \frac{1}{2}$ then*

$$\mathbb{P}(W < \infty) < 1.$$

*More precisely,*

$$\mathbb{P}(W < \infty) = 1 - |p - q| = 1 - |2p - 1|.$$

***Proof*** For reasons of symmetry, it is sufficient to consider the case $p > \frac{1}{2}$. We decompose the event $\{W < \infty\}$ according to the two possible outcomes for the first step of the random walk and use the law of total probability (7.4), according to which

$$\mathbb{P}(W < \infty) = p \cdot \mathbb{P}(W < \infty | X_1 = 1) + q \cdot \mathbb{P}(W < \infty | X_1 = -1).$$

Given $X_1 = 1$, the event $\{W < \infty\}$ occurs if and only if the random walk, starting from the point $(1, 1)$, eventually hits the $x$-axis. This situation is equivalent to a random walk, starting at the origin, which must descend to the level $-1$ in finite time, where the probability for a downward step is $q$. According to part (b) of Theorem 4.1, if we swap the roles of $p$ and $q$, $\mathbb{P}(W < \infty | X_1 = 1) = \frac{q}{p}$. For the event $\{W < \infty\}$ to occur when $X_1 = -1$, the random walk must ascend one step in finite time, starting from the point $(1, -1)$. This is equivalent to a random walk starting at the origin and reaching a height of 1 in finite time. Since $p > \frac{1}{2}$, according to Theorem 4.1 (b), this probability is one. We thus obtain

$$\mathbb{P}(W < \infty) = p \cdot \frac{q}{p} + q \cdot 1 = 2q = 2\min(p, q) = 1 - |p - q| < 1.$$

∎

The result states that an asymmetric random walk starting at zero will return to its starting point with probability $1 - |p - q|$. We will soon see that, in this case, the number of visits to zero follows a geometric distribution. First, however, we are interested in the distribution of first return time $W$. Generalizing Theorem 2.9 (a), the following holds:

**Theorem 4.6 (Distribution of the First Return Time)** *Let $W$ be the first return time to zero. Then*

$$\mathbb{P}(W = 2n) = \frac{2}{n}\binom{2n-2}{n-1}(pq)^n, \qquad n \geq 1.$$

***Proof*** We decompose the event $\{W = 2n\}$ based on the first step of the random walk and apply the rule (7.4) of total probability, along with Theorem 4.1 (a). This yields

$$\mathbb{P}(W = 2n) = \mathbb{P}(W = 2n | X_1 = 1) \cdot p + \mathbb{P}(W = 2n | X_1 = -1) \cdot q. \qquad (4.15)$$

Given that $X_1 = 1$, the event $\{W = 2n\}$ is equivalent to the first-passage time $V_{-1}$ taking the value $2n - 1$. According to Theorem 4.1 (a), swapping the roles of $p$ and $q$, we have

$$\mathbb{P}(W = 2n | X_1 = 1) = \mathbb{P}(V_{-1} = 2n - 1) = \frac{1}{pn}\binom{2n-2}{n-1}(pq)^n.$$

## 4.2 Visits to Zero

Similarly, for $X_1 = -1$, we obtain

$$\mathbb{P}(W = 2n | X_1 = -1) = \mathbb{P}(V_1 = 2n - 1) = \frac{1}{qn}\binom{2n-2}{n-1}(pq)^n.$$

Substituting these conditional probabilities into (4.15), the result follows. ∎

The first return time $W$ has the generating function

$$g_W(t) = \sum_{n=1}^{\infty} \mathbb{P}(W = 2n) t^{2n} = 1 - \sqrt{1 - 4pqt^2}, \quad |t| \leq 1 \quad (4.16)$$

(see Exercise 4.10). Since $g_W(1) = \mathbb{P}(W < \infty) = 1 - \sqrt{1 - 4pq} = 1 - |p - q|$, this also confirms the result of Theorem 4.5.

In generalizing these ideas, just as in the case of the symmetric random walk (cf. Theorem 2.28), we can determine the distribution of the time of the $k$-th return to the starting point 0, denoted by

$$W_k := \inf\left\{2j : j \in \mathbb{N} \text{ and } \sum_{m=1}^{j} \mathbf{1}\{S_{2m} = 0\} = k\right\}. \quad (4.17)$$

In this context, $W = W_1$. The probabilities $\mathbb{P}(W_k = 2n)$ for $n \geq k$ can be derived using generating functions. Due to the memoryless property of the random walk, the generating function of $W_k$ is the $k$-th power of $g_W$ from Eq. (4.16). One method for obtaining $\mathbb{P}(W_k = 2n)$ via a recursion formula is described on p. 275 in [FE1]. However, it is much simpler to use part (a) of Theorem 3.1 on the distribution of the number of returns to zero in a bridge. Since the resulting denominator $\binom{2n}{n}$ corresponds to the total number of $2n$-bridges, replacing $k$ with $k-1$ gives

$$\frac{2^k k}{n}\binom{2n-k-1}{n-1}$$

as the number of all $2n$-bridges that return to zero $k-1$ times. Furthermore, because

$$\frac{2^k k}{n}\binom{2n-k-1}{n-1} = \frac{2^k k}{2n-k}\binom{2n-k}{n},$$

the following result, stated without proof on p. 275 in [FE1], holds:

**Theorem 4.7 (Distribution of the $k$-th Return Time)** *Let $W_k$ be the time of the $k$-th return to 0, as defined in (4.17). Then*

$$\mathbb{P}(W_k = 2n) = \frac{2^k k}{2n-k}\binom{2n-k}{n}(pq)^n, \quad n = k, k+1, \ldots$$

**Proof** The event $\{W_k = 2n\}$ means that the random walk takes $n$ upward and downward steps up to time $2n$, and returns to zero exactly $k - 1$ times before that point. Each specific path from $(0, 0)$ to $(2n, 0)$ has the same probability $(pq)^n$. Since the coefficient in the theorem indicates the number of such paths, the result follows. ∎

Note that Theorem 4.7 generalizes Theorem 2.28. We now examine the number of returns to the origin for an asymmetric random walk, denoted by

$$N := \sum_{k=1}^{\infty} \mathbf{1}\{S_{2k} = 0\}.$$

If the random walk is symmetric, then the random variable $N$ takes the value $\infty$ with probability one due to the recurrence property. Additionally, as discussed in Sect. 2.4, the number

$$N_{2n} := \sum_{k=1}^{2n} \mathbf{1}\{S_{2k} = 0\}$$

of returns to zero up to time $2n$ (or simply, number of zeros) is of order $\sqrt{2n}$. For the expectation of $N$, the following holds:

**Theorem 4.8 (Expectation of the Number of Zeros)** *If $N$ denotes the number of zeros in an asymmetric random walk on $\mathbb{Z}$ starting at 0, then*

$$\mathbb{E}(N) = \frac{1}{|p - q|} - 1.$$

**Proof** According to Eq. (7.14) from Sect. 7.6,

$$\mathbb{E}(N) = \lim_{n \to \infty} \mathbb{E}(N_{2n}).$$

Because $\mathbb{E}(N_{2n}) = \sum_{k=1}^{n} \mathbb{P}(S_{2k} = 0)$ and $\mathbb{P}(S_{2k} = 0) = \binom{2k}{k}(pq)^k$, where

$$\binom{2k}{k} = (-1)^k \binom{-\frac{1}{2}}{k} 2^{2k},$$

it follows that

$$\mathbb{E}(N) = \sum_{k=1}^{\infty} \mathbb{P}(S_{2k} = 0) = \sum_{k=1}^{\infty} \binom{2k}{k}(pq)^k = \sum_{k=1}^{\infty} \binom{-\frac{1}{2}}{k}(-4pq)^k$$

## 4.2 Visits to Zero

$$= \sum_{k=0}^{\infty} \binom{-\frac{1}{2}}{k}(-4pq)^k - 1 = (1-4pq)^{-1/2} - 1$$

$$= \frac{1}{|p-q|} - 1.$$

The penultimate equality used the binomial series from Sect. 7.11. ∎

The above theorem states, among other things, that an asymmetric random walk with $p = 0.51$ visits zero, on average, 49 times. Figure 4.3 illustrates the dependence of the expected number of visits to zero as a function of $p$.

**Theorem 4.9 (Distribution of the Number of Visits to Zero)** *The number $N$ of visits to zero of an asymmetric walk on $\mathbb{Z}$ has a geometric distribution with parameter $|p-q|$. Specifically,*

$$\mathbb{P}(N=k) = |p-q| \cdot (1-|p-q|)^k, \qquad k \geq 0.$$

**Proof** The assertion can be understood by defining $W$ as in Eq. (4.14). The event $\{W = \infty\}$, where the random walk *does not* return to zero, is considered a *hit*, while the event $\{W < \infty\}$ represents a *miss*. The event $\{N = k\}$ indicates that exactly $k$ misses occur before the first hit. For this to happen, the random walk must return to zero exactly $k$ times starting from the origin, and after the $k$-th return, it must never return to zero again. Due to the memoryless property of the random walk, when a return to zero occurs, the likelihood of another visit to zero is the same as if the random walk were starting again from the origin. Therefore, we have a sequence of similar and mutually independent trials (a so-called Bernoulli sequence), where each trial results in a miss with probability $1 - |p-q|$ or a hit with probability $|p-q|$. This leads to the conclusion that $N$ follows a geometric distribution, as stated. ∎

**Fig. 4.3** Expectation of $N$ as a function of $p$

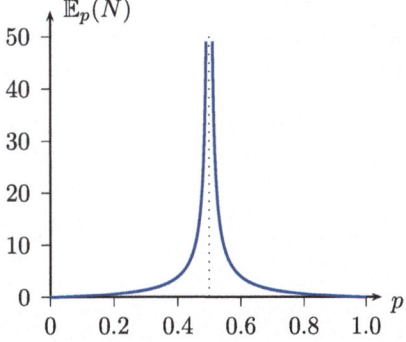

From the distribution of $N$, one can naturally derive the expectation and the variance of $N$. The variance is given by

$$\mathbb{V}(N) = \frac{1 - |p - q|}{|p - q|^2}.$$

**SAQ 5**  Can you derive this equation yourself?

## 4.3 Random Walks with Absorbing Boundaries: The Gambler's Ruin Problem

In this section, we examine a random walk on $\mathbb{Z}$ that starts at the origin and is not necessarily asymmetric. This walk ends when it first hits either the height $-a$ or the height $b$, where $a$ and $b$ are arbitrary positive integers. The lines $y = -a$ and $y = b$ thus represent *absorbing boundaries*. This scenario raises several natural questions:

- How long does it take for one of the absorbing boundaries to be reached?
- What is the probability that absorption occurs at the upper boundary $y = b$ or at the lower boundary $y = -a$?

If we interpret $a$ and $b$ as the capital assets (in euros) of two players A and B, who repeatedly play a game in independent sequence—where A and B win with probabilities $p$ and $q$, respectively, and receive one euro from their opponent in case of a win—then absorption of the random walk at height $b$ or $-a$ corresponds to player B or A becoming bankrupt. In this context, an upward or downward step of the random walk indicates that A or B wins a single game (see Fig. 4.4).

We model this situation using independent, identically distributed random variables $X_1, X_2, \ldots$, with $\mathbb{P}(X_j = 1) = p = 1 - \mathbb{P}(X_j = -1)$ for $j \geq 1$,

**Fig. 4.4** Illustrating the gambler's ruin problem

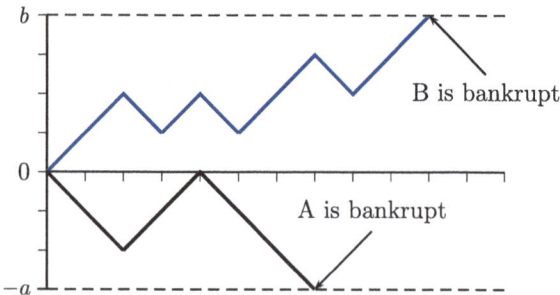

## 4.3 Random Walks with Absorbing Boundaries: The Gambler's Ruin Problem

where $0 < p < 1$. Letting $S_0 := 0$ and $S_n := X_1 + \ldots + X_n$ for $n \geq 1$, we define the random variable $T$ as

$$T := \inf\{n \geq 0 : S_n \in \{-a, b\}\}. \tag{4.18}$$

This formalizes the stopping time when the random walk reaches either absorbing boundary. We already know that $\mathbb{P}(T < \infty) = 1$, meaning that, with certainty, one of the two boundaries will be hit in finite time. According to Theorem 4.1, if $p \geq \frac{1}{2}$, the random walk will almost surely reach the height $b$ in finite time. Conversely, if $p < \frac{1}{2}$, it will almost surely reach the height $-a$ in finite time. This implies that the random variable $S_T : \Omega \to \mathbb{N}_0$, defined by

$$S_T(\omega) := \begin{cases} S_{T(\omega)}(\omega), & \text{if } T(\omega) < \infty, \\ 0, & \text{if } T(\omega) = \infty, \end{cases} \quad \omega \in \Omega,$$

will almost surely take one of the two values $-a$ or $b$. We use the canonical probability space for infinitely long random walks as specified in Sect. 7.2. The probability that absorption occurs at the upper boundary $y = b$, and thus player A wins, is given by $\mathbb{P}(S_T = b)$.

**Theorem 4.10 (Absorption or Ruin Probabilities)** *For the absorption probabilities $\mathbb{P}(S_T = b)$ and $\mathbb{P}(S_T = -a)$, the following hold:*

(a) $\mathbb{P}(S_T = b) = \begin{cases} \dfrac{a}{a+b}, & \text{if } p = \dfrac{1}{2}, \\[2mm] \dfrac{1 - \left(\frac{q}{p}\right)^a}{1 - \left(\frac{q}{p}\right)^{a+b}}, & \text{if } p \neq \dfrac{1}{2}. \end{cases}$

(b) $\mathbb{P}(S_T = -a) = 1 - \mathbb{P}(S_T = b),$

*where $q := 1 - p$.*

**Proof** The problem can be reformulated by considering the random walk starting at height $y = a$ and shifting the absorbing boundaries upward by $a$ units. This means the new absorbing lines are at $y = a + b$ and at the $x$-axis. Player B is ruined if this random walk first reaches the height $a + b$ without hitting the $x$-axis beforehand. To derive the ruin probability $\mathbb{P}(S_T = b)$, we treat the initial height of this random walk as a parameter $k$ and study the probability that absorption occurs at height $a + b$ as a function of $k$. Let this probability be denoted as $P_k$. For simplicity, let $r := a + b$. Since absorption occurs immediately if $k = 0$ or $k = r$, it follows that

$$P_0 = 0, \qquad P_r = 1. \tag{4.19}$$

For $1 \leq k \leq r-1$, the random walk, after one step, is either at height $k+1$ or at height $k-1$. Since the steps are independent of each other, the situation after the first step is the same as at the beginning, only with the parameter $k$ changed. Using the law of total probability (7.4), we get

$$P_k = p \cdot P_{k+1} + q \cdot P_{k-1}, \qquad k = 1, 2, \ldots, r-1. \tag{4.20}$$

This leads to the recursion formula

$$d_k = d_{k-1} \cdot \frac{q}{p}, \qquad k = 1, \ldots, r-1, \tag{4.21}$$

for the differences $d_k := P_{k+1} - P_k$. If $p = q = \frac{1}{2}$, then $d_0, \ldots, d_{r-1}$ are equal, resulting in $P_k = \frac{k}{r}$. Thus, $\mathbb{P}(S_T = b) = P_a = \frac{a}{a+b}$, if $p = \frac{1}{2}$. For $p \neq \frac{1}{2}$, Eq. (4.21) inductively gives $d_j = (q/p)^j \cdot d_0$ for $j = 1, \ldots, r-1$. Therefore,

$$P_k = P_k - P_0 = \sum_{j=0}^{k-1} d_j = d_0 \sum_{j=0}^{k-1} \left(\frac{q}{p}\right)^j = d_0 \cdot \frac{1 - \left(\frac{q}{p}\right)^k}{1 - \frac{q}{p}}.$$

Setting $k = r$, we get $P_r = 1$, so

$$d_0 = \frac{1 - \frac{q}{p}}{1 - \left(\frac{q}{p}\right)^r}.$$

Thus,

$$P_k = \frac{1 - \left(\frac{q}{p}\right)^k}{1 - \left(\frac{q}{p}\right)^r}, \qquad \text{if } p \neq \frac{1}{2}.$$

In particular, we have $\mathbb{P}(S_T = b) = P_a$, as stated in part (a). Since $\mathbb{P}(S_T \in \{b, -a\}) = 1$, it follows that $\mathbb{P}(S_T = -a) = 1 - \mathbb{P}(S_T = b)$. ∎

Figure 4.5 illustrates the absorption probability at the upper boundary, which corresponds to the ruin probability for player B, depending on the success probability $p$ for player A. Two scenarios are shown: one where each player starts with 3 euros (left), and another where each starts with 10 euros (right). It is noteworthy how the larger starting capital affects the ruin probability. When both players start with 3 euros, player B goes bankrupt with a probability of about 0.65 when the success probability for player A is $p = 0.55$. However, if both players start with 10 euros, the higher success probability of A over B becomes more pronounced over a longer series of games. This is reflected in the increased ruin probability of 0.88 for player

### 4.3 Random Walks with Absorbing Boundaries: The Gambler's Ruin Problem

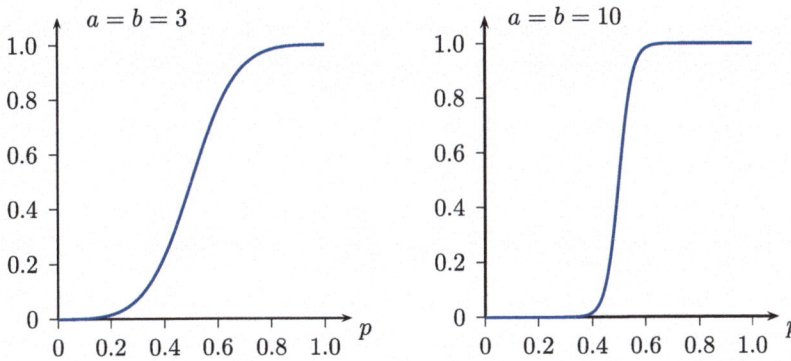

**Fig. 4.5** $\mathbb{P}(S_T = b)$ as a function of the success probability $p$ for player A for the cases where $a = b = 3$ (left) and $a = b = 10$ (right)

B. Additionally, note that for $p > \frac{1}{2}$, the probability of ruin for player B is always at least $1 - (q/p)^a$, regardless of his starting capital $b$.

Next, we consider the expectation of the time span $T$ until absorption, as defined in (4.18). Before determining this expectation, we will first prove its existence.

**Theorem 4.11 (Existence of the Expectation of $T$)** *For the time span $T$ until absorption, the following hold:*

(a) *There exist constants $c > 0$ and $\varrho$ with $0 < \varrho < 1$, such that*

$$\mathbb{P}(T \geq n) \leq c\varrho^n, \qquad n \geq 1.$$

(b) $\mathbb{E}(T) < \infty$.

**Proof** (a) Let $r := a+b$, and choose a positive integer $m > r$. For $k \geq 0$, if the event $\{X_{k+1} = 1, \ldots, X_{k+m} = 1\}$ occurs (which happens with probability $p^m$), then it follows that $S_{k+m} - S_k > r$; thus, $\mathbb{P}(S_{k+m} - S_k > r) \geq p^m$. If $T \geq km+1$ holds for an integer $k$, then this inequality implies the occurrence of all events $\{S_m - S_0 \leq r\}$, $\{S_{2m} - S_m \leq r\}, \ldots, \{S_{km} - S_{(k-1)m} \leq r\}$.

> **SAQ 6** Why does the last statement hold?

Since these events are formed by pairwise disjoint sets of the random variables $X_j$, $j \geq 1$, they are independent. It follows that

$$\mathbb{P}(T \geq mk + 1) \leq (1 - p^m)^k, \qquad k \in \mathbb{N}_0. \tag{4.22}$$

Now, set $c := (1 - p^m)^{-1}$ and $\varrho := (1 - p^m)^{1/m}$. For any $n \geq 1$, there is a non-negative integer $k$ such that $km < n \leq (k + 1)m$. Using (4.22), we get

$$\mathbb{P}(T \geq n) \leq \mathbb{P}(T \geq km + 1) \leq \left(1 - p^m\right)^k = c \left(1 - p^m\right)^{k+1}$$
$$= c \varrho^{m(k+1)} \leq c \varrho^n,$$

which proves (a). Since the series $\sum_{n=1}^{\infty} \varrho^n$ converges, the expectation of $T$ exists according to Sect. 7.6. ∎

**Note** We have already shown the existence of the expectation of $T$ for the case $p \neq \frac{1}{2}$ in Theorem 4.1. In fact, $T \leq \widetilde{T} := \inf\{n \geq 1 : S_n = b\}$, and $\widetilde{T}$ has the same distribution as the sum $V_{1,1} + \ldots + V_{1,b}$, where $V_{1,1}, \ldots, V_{1,b}$ are independent and have the same distribution as the first-passage time $V_1$. According to Theorem 4.1, if $p > \frac{1}{2}$, then $\mathbb{E}(V_1) < \infty$ and thus $\mathbb{E}(\widetilde{T}) < \infty$, so $\mathbb{E}(T) < \infty$ as well. The same argument applies in the case where $p < \frac{1}{2}$.

**Theorem 4.12 (Expected Duration Until Absorption)** *For the time $T$ until absorption,*

$$\mathbb{E}(T) = \begin{cases} \dfrac{1}{p - q} \cdot \dfrac{b\left\{1 - \left(\frac{p}{q}\right)^a\right\} - a\left\{\left(\frac{q}{p}\right)^b - 1\right\}}{\left(\frac{q}{p}\right)^b - \left(\frac{p}{q}\right)^a}, & \text{if } p \neq \dfrac{1}{2}, \\ a \cdot b, & \text{if } p = \dfrac{1}{2}. \end{cases}$$

**Proof** As in the proof of Theorem 4.10, we choose the absorbing boundaries $y = r$ (where $r = a + b$) and $y = 0$. The random walk starts at height $k$, with $k \in \{0, 1, \ldots, r\}$ considered as a parameter. Let $m_k$ denote the expectation of $T$ (where $T = \inf\{n \geq 0 : S_n \in \{0, r\}\}$) under the condition $S_0 = k$. Then, the expectation we seek is clearly $m_a$. Since $T$ takes the value 0 with probability one when starting at height 0 or $r$, it follows that

$$m_0 = m_r = 0. \tag{4.23}$$

In the case $1 \leq k \leq r - 1$, the random walk is at height $k + 1$ with probability $p$ and at height $k - 1$ with probability $q$ after the first step. Due to the memoryless property of the random walk, the situation after the first step is the same as at the beginning, but with different parameter values. It is also important to note that one time step has been used in relation to reaching one of the two boundaries. Therefore, we have

$$m_k = p(1 + m_{k+1}) + q(1 + m_{k-1})$$
$$= 1 + p\, m_{k+1} + q\, m_{k-1}, \quad k = 1, \ldots, r - 1.$$

## 4.3 Random Walks with Absorbing Boundaries: The Gambler's Ruin Problem

For the differences $d_j := m_{j+1} - m_j$, where $j = 0, 1, \ldots, r-1$, we obtain the recursion

$$d_j = \frac{1}{p}(q\, d_{j-1} - 1), \qquad j = 1, \ldots, r-1. \tag{4.24}$$

According to (4.23), $d_0 = m_1$. Using the sum formula for the geometric series in the case $p \neq q$, we get

$$d_j = \left(\frac{q}{p}\right)^j m_1 - \frac{1 - \left(\frac{q}{p}\right)^j}{p - q}, \qquad j = 0, 1, \ldots, r-1. \tag{4.25}$$

Because $m_r = 0$, $0 = d_0 + \ldots + d_{r-1}$. From the above representation, $m_1$ is given by

$$m_1 = \frac{1}{p-q}\left(\frac{r\left(1 - \frac{q}{p}\right)}{1 - \left(\frac{q}{p}\right)^r} - 1\right). \tag{4.26}$$

With some effort, and using $m_a = m_1 + d_1 + \ldots + d_{a-1}$, we arrive at the assertion of the theorem for the case $p \neq q$ (see Exercise 4.13). If $p = q$, Eq. (4.24) takes the form $d_j = m_1 - 2j$. From this, it follows as before that $m_1 = r - 1$ and $m_a = a(r-a) = ab$. ∎

Figure 4.6 shows the expectation of $T$ as a function of $p$ for the cases where $a = b = 3$ (left) and $a = 5, b = 10$ (right). Interpreting the random walk as a repeated game, where player A has a probability of success $p$ and the initial capital amounts are $a$ and $b$ euros for players A and B, respectively, the two graphs illustrate the

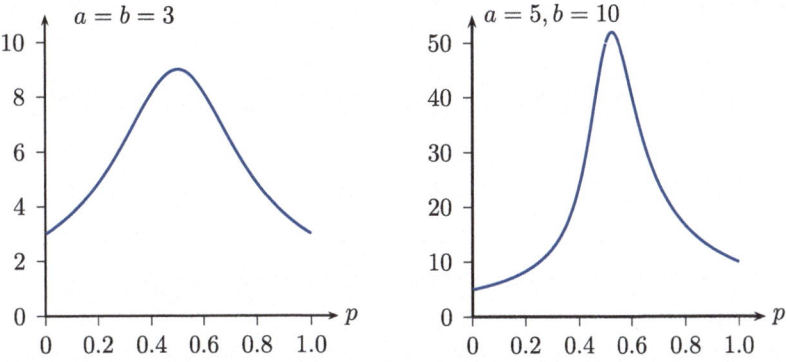

**Fig. 4.6** Expectation of the game duration until one of the two players is ruined in the case $a = b = 3$ (left) and $a = 5, b = 10$ (right)

expected duration of the game until one of the two players is ruined. In the left figure, the graph is symmetric about the line $x = \frac{1}{2}$, due to the equal initial capital amounts. For very small or large values of $p$, the expected value is barely larger than 3, which is to be expected. The longest average game duration, 9 games, is attained in the case of a fair game. If player A starts with 5 euros and player B with 10 euros, it is not surprising that with very small $p$, hardly more than 5 games are played before player A is presumably bankrupt. With very large $p$, player A will win almost all games. Since player B started the game with 10 euros, slightly more than 10 games must be played before one of the players (presumably player B) is bankrupt.

Interestingly, the expected game duration in the right figure does not attain its maximum value at $p = \frac{1}{2}$, but at a slightly larger value. This is hardly surprising because if player A has a slightly higher chance of winning each game than player B, it could compensate for A's smaller capital, balancing the ruin probabilities for both players and generally extending the game duration. Setting the ruin probability of player B from Theorem 4.10 for $a = 5$ and $b = 10$ equal to $\frac{1}{2}$, we get $x := (q/p)^5$. The resulting cubic equation $1 - 2x + x^3 = 0$ has a solution $x_0 \approx 0.61803$ in the interval $(0, 1)$, leading to $p \approx 0.524$. At this success probability for player A, it takes on average slightly more than 52 individual games until of one of the players is ruined.

Finally, we return to the random variable

$$B_k = \sum_{j=1}^{\infty} \mathbf{1}\{S_j = k \text{ and } W > j\},$$

defined in (2.51), which represents the number of times a *symmetric* random walk stays at height $k \neq 0$ *before hitting zero for the first time*. The following surprising result holds:

**Theorem 4.13** *Let $k$ be any integer different from zero. Then*

$$\mathbb{E}(B_k) = 1.$$

**Proof** For reasons of symmetry, we assume $k > 0$. Since $\mathbb{P}(B_k \geq 1 | X_1 = -1) = 0$, it follows that

$$\mathbb{P}(B_k \geq 1) = \frac{1}{2}\mathbb{P}(B_k \geq 1 | X_1 = 1).$$

Under the condition $X_1 = 1$, we have a random walk starting at height 1. According to Theorem 4.10, the probability that it reaches height $k$ before hitting zero (i.e., being absorbed at height 0) is $\frac{1}{k}$, as this corresponds to a gambler's ruin problem with $b = k - 1$ and $a = 1$. Thus, $\mathbb{P}(B_k \geq 1) = \frac{1}{2k}$. Given $B_k \geq 1$, the random walk reaches height $k$ for the first time before hitting zero. We need to determine the probability that this stay at $k$ is the last before hitting zero, i.e., $\mathbb{P}(B_k = 1 | B_k \geq 1)$. If the next step of the random walk is upward (which happens with probability $\frac{1}{2}$), this conditional probability is zero. Otherwise, the random walk moves to height $k - 1$, and the probability of being absorbed at height 0 before hitting $k$ again is $\frac{1}{k}$, as this again corresponds to the gambler's ruin problem, with $b = k - 1$ and $a = 1$. Thus, $\mathbb{P}(B_k = 1 | B_k \geq 1) = \frac{1}{2k}$. Therefore, the probability of having another stay at height $k$ before hitting zero is $1 - \frac{1}{2k}$. Due to the memoryless property, the number of additional stays at height $k$ (after the first stay) follows a geometric distribution, shifted by 1. Hence, $\mathbb{E}(B_k | B_k \geq 1) = 2k$. Overall, we obtain the surprising result

$$\mathbb{E}(B_k) = \mathbb{P}(B_k = 0) \cdot \mathbb{E}(B_k | B_k = 0) + \mathbb{P}(B_k \geq 1) \cdot \mathbb{E}(B_k | B_k \geq 1) = \frac{1}{2k} \cdot 2k$$

$$= 1.$$

∎

## 4.4 Longest Upward and Downward Runs

In this section, we consider a random walk of length $n$ on $\mathbb{Z}$, which can be symmetric or asymmetric. We assume that the random variables $X_1, \ldots, X_n$ are independent, where $\mathbb{P}(X_j = 1) = p = 1 - \mathbb{P}(X_j = -1)$ for each $j = 1, \ldots, n$, with $0 < p < 1$. If we interpret an upward step as a *hit* and a downward step as a *miss*, associating hits with "luck" and misses with "bad luck", we seek to determine the distribution of the longest series of consecutive hits—i.e., the longest streak of luck in a Bernoulli sequence of length $n$ with hit probability $p$.

In more neutral terms, we refer to a series of maximum length of consecutive hits as an *upward run* and a series of maximum length of consecutive misses as a *downward run*. Figure 4.7 illustrates the result of a simulated random walk of length $n = 20$ with probability $p = \frac{1}{2}$. The walk begins with an upward run of length 2, followed by a downward run of length 4, and then an upward run of length 5. This is followed by, in order, a downward run of length 3, an upward run of length 1,

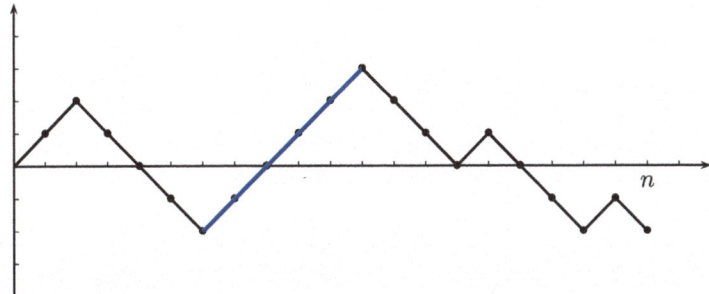

**Fig. 4.7** Random walk of length 20 with longest upward run of length 5

another downward run of length 3, and finally one upward and one downward run, each of length 1. Here, the unusually long streak of luck—5 consecutive upward steps—stands out.

The following theorem provides information on the distribution of the length of the longest upward run (see, e.g., [PM]):

**Theorem 4.14 (Distribution of the Length of the Longest Upward Run)** *Let*

$$R_n := \max\left\{m \in \mathbb{N} : \exists \ell \in \{1, \ldots, n-m+1\} \text{ with } X_\ell = 1, \ldots, X_{\ell+m-1} = 1\right\}$$

*denote the length of the longest series of consecutive hits in a Bernoulli sequence of length n with hit probability p, where $0 < p < 1$ (i.e., the length of the longest upward run). Then, for every $k \in \{1, \ldots, n\}$,*

$$\mathbb{P}(R_n \geq k) = \sum_{r=1}^{\lfloor \frac{n+1}{k+1} \rfloor} (-1)^{r-1} q^{r-1} p^{rk} \left( \binom{n-rk}{r-1} + q \binom{n-rk}{r} \right). \quad (4.27)$$

***Proof*** Let

$$A_1 := \{X_1 = 1, \ldots, X_k = 1\},$$
$$A_j := \{X_{j-1} = -1, X_j = 1, \ldots, X_{j+k-1} = 1\}, \quad j = 2, \ldots, n-k+1,$$

denote the events that a series of *at least* $k$ consecutive hits begins at the $j$-th trial. In the case where $j \geq 2$, the $(j-1)$-th trial must not result in a hit. With this setup, the event $\{R_n \geq k\}$ can be expressed as

$$\{R_n \geq k\} = \bigcup_{j=1}^{n-k+1} A_j,$$

which suggests using the inclusion-exclusion formula (see, e.g., [DU], p. 35). According to this formula,

$$\mathbb{P}(R_n \geq k) = \sum_{r=1}^{n-k+1} (-1)^{r-1} \sum_{1 \leq i_1 < \ldots < i_r \leq n-k+1} \mathbb{P}(A_{i_1} \cap \ldots \cap A_{i_r}). \quad (4.28)$$

This means we need to determine the probabilities of the events $A_i$, as well as the probabilities of the intersections of these events. Clearly,

$$\mathbb{P}(A_1) = p^k \quad (4.29)$$

## 4.4 Longest Upward and Downward Runs

and

$$\mathbb{P}(A_j) = q\, p^k, \qquad j = 2, 3, \ldots, n-k+1. \tag{4.30}$$

For $r \geq 2$, the probabilities $\mathbb{P}(A_{i_1} \cap \ldots \cap A_{i_r})$ depend on $i_1, \ldots, i_r$ with $1 \leq i_1 < \ldots < i_r \leq n-k+1$ as follows: If $i_{s+1} - i_s \leq k$ for some $s \in \{1, \ldots, r-1\}$, then the events $A_{i_s}$ and $A_{i_{s+1}}$ mutually exclude each other because $X_{i_{s+1}-1} = 1$ and $X_{i_{s+1}-1} = -1$ must hold simultaneously. Consequently, $\mathbb{P}(A_{i_1} \cap \ldots \cap A_{i_r}) = 0$. In the opposite case, where $i_{s+1} - i_s \geq k+1$ for every $s \in \{1, \ldots, r-1\}$, the events $A_{i_1}, \ldots, A_{i_r}$ are independent. Therefore, the probability $\mathbb{P}(A_{i_1} \cap \ldots \cap A_{i_r})$ equals $q^r p^{kr}$ or $q^{r-1} p^{kr}$, depending on whether $i_1 \geq 2$ or $i_1 = 1$.

The number of $r$-tuples $(i_1, \ldots, i_r)$ with $2 \leq i_1 < \ldots < i_r \leq n-k+1$ and $i_{s+1} - i_s \geq k+1$ for every $s = 1, \ldots, r-1$ equals the binomial coefficient $\binom{n-kr}{r}$, since the set of these tuples can be bijectively mapped to the set of all tuples $(j_1, \ldots, j_r)$ with $2 \leq j_1 < \ldots < j_r \leq n-kr+1$ (see Exercise 4.18). Similarly, the number of $r$-tuples $(i_1, \ldots, i_r)$ with $1 = i_1 < \ldots < i_r \leq n-k+1$ and $i_{s+1} - i_s \geq k+1$ for every $s = 1, \ldots, r-1$ equals the binomial coefficient $\binom{n-kr}{r-1}$. From (4.29) and (4.30) together with (4.28), the assertion follows. Note that the upper summation index in (4.27) results from the inequality $n - kr \geq r - 1$. ∎

Since $\mathbb{P}(R_n = k)$ can be written as the difference between the probabilities $\mathbb{P}(R_n \geq k)$ and $\mathbb{P}(R_n \geq k+1)$, Theorem 4.14 provides an explicit formula for numerically determining the distribution of $R_n$, as long as $n$ is not too large. Figure 4.8 displays bar charts of the distribution of $R_n$ for the case where $p = \frac{18}{37}$ with $n = 50$ and $n = 100$. The value $\frac{18}{37}$ for $p$ was chosen because it corresponds to the probability of winning in roulette on a so-called *simple chance*. A simple chance is characterized by the fact that 18 of the 37 fields of the roulette wheel—i.e., 18 of the numbers from 0 to 36—are favorable for the player. An example of a simple chance is betting on *Impair*, which involves wagering on the occurrence of an odd number.

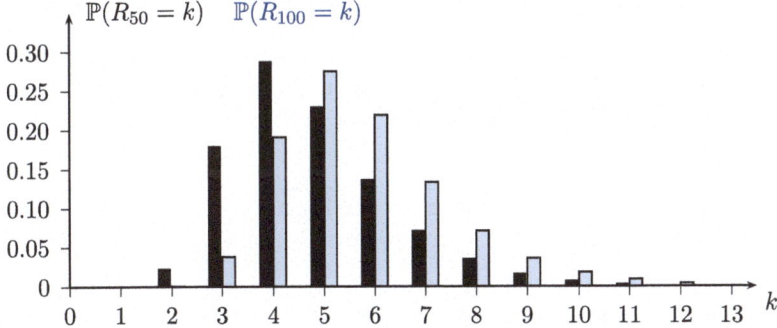

**Fig. 4.8** Bar charts of the distribution of $R_{50}$ (black) and $R_{100}$ (blue) for the case where $p = \frac{18}{37}$ (simple chance in roulette)

When examining Fig. 4.8, it is noticeable that—if we interpret an upward run as a lucky streak in roulette when betting on a simple chance—a lucky streak of at least 7 is not unlikely in 100 consecutive games. However, "luck" can also be interpreted as "bad luck", meaning that a player betting on a simple chance must also expect longer streaks of losses. Many players overlook the independence of individual game outcomes and may double their bets after their second or third consecutive loss.

It is suspected that the probability $\mathbb{P}(R_n \geq k)$ appearing in Eq. (4.27) increases monotonically with $p$. The following results confirms this conjecture. The purely probabilistic proof avoids the usual analytical tools, such as examining the derivative. Readers are encouraged to prove the monotonicity of the function $p \to \mathbb{P}_p(R_n \geq k)$, $0 < p < 1$, using a different approach. In doing so, we emphasize the dependence of the probability $\mathbb{P}(R_n \geq k)$ on $p$ by indexing it with $p$.

**Theorem 4.15 (Monotonicity of $\mathbb{P}_p(R_n \geq k)$ in $p$)** *Under the conditions of Theorem 4.14, the exceedance probability $\mathbb{P}_p(R_n \geq k)$ is monotonically increasing in $p$ for fixed n and k.*

*Proof* The idea behind the proof is to express the random variables $X_1, \ldots, X_n$ as functions of independent and uniformly distributed random variables $U_1, \ldots, U_n$ on the unit interval $(0, 1)$. Specifically, we define

$$X_j(\omega) := 2\,\mathbf{1}\{U_j(\omega) \leq p\} - 1, \quad \omega \in \Omega, \qquad j = 1, \ldots, n. \tag{4.31}$$

Here, $U_1, \ldots, U_n$ are defined on a sample space $\Omega$, which does not need to be further specified. With this construction, $X_1, \ldots, X_n$ are independent and identically distributed, and since $\mathbb{P}(U_j \leq p) = p$, it follows that $\mathbb{P}(X_j = 1) = p = 1 - \mathbb{P}(X_j = -1)$, satisfying the assumptions of Theorem 4.14. The advantage of this construction is that, if $p_1 < p_2$ in Eq. (4.31), it implies that $\mathbf{1}\{U_j(\omega) \leq p_1\} \leq \mathbf{1}\{U_j(\omega) \leq p_2\}$. Therefore, when transitioning from $p_1$ to $p_2$, some values of $X_j(\omega)$ can change from $-1$ to $1$. As a result, the value of $R_n$, which depends on $X_1(\omega), \ldots, X_n(\omega)$, can only increase. ∎

The behavior of the length $R_n$ of the longest run in a Bernoulli sequence of length $n$ as $n$ increases has generated extensive literature. P. Erdös[1] and P. Révész[2] ([ER]) proved the strong law of large numbers:

$$\lim_{n \to \infty} \frac{R_n}{\log n} = \frac{1}{-\log p}$$

---

[1] Paul Erdös (1913–1996), one of the most significant mathematicians of the twentieth century, published about 1500 mathematical works. This large number of publications led to the creation of the so-called *Erdös number*. The 509 mathematicians who co-authored a publication with him are assigned an Erdös number of 1. Those who co-authored a publication with a mathematician with an Erdös number of 1 have an Erdös number of 2, and so on. Main areas of work were number theory and combinatorics.

[2] Pál Révész (1924-2022) was a Hungarian mathematician. His primary areas of research were probability theory and mathematical statistics.

with probability one. From this, it follows that $R_n$ grows logarithmically with $n$. A good overview of various results on the behavior of $R_n$ as $n \to \infty$ is provided in Section 8.5 of [EKM].

We also note that in an *infinitely long* random walk—whether symmetric or asymmetric—an upward run of a given fixed length occurs infinitely often with probability one. In fact, let $X_1, X_2, \ldots$ be independent random variables on a suitable probability space $(\Omega, \mathcal{A}, \mathbb{P})$ (see Sect. 7.2) with $\mathbb{P}(X_j = 1) = p = 1 - \mathbb{P}(X_j = -1)$ for each $j \geq 1$. Here, $p \in (0, 1)$ is arbitrary. For a fixed positive integer $k$, let

$$A_1 := \{X_1 = 1, \ldots, X_k = 1, X_{k+1} = -1\},$$
$$A_j := \{X_{j-1} = -1, X_j = 1, \ldots, X_{j+k-1} = 1, X_{j+k} = -1\}, \qquad j \geq 2,$$

denote the events that a series of *exactly* $k$ consecutive hits begins at the $j$-th trial. We set

$$A := \bigcap_{n=1}^{\infty} \left( \bigcup_{j=n}^{\infty} A_j \right),$$

so that $\omega \in \Omega$ belongs to $A$ if and only if it is an element of infinitely many of the events $A_1, A_2, \ldots$. Therefore, the event $A$ occurs if and only if infinitely many of the $A_j$ occur. To show $\mathbb{P}(A) = 1$, we choose the subsequence $(A_{rk+2r+1})_{r \geq 0}$ from $A_1, A_2, \ldots$. Since the events $A_1, A_{k+3}, A_{2k+5}, A_{3k+7}, \ldots$ are formed by pairwise disjoint sets of the random variables $X_1, X_2, \ldots$, the events $A_{rk+2r+1}$, $r \geq 1$, are independent. Because $\mathbb{P}(A_j) = p^k(1-p)^2 > 0$ for $k \geq 2$, it follows that $\sum_{r=1}^{\infty} \mathbb{P}(A_{rk+2r+1}) = \infty$. From part (b) of the Borel–Cantelli lemma in Sect. 7.13, we deduce

$$\mathbb{P}\left( \bigcap_{n=1}^{\infty} \left( \bigcup_{r=n}^{\infty} A_{rk+2r+1} \right) \right) = 1.$$

Since this event, which has probability one, is a subset of $A$, the assertion follows.

Clearly, the above result also applies if, instead of a sequence of directly consecutive ones, any other sequence of 1's and $-1$'s—no matter how long—is given. The probability that this sequence occurs infinitely often in an infinitely long random walk is one!

## 4.5 The Galton–Watson Process

In 1873, Francis Galton posed the following problem: What is the probability that the male line of a man's descendants will die out if he and each of his sons, grandsons, etc., independently of one another, have exactly $k$ sons with the same probability $p_k$, where $k \in \{0, 1, 2, \ldots\}$?

In a more neutral setting, we imagine a population of individuals who all have a lifespan of one time unit and reproduce asexually. All individuals in a generation are born simultaneously and die at the same time. Let $M_n$ denote the size of the population at time $n \geq 1$, with $M_0 := 1$, assuming that the population dynamic process starts with one ancestor in the zeroth generation.

To model the stochastic development of the population, we specify a so-called *offspring distribution*. This is a probability distribution on the set $\mathbb{N}_0$, where the number $k$ occurs with probability $p_k$. The generating function of this distribution is given by

$$g(t) := \sum_{k=0}^{\infty} p_k t^k, \qquad |t| \leq 1,$$

which will play a central role in the following analysis.

We assume that each individual in each generation reproduces independently of all others, according to this offspring distribution. This assumption leads to the *reproduction equation*

$$M_{n+1} = \sum_{j=1}^{M_n} X_{n+1}^{(j)}. \tag{4.32}$$

Here, $X_n^{(j)}$, for $n, j \geq 1$, are independent $\mathbb{N}_0$-valued random variables with the given offspring distribution, and $X_{n+1}^{(j)}$ denotes the number of offspring of the $j$-th individual in the $n$-th generation. The sequence $(M_n)_{n \geq 0}$, defined recursively by Eq. (4.32), is called the (simple) *Galton–Watson[3]-process* (GW process for short).

We can view the development of the population sizes $M_n$ as a planar random walk starting at the point $(0, 1)$ and being at the point $(n, M_n)$ at time $n$ (the $n$-th generation). Regardless of its previous behavior, the process moves from there to the point $(n+1, M_{n+1})$. The randomness generates the new height $M_{n+1}$ according to Eq. (4.32), based on the previous height $M_n$. This is done by independently observing the realization of a random variable with the offspring distribution $(p_k)_{k \geq 0}$ a total of $M_n$ times and summing the resulting values. Figure 4.9 shows the result of such a random walk using a binomial distribution Bin$(3, \frac{1}{3})$, as the offspring distribution, with probabilities

$$p_0 = \frac{8}{27}, \quad p_1 = \frac{12}{27}, \quad p_2 = \frac{6}{27}, \quad p_3 = \frac{1}{27}.$$

---

[3] Henry William Watson (1827–1903), clergyman, mathematician, and mountaineer, authored his most important book, *A Treatise on the Kinetic Theory of Gases* (1876). His correspondence with Francis Galton in 1873 laid the foundation for the theory of branching processes.

## 4.5 The Galton–Watson Process

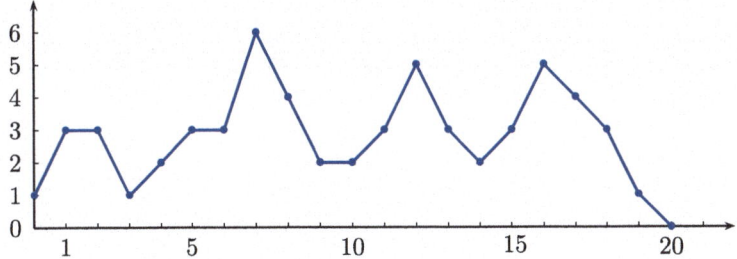

**Fig. 4.9** Realization of a Galton–Watson process for the offspring distribution $\mathrm{Bin}(3, \frac{1}{3})$ as a polygonal chain

In the simulation shown, the population dies out in the 20th generation. Note that this offspring distribution has an expected value of 1, leading to a population that eventually dies out with probability one, as we will see later.

In the following, let

$$\mu := \sum_{k=1}^{\infty} k p_k = g'(1)$$

denote the expectation of the offspring distribution, assuming it exists. When $\mu > 1$, each individual has, on average, more than one offspring; this is referred to as the *supercritical case*. The *critical case* occurs when each individual is followed, on average, by exactly one individual, i.e., $\mu = 1$. The remaining case, $\mu < 1$, is known as the *subcritical case*.

The probability that the process $(M_n)_{n \geq 0}$ will eventually die out is denoted by

$$w := \mathbb{P}\left(\bigcup_{n=1}^{\infty} \{M_n = 0\}\right).$$

This probability depends solely on the offspring distribution, excluding trivial cases like $p_1 = 1$ or $p_0 = 0$. Therefore, we make the fundamental assumption $0 < p_0 < 1$ for all further considerations.

> **SAQ 7** Why are the cases $p_1 = 0$ and $p_1 = 1$ trivial?

The key to determining the extinction probability is provided by the following lemma, which concerns the generating function of a sum of "randomly many" independent and identically distributed random variables, as in Eq. (4.32):

**Lemma 4.16 (Generating Function of a Randomized Sum)** *Let $N, X_1, X_2, \ldots$ be independent $\mathbb{N}_0$-valued random variables defined on a common probability space, where $X_1, X_2, \ldots$ have the same distribution and therefore share the same generating function*

$$g(t) := \sum_{n=0}^{\infty} \mathbb{P}(X_1 = n)\, t^n.$$

*The generating function of $N$ is denoted by*

$$\varphi(t) := \sum_{k=0}^{\infty} \mathbb{P}(N = k)\, t^k.$$

*Then the generating function of randomized sum $S_N$, defined by $S_0 := 0$, $S_k := X_1 + \cdots + X_k$ for $k \geq 1$, and*

$$S_N(\omega) := S_{N(\omega)}(\omega) \quad \text{for } \omega \in \Omega,$$

*is given by*

$$h(t) = \varphi(g(t)). \tag{4.33}$$

*Proof* By decomposing the event $\{S_N = j\}$ according to the realized value of $N$ and using the independence of $N$ and $S_k$,

$$\mathbb{P}(S_N = j) = \sum_{k=0}^{\infty} \mathbb{P}(N = k, S_k = j) = \sum_{k=0}^{\infty} \mathbb{P}(N = k)\, \mathbb{P}(S_k = j).$$

According to Sect. 7.9, $S_k$ has the generating function $g(t)^k$. Therefore, Eq. (4.33) follows from

$$h(t) = \sum_{j=0}^{\infty} \mathbb{P}(S_N = j)\, t^j$$

$$= \sum_{k=0}^{\infty} \mathbb{P}(N = k) \left( \sum_{j=0}^{\infty} \mathbb{P}(S_k = j)\, t^j \right)$$

$$= \sum_{k=0}^{\infty} \mathbb{P}(N = k)\, (g(t))^k.$$

■

## 4.5 The Galton–Watson Process

**Corollary 4.17 (Expectation and Variance of $S_N$)**

(a) *In the situation of Lemma 4.16, suppose $\mathbb{E}|N| < \infty$ and $\mathbb{E}|X_1| < \infty$. Then*

$$\mathbb{E}(S_N) = \mathbb{E}(N)\,\mathbb{E}(X_1).$$

(b) *In the situation of Lemma 4.16, suppose $\mathbb{E}(N^2) < \infty$ and $\mathbb{E}(X_1^2) < \infty$. Then*

$$\mathbb{V}(S_N) = \mathbb{V}(N)\,(\mathbb{E}\,X_1)^2 + \mathbb{E}(N)\,\mathbb{V}(X_1).$$

**Proof** Using (4.33), we obtain $h'(t) = \varphi'(g(t))\,g'(t)$ and

$$h''(t) = \varphi''(g(t))\,(g'(t))^2 + \varphi'(g(t))\,g''(t), \qquad |t| < 1. \tag{4.34}$$

Under the given assumptions, the (left-sided) derivatives exist: $\varphi'(1) = \mathbb{E}(N)$ and $g'(1) = \mathbb{E}(X_1)$ (for part a)) or $\varphi''(1) = \mathbb{E}(N(N-1))$ and $g''(1) = \mathbb{E}(X_1(X_1-1))$ (for part b)), see Sect. 7.9. Therefore, the (left-sided) derivatives $h'(1) = \mathbb{E}(S_N)$ and $h''(1) = \mathbb{E}(S_N(S_N-1))$ also exist. Thus, assertion (a) is straightforward, and part (b) follows by direct calculation (Exercise 4.19). ∎

Corollary 4.17 provides the following statement regarding the expectation and variance of the population size at time $n$.

**Theorem 4.18 (Expectation and Variance of $M_n$)** *Let $(M_n)_{n\geq 0}$ with $M_0 := 1$ be a GW process whose offspring distribution $(p_k)_{k\geq 0}$ has both a finite expectation $\mu$ and a finite variance $\sigma^2$. Then the following hold:*

(a) $\mathbb{E}(M_n) = \mu^n$,
(b)

$$\mathbb{V}(M_n) = \begin{cases} \dfrac{\sigma^2 \mu^{n-1}(\mu^n - 1)}{\mu - 1}, & \text{if } \mu \neq 1, \\ n\sigma^2, & \text{if } \mu = 1. \end{cases}$$

**Proof** From Eq. (4.32) and Corollary 4.17, for each $n \geq 0$:

$$\mathbb{E}(M_{n+1}) = \mathbb{E}(M_n)\,\mu, \qquad \mathbb{V}(M_{n+1}) = \mathbb{V}(M_n)\,\mu^2 + \mathbb{E}(M_n)\,\sigma^2. \tag{4.35}$$

The assertion now follows by induction on $n$ (see Exercise 4.20). ∎

Since $\mathbb{E}(M_n) \to 0$ when $\mu < 1$, it is expected that the population will eventually die out. In contrast, the long-term behavior of $M_n$ in the critical case $\mu = 1$ is less clear. When $\mu > 1$, one might expect the population size $M_n$ to grow geometrically with $n$, because $\lim_{n\to\infty} \mathbb{E}(M_n) = \infty$. However, there is a positive probability that the ancestor in the zeroth generation will die without offspring. For this reason, one might suspect that a population with supercritical offspring distribution, if it has not died out by a large time $n$, is typically very large. We will explore the validity of these assumptions, beginning with the probability that a GW process will die out in finite time.

**Theorem 4.19 (Criterion for the Extinction Probability)** *Let $(M_n)_{n \geq 0}$ with $M_0 := 1$ be a GW process whose offspring distribution $(p_k)_{k \geq 0}$ has the generating function $g$. Then the following hold:*

(a) *The extinction probability $w$ of $(M_n)$ is equal to the smallest non-negative solution $t$ of the fixed-point equation $g(t) = t$.*
(b) *If $\mu \leq 1$, then $w = 1$.*
(c) *If $\mu > 1$, then $w < 1$.*

*Proof*

(a) Let $\varphi_n$ denote the generating function of $M_n$. Using the recursive relationship (4.32) and Lemma 4.16, it follows that

$$\varphi_{n+1}(t) = \varphi_n(g(t))$$

and thus, because $\varphi_1(t) = g(t)$,

$$\varphi_n(t) = (g \circ \ldots \circ g)(t) \quad (n - \text{fold iteration of} g).$$

Since the events $\{M_n = 0\}$, $n \geq 1$, form an ascending sequence of sets, it follows from the definition of the extinction probability and the fact that $\mathbb{P}$, as a probability measure, is continuous from below (see (7.1)), that

$$w = \lim_{n \to \infty} \mathbb{P}(M_n = 0) = \lim_{n \to \infty} \varphi_n(0).$$

Furthermore, since

$$g(w) = g\big(\lim \varphi_n(0)\big) = \lim g(\varphi_n(0)) = \lim \varphi_{n+1}(0) = w,$$

the extinction probability $w$ is a fixed point of $g$. If $x \geq 0$ is any fixed point of $g$, then $x = g(x) \geq g(0) = \varphi_1(0)$, and thus, by induction, $x \geq \varphi_n(0)$ for every $n \geq 1$, so $x \geq w = \lim \varphi_n(0)$.

(b) We distinguish between the cases where $p_0 + p_1 = 1$ and $p_0 + p_1 < 1$. In the first case, it follows that $\mathbb{P}(M_n = 0) = 1 - p_1^n$ and thus $w = 1$. In the second case, the argument proceeds as in part (c). The equation $g(t) = t$ has the unique fixed point $w = 1$ (see Fig. 4.10, left).

(c) In the case where $\mu = g'(1) > 1$, it follows that $p_0 + p_1 < 1$, and the derivative $g'(t) = \sum_{k=1}^{\infty} k p_k t^{k-1}$ is strictly increasing on $[0, 1]$, indicating that $g$ is strictly

## 4.5 The Galton–Watson Process

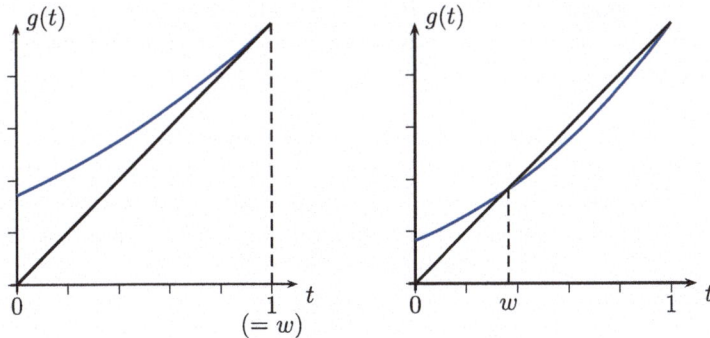

**Fig. 4.10** Illustrating the extinction probability of a GW process

convex on this interval. Consequently, $g(t)$ has exactly one additional fixed point $w$ in the interval $[0, 1]$, besides the trivial fixed point $t = 1$, as shown in Fig. 4.10 (right).

∎

The above result states that the process dies out with probability one in the subcritical or critical case—i.e., when, on average, at most one offspring is born. Otherwise, the extinction probability is less than one.

**Example 4.20 (Geometric Offspring Distribution)** The geometric offspring distribution

$$p_k = \frac{1}{\mu + 1} \left( \frac{\mu}{\mu + 1} \right)^k, \quad k \in \mathbb{N}_0, \tag{4.36}$$

has expectation $\mu > 1$ and the generating function

$$g(t) = \frac{1}{\mu + 1 - \mu t}, \quad |t| \leq 1.$$

The fixed-point equation $g(t) = t$ leads to the quadratic equation

$$\mu t^2 - (\mu + 1)t + 1 = 0.$$

Besides the trivial solution $t = 1$, this equation has a second solution $\frac{1}{\mu} < 1$. Therefore, the GW process with the offspring distribution (4.36) dies out with probability $\frac{1}{\mu}$.

**Example 4.21 (Family Planning)** If each individual can have *at most two offspring*, then $g(t) = p_0 + p_1 t + p_2 t^2$. Since $g'(t) = p_1 + 2t p_2$ and $1 = p_0 + p_1 + p_2$, it follows that

$$\mu = g'(1) = p_1 + 2p_2 = 1 + p_2 - p_0.$$

Thus, $\mu > 1$ if and only if $p_2 > p_0$. In this supercritical case, the extinction probability is equal to the unique solution $w \in (0, 1)$ of the quadratic equation $t = p_0 + p_1 t + p_2 t^2$. After a direct calculation, this results in $w = p_0/p_2$, which does not depend on $p_1$.

Note that in the special case $p_1 = 0$, we have a random walk that starts at height 1 and takes a step up or down with probabilities $p_2$ and $p_0$, respectively. The extinction probability can thus also be obtained using Theorem 4.1.

**SAQ 8** Why is the last statement true?

**Example 4.22 (Offspring Distribution with Infinite Expectation)** For the offspring distribution $(p_j)$ with $p_j := \frac{1}{(j+1)(j+2)}$, $j \geq 0$, it follows that $\frac{1}{(j+1)(j+2)} = \frac{1}{j+1} - \frac{1}{j+2}$. Using the power series expansion of the logarithm function, the generating function is given by

$$g(t) = \frac{1}{t}\left(1 + \left(\frac{1}{t} - 1\right)\log(1 - t)\right), \qquad |t| < 1,\ t \neq 0,$$

as well as $g(0) = \frac{1}{2}$ (see Exercise 4.21). Since the harmonic series diverges, this offspring distribution has an infinite expectation. The equation $w = g(w)$ for determining the extinction probability $w$ leads to the solution $w = 0.683803\ldots$, obtained by iteration.

The following result clarifies the previously mentioned assumption that a supercritical GW process exhibits highly unstable long-term behavior: it either becomes extinct or "explodes," resulting in a very large population size (the so-called *extinction-explosion dichotomy*).

**Theorem 4.23 (Extinction-Explosion Dichotomy)** *For a Galton–Watson process $(M_n)_{n\geq 0}$ with $M_0 = 1$ and extinction probability $w = \mathbb{P}(\cup_{n=1}^\infty \{M_n = 0\})$,*

$$\mathbb{P}\left(\lim_{n\to\infty} M_n = 0\right) = w, \qquad \mathbb{P}\left(\lim_{n\to\infty} M_n = \infty\right) = 1 - w.$$

**Proof** For any positive integer $k$, let $T_k := \sup\{n \geq 1 : M_n = k\}$ denote the last time at which the population size takes the value $k$ (where $T_k = \infty$ is possible). Due

## 4.5 The Galton–Watson Process

to our fundamental assumption $p_0 > 0$ for the offspring distribution, it follows for any positive integer $n$ that

$$\mathbb{P}(T_k = n) = \mathbb{P}(M_n = k) \, \mathbb{P}\left(\bigcap_{j=1}^{\infty} \{M_{n+j} \neq k\} \,\Big|\, M_n = k\right)$$
$$\geq \mathbb{P}(M_n = k) \, \mathbb{P}(M_{n+1} = 0 | M_n = k) \qquad (4.37)$$
$$= \mathbb{P}(M_n = k) \, p_0^k.$$

**SAQ 9** Why does inequality (4.37) hold?

Summing over $n$ gives

$$1 \geq \sum_{n=1}^{\infty} \mathbb{P}(T_k = n) \geq p_0^k \sum_{n=1}^{\infty} \mathbb{P}(M_n = k),$$

and thus, in particular, $\sum_{n=1}^{\infty} \mathbb{P}(M_n = k) < \infty$. From this, we not only obtain

$$\lim_{n \to \infty} \mathbb{P}(M_n = k) = 0 \quad \text{for every } k \geq 1,$$

but also

$$\mathbb{P}\left(\bigcap_{\ell=1}^{\infty} \bigcup_{n=\ell}^{\infty} \{M_n = k\}\right) = 0 \quad \text{for every } k \geq 1,$$

according to the Borel-Cantelli lemma from Sect. 7.13. Considering the complementary event and noting that the intersection of countably many sets, each with probability one, also has probability one, it follows that

$$\mathbb{P}\left(\bigcap_{k=1}^{\infty} \bigcup_{\ell=1}^{\infty} \bigcap_{n=\ell}^{\infty} \{M_n \neq k\}\right) = 1. \qquad (4.38)$$

Let us denote the event occurring here briefly as $\Omega_0$. For each $\omega \in \Omega_0$, the sequence $(M_n(\omega))_{n \geq 1}$ from $\mathbb{N}_0$ has the property that each positive integer appears only finitely often as a sequence element. Thus, we have the subset relation

$$\Omega_0 \subset \left\{\lim_{n \to \infty} M_n = 0\right\} \uplus \left\{\lim_{n \to \infty} M_n = \infty\right\}, \qquad (4.39)$$

because for each $\omega \in \Omega_0$, there is either an $n \geq 1$ with $M_n(\omega) = 0$ (in which case $M_k(\omega) = k$ for every $k \geq n$, so $\lim_{n \to \infty} M_n(\omega) = 0$), or $M_n(\omega) \neq 0$ for every

$n \in \mathbb{N}$. In this case, $\lim_{n\to\infty} M_n(\omega) = \infty$, because otherwise at least one positive integer would appear infinitely often as a sequence element. From (4.38) and (4.39), the assertion follows. ∎

For a supercritical Galton–Watson process, one would expect the process to grow exponentially at a rate of $\mu^n$ on the set $\{\omega \in \Omega : \lim_{n\to\infty} M_n(\omega) = \infty\}$, which has positive probability $1 - w$. Specifically, if we consider the rescaled process

$$Z_n := \frac{M_n}{\mu^n}, \qquad n \geq 0,$$

then, according to Theorem 4.18 and the rules $\mathbb{E}(aX) = a\mathbb{E}(X)$ and $\mathbb{V}(aX) = a^2\mathbb{V}(X)$ for the expectation and variance of a random variable $X$ multiplied by the constant $a$,

$$\mathbb{E}(Z_n) = 1, \qquad \mathbb{V}(Z_n) = \sigma^2 \cdot \frac{\mu^{n-1}(\mu^n - 1)}{\mu^{2n}(\mu - 1)} \to \frac{\sigma^2}{\mu(\mu - 1)} \text{ as } n \to \infty.$$

Here, we have assumed that the offspring distribution has a finite variance $\sigma^2$.

The key to more detailed investigations of $Z_n$ lies in the observation that the process $(Z_n)$ exhibits the characteristic *martingale property*:

$$\mathbb{E}(Z_{n+1}|Z_n) = \mathbb{E}\left[\frac{M_{n+1}}{\mu^{n+1}}\bigg|Z_n\right] = \frac{1}{\mu^{n+1}} \cdot \mathbb{E}[M_{n+1}|M_n] = \frac{1}{\mu^{n+1}} \cdot \mu \cdot M_n$$
$$= Z_n.$$

With this, it can be shown that $Z_n$ converges to a random variable $Z$ with probability one as $n \to \infty$, where $Z$ has an expectation of 1 and a variance of $\sigma^2/(\mu(\mu - 1))$. For more details, interested readers are referred to further literature, such as [HAR] or [AN].

### Answers to the Self-Assessment Questions

**Answer 1** The event $\{S_n = j\}$ occurs if and only if $\frac{n+j}{2}$ of the random variables $X_1, \ldots, X_n$ take the value $+1$ and $\frac{n-j}{2}$ take the value $-1$. The binomial coefficient gives the number of ways to make such a selection, and each specific selection has the probability shown to the right of the binomial coefficient.

**Answer 2** Note that $p > q$ and $1 - 4pq = (p + q)^2 - 4pq = p^2 - 2pq + q^2 = (p - q)^2$.

**Answer 3** If $\ell \geq r$, then $S_n = \ell$ implies $M_n \geq r$.

**Answer 4** Here, $p$ stands for $\mathbb{P}(X_n = 1)$. The conditional probability $\mathbb{P}(M_{n-1} = S_{n-1} = k - 1, S_n = k | X_n = 1)$ equals $\mathbb{P}(M_{n-1} = S_{n-1} = k - 1, S_{n-1} = k - 1)$.

## 4.5 The Galton–Watson Process

For any event $A$,

$$\{M_{n-1} = k-1\} \cap A \uplus \{M_{n-1} \geq k\} \cap A = \{M_{n-1} \geq k-1\} \cap A.$$

**Answer 5** Setting $w := |p-q|$, we have

$$\mathbb{E}[N(N-1)] = w(1-w)^2 \sum_{k=2}^{\infty} k(k-1)(1-w)^{k-2}$$

$$= w(1-w)^2 \frac{d^2}{dt^2} \frac{1}{1-t}\bigg|_{t=1-w} = \frac{2(1-w)^2}{w^2},$$

and thus

$$\mathbb{V}(N) = \mathbb{E}[N(N-1)] + \mathbb{E}(N) - (\mathbb{E} N)^2 = \frac{2(1-w)^2}{w^2}$$

$$+ \frac{1}{w} - 1 - \left(\frac{1}{w} - 1\right)^2 = \frac{1-w}{w^2}.$$

**Answer 6** The difference between the two absorbing boundaries is $r = a + b$. Suppose that $S_{\ell m} - S_{(\ell-1)m} > r$ for some $\ell \in \{1, \ldots, k\}$. Then $T \leq km$, because absorption (at the upper boundary) would have already occurred by time $km$.

**Answer 7** In the first case, the population is constant, and in the second case, at least one offspring is always born. In both of these cases, $w = 0$.

**Answer 8** The GW process dies out if and only if the random walk starting at height 1 eventually reaches height 0. This is equivalent to a random walk starting at 0 eventually reaching height $-1$. The probability of this happening is $\mathbb{P}(V_{-1} < \infty)$. According to Theorem 4.1 (b), $\mathbb{P}(V_{-1} = \infty) = 1 - p_0/p_2$.

**Answer 9** Given that $M_n = k$, if $M_{n+1} = 0$ the population is extinct at time $n+1$. This implies that $M_{n+j} = 0 \neq k$ for each $j \geq 1$, since $k \geq 1$.

## Exercises

**Exercise 4.1** As at the beginning of this chapter, assume an asymmetric random walk on $\mathbb{Z}$ starting at 0, where $p = \frac{1}{4}$. Show that

$$\mathbb{P}\left(\max_{j \in \mathbb{N}_0} S_j = 0\right) = \frac{2}{3}.$$

**Exercise 4.2** Using (4.4), derive the equation

$$g(t) = qtg^2(t), \quad |t| \leq 1,$$

for the generating function $g$ of the first-passage time $V_1$.

**Exercise 4.3** Show that

$$\frac{1 - \sqrt{1 - 4pqt^2}}{2qt} = \sum_{n=0}^{\infty} b_n t^n$$

for every $t \in \mathbb{R}$ with $|t| \leq 1$ and $t \neq 0$, where

$$b_{2k-1} = \frac{1}{kq} \binom{2k-2}{k-1} (pq)^k, \quad k \geq 1,$$

and $b_{2k} = 0$ for every $k \geq 0$.

**Exercise 4.4** Show that

$$\mathbb{V}(V_1) = \frac{4pq}{(p-q)^3}.$$

**Exercise 4.5** Show that the generating function $g_2(t) = \mathbb{E}(t^{V_2})$ of the first-passage time $V_2$ to height 2 is given by

$$g_2(t) = \sum_{j=1}^{\infty} \frac{1}{j+1} \binom{2j}{j} p^{j+1} q^{j-1} t^{2j}, \quad |t| \leq 1.$$

This directly results in Eq. (4.10).

**Exercise 4.6** Assume an asymmetric random walk on $\mathbb{Z}$ starting at 0 with $p < \frac{1}{2}$, and let $V_k$ denote the first-passage time to height $k$. Prove that

$$\mathbb{P}(V_k < \infty) = \left(\frac{p}{q}\right)^k, \quad k \geq 1. \tag{4.40}$$

**Hint:** You can proceed inductively, using the fact that for any choice of $\ell \geq k+1$ and $m \geq 1$, the events $\{V_{k+1} = \ell\}$ and $\{V_{k+1} - V_k = m\}$ are independent, and that $V_{k+1} - V_k$ has the same distribution as $V_1$.

**Exercise 4.7** Let $X_1, X_2, \ldots$ independent random variables with $\mathbb{P}(X_j = 1) = p$ and $\mathbb{P}(X_j = -1) = q = 1 - p$ for each $j$, where $0 < p < \frac{1}{2}$. Let $(S_n)_{n \geq 0}$ be an asymmetric random walk with $S_0 = 0$ and $S_\ell = X_1 + \ldots + X_\ell$ for $\ell \geq 1$. Show that:

(a) $\mathbb{P}\left(\max_{j \geq 0} S_j = k\right) = \left(\frac{p}{q}\right)^k \left(1 - \frac{p}{q}\right), \quad k \in \mathbb{N}_0,$

## 4.5 The Galton–Watson Process

(b) $\mathbb{E}\left(\max_{j\geq 0} S_j\right) = \dfrac{q}{p-q}.$

**Hint:** Exercise 4.6 may be helpful.

**Exercise 4.8** Verify that Eq. (4.10) agrees with the statement of Theorem 2.24 in the special case $p = \frac{1}{2}$.

**Exercise 4.9** In the situation described in Sect. 4.1, let

$$T := \inf\{2j + 1 : j \in \mathbb{N} \text{ and } S_{2j+1} S_{2j-1} = -1\}$$

be the time of the first change of sign. Show that

$$\mathbb{P}(T = 2k + 1) = \frac{1}{k+1}\binom{2k}{k}(pq)^k, \qquad k \geq 1.$$

**Exercise 4.10** Show that the generating function of the time $W$ of the first return to 0 is given by

$$g_W(t) = \sum_{n=1}^{\infty} \mathbb{P}(W = 2n) t^{2n} = 1 - \sqrt{1 - 4pqt^2}, \qquad |t| \leq 1.$$

**Exercise 4.11** Assume an asymmetric random walk on $\mathbb{Z}$ starting at 0.

(a) Show that there is a constant $\varrho \in (0, 1)$ such that

$$\mathbb{P}(S_{2n} = 0) \sim \frac{1}{\sqrt{\pi n}} \cdot \varrho^n \quad \text{as } n \to \infty,$$

and determine $\varrho$.

(b) Conclude that, with probability 1, the random walk returns to 0 only finitely often.

**Exercise 4.12** In the gambler's ruin problem, A and B start with 10 euros and 100 euros, respectively. Player A can choose the amount to bet per game, with options of 1 euro, 2 euros, 5 euros, or 10 euros. Which bet should A choose to maximize the probability of B's ruin if A's winning probability $p$ per game is:

(a) $p = 0.5$,
(b) $p = 0.4$,
(c) $p = 0.8$?

What is the probability of B's ruin in each case?

**Exercise 4.13** Prove the statement of Theorem 4.12 in the case where $p \neq \frac{1}{2}$.
**Hint:** Start with (4.25) and (4.26).

**Exercise 4.14** In the context of the gambler's ruin problem, let $a + b = 100$. For which value of $a$ is the expected duration of the game maximized or minimized in the following cases:

(a) $p = 0.3$,
(b) $p = 0.8$?

**Exercise 4.15** In the gambler's ruin problem, suppose player B has infinite capital. What is the probability of player A will go bankrupt after exactly $a + 2k$ games, where $k \in \mathbb{N}_0$, if his starting capital is $a$ euros and he wins a single game with probability $p$?

**Exercise 4.16** Let $r$ be any integer greater than or equal to 2. A symmetric random walk on $\mathbb{Z}$ starts at $k$, where $1 \leq k \leq r - 1$. The walk ends when it reaches the height $r$, which is an absorbing boundary. If it reaches height 0, it is reflected there, meaning the next step is automatically an upward step. The random variable $Y$ denotes the number of visits of the random walk to 0 before absorption. Show that:

(a) $\mathbb{P}(Y = 0) = \dfrac{k}{r}$,

(b) $\mathbb{P}(Y = \ell) = \dfrac{1}{r}\left(1 - \dfrac{k}{r}\right)\left(1 - \dfrac{1}{r}\right)^{\ell-1}$, $\quad \ell \geq 1$.

**Hint:** Consider the gambler's ruin problem.

**Exercise 4.17** Let $X_1, X_2, \ldots$ be independent random variables with $\mathbb{P}(X_j = 1) = p$ and $\mathbb{P}(X_j = -1) = 1 - p =: q$ for each $j \geq 1$, where $0 < p < 1$. As in Sect. 4.4, a *run* is defined as a sequence of maximum length consisting of consecutive symbols 1 (for *hits*) or $-1$ (for *misses*). For example, the initial sequence of length 20 depicted as a random walk in Fig. 4.7 starts with a first run of length 2, followed by a second run of length 4. Generally, $R_1$ and $R_2$ denote the lengths of the first and second runs, respectively. Show that:

(a) $\mathbb{P}(R_1 = k) = p^k q + q^k p$, $\quad k \geq 1$,
(b) $\mathbb{E}(R_1) = \dfrac{1}{pq} - 2$,
(c) $\mathbb{P}(R_2 = k) = p^2 q^{k-1} + q^2 p^{k-1}$, $\quad k \geq 1$,
(d) $\mathbb{E}(R_2) = 2$.

**Exercise 4.18** Let $M_1$ be the set of all $r$-tuples $(i_1, \ldots, i_r)$ with $2 \leq i_1 < \ldots < i_r \leq n - k + 1$ and $i_{s+1} - i_s \geq k + 1$ for each $s = 1, \ldots, r - 1$. Furthermore, let $M_2$

## 4.5 The Galton–Watson Process

be the set of all tuples $(j_1, \ldots, j_r)$ with $2 \leq j_1 < \ldots < j_r \leq n - kr + 1$. Construct a bijective mapping $f : M_1 \to M_2$.

**Exercise 4.19** In the situation of Corollary 4.17, prove that

$$\mathbb{V}(S_N) = \mathbb{V}(N) \, (\mathbb{E} \, X_1)^2 + \mathbb{E}(N) \, \mathbb{V}(X_1).$$

**Hint:** Use Eq. (4.34).

**Exercise 4.20** Use Theorem 4.18 to prove the variance formula

$$\mathbb{V}(M_n) = \begin{cases} \dfrac{\sigma^2 \mu^{n-1}(\mu^n - 1)}{\mu - 1}, & \text{if } \mu \neq 1, \\ n\sigma^2, & \text{if } \mu = 1, \end{cases}$$

by mathematical induction on $n$.

**Exercise 4.21** Show that the generating function of the offspring distribution in Example 4.22 is given by

$$g(s) = \frac{1}{s}\left(1 + \left(\frac{1}{s} - 1\right)\log(1-s)\right), \qquad |s| < 1, \, s \neq 0,$$

and $g(0) = \frac{1}{2}$.

**Exercise 4.22** Let $(M_n)_{n \geq 0}$ with $M_0 := 1$ be a supercritical GW process, and suppose the offspring distribution has a finite variance $\sigma^2$. Show: If $\mu$ denotes the expectation of the offspring distribution, then the extinction probability $w$ satisfies

$$w \leq 1 - \frac{\mu - 1}{\sigma^2 + \mu^2 - \mu}.$$

**Hint:** The first and second derivative of the generating function $g$ of the offspring distribution are monotonically increasing functions, and $g''(1) = \sigma^2 + \mu^2 - \mu$.

**Exercise 4.23** Show that, in the case of the geometric offspring distribution from Example 4.20, the inequality from Exercise 4.22 is equivalent to

$$\frac{3}{2} \leq \mu + \frac{1}{2\mu}.$$

**Exercise 4.24** Suppose the offspring distribution of a Galton–Watson process follows the Poisson distribution $\text{Po}(\lambda)$ with the generating function $g(t) = \exp(\mu(t-1))$, $t \in \mathbb{R}$, where $\mu > 0$. Show that the extinction probability $w$ is less than $\frac{1}{\mu}$.

**Hint:** Use the inequality $e^t > 1 + t$, which holds for each $t \neq 0$.

# Random Walks on the Integer Lattice $\mathbb{Z}^d$ 5

In this chapter, we consider a random walk on the integer lattice $\mathbb{Z}^d$, defined as $\{k := (k_1, \ldots, k_d) : k_j \in \mathbb{Z} \text{ for } j = 1, \ldots, d\}$. This walk starts at the origin $\mathbf{0} := (0, \ldots, 0)$ in $\mathbb{R}^d$ and moves memorylessly, such that at each step, it moves from point k to one of the $2d$ possible neighboring points of k with probability $\frac{1}{2d}$. Here, $j := (j_1, \ldots, j_d) \in \mathbb{Z}^d$ is called a *neighboring point of* k if $\sum_{m=1}^{d} |k_m - j_m| = 1$. A particle performing such a random walk thus chooses one of the coordinate axes with equal probability $\frac{1}{d}$ and then moves randomly by one unit along the chosen axis, either forward or backward. We initially assume that this random walk is symmetric. The consequences of deviating from this assumption of a uniform distribution will be discussed later. In what follows, we will use the words *site* or *position* synonymously with *point*.

Figure 5.1 shows the first 21 steps of such a random walk in the case where $d = 2$. The path of this random walk can be uniquely reconstructed from the image, given the information that exactly 21 steps were taken. For example, the 5-th step goes from $(-1, 1)$ to $(-1, 2)$, and the 14-th step goes from $(-5, 4)$ to $(-5, 3)$. In the case $d = 3$, one can imagine a random walk as a journey on a cubic climbing frame.

We model the symmetric random walk on $\mathbb{Z}^d$ by a sequence of independent, identically distributed random vectors $X_1, X_2, \ldots$, each having a uniform distribution on the set $E_d := \{e_1, -e_1, \ldots, e_d, -e_d\}$. Here, for each $j \in \{1, \ldots, d\}$, $e_j := (0, \ldots, 0, 1, 0, \ldots, 0)$ denotes the $j$-th canonical unit vector of $\mathbb{R}^d$, with a one in the $j$-th component, and $-e_j$ its reflection at the origin. Setting $S_0 := \mathbf{0}$ and $S_n := X_1 + \ldots + X_n$ for $n \geq 1$, the random vector $S_n$ models the position of the random walk at time $n$. The expectation $\mathbb{E}(X_1)$ of $X_1$ is, by definition, the vector of the expectations of the $d$ components of $X_1$, and thus the zero vector $\mathbf{0}$ in $\mathbb{R}^d$.

**Fig. 5.1** The first 21 steps of a random walk on $\mathbb{Z}^2$

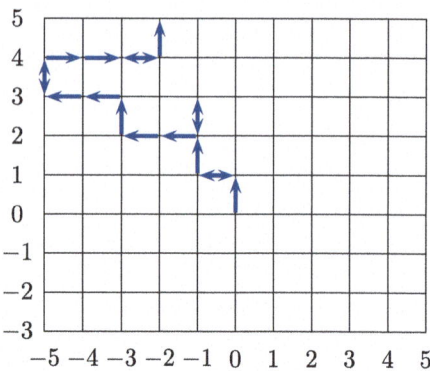

**SAQ 1** Why is $\mathbb{E}(X_1) = \mathbf{0}$?

Thus, it follows that $\mathbb{E}(S_n) = \mathbf{0}$ for every $n \geq 1$. For the square $\|S_n\|^2$ of the Euclidean norm of $S_n$, $\mathbb{E}\|S_n\|^2 = n$ (Exercise 5.3).

As defined in Sect. 2.5, let

$$T := \inf\{2k : k \in \mathbb{N} \text{ and } S_{2k} = \mathbf{0}\} \tag{5.1}$$

be the first return time to the starting point $\mathbf{0}$. The question then arises whether the symmetric random walk on $\mathbb{Z}^d$ is recurrent for general $d$, that is, whether it returns to the starting point with probability one in finite time, as is the case for $d = 1$. The next section is dedicated to this problem. Section 5.2 investigates how many distinct lattice points a random walk visits after $n$ steps. Section 5.3 addresses a far-reaching generalization of the gambler's ruin problem.

## 5.1 Recurrence and Transience

The following central theorem, attributed to George Pólya[1] [PO], states that the recurrence property of the symmetric random walk on $\mathbb{Z}$ is preserved in the case $d = 2$ (the planar walk), but is lost for $d \geq 3$.

---

[1] George Pólya (1887–1985), taught at ETH Zurich from 1914 to 1940 (appointed professor in 1928), held a visiting professorship at Brown University from 1940 to 1942, and was a professor at Stanford University from 1942 to 1953. Pólya was primarily an analyst, he made significant contributions to analysis, probability theory, mathematical physics, combinatorics, and number theory.

## 5.1 Recurrence and Transience

**Theorem 5.1 (Recurrence/Transience of the Symmetric Random Walk on $\mathbb{Z}^d$)**
*For the first return time of a symmetric random walk on $\mathbb{Z}^d$ as defined in (5.1), the following holds:*

(a) $\mathbb{P}(T < \infty) = 1$, if $d \leq 2$.
(b) $\mathbb{P}(T < \infty) < 1$, if $d \geq 3$.

*Proof*

(a) Let $N := \sum_{n=0}^{\infty} \mathbf{1}\{S_{2n} = \mathbf{0}\}$ be the number of visits of the random walk to the origin (including the initial visit), and let

$$L := \sup\{2n : n \geq 0 \text{ and } S_{2n} = \mathbf{0}\}, \qquad 0 \leq L \leq \infty,$$

be the random time of the *last* visit of the walk to $\mathbf{0}$. According to (7.14),

$$\mathbb{E}(N) = \sum_{n=0}^{\infty} \mathbb{P}(S_{2n} = \mathbf{0}). \qquad (5.2)$$

Due to the memoryless property of the walk, for each $n \geq 0$, we have

$$\mathbb{P}(L = 2n) = \mathbb{P}(S_{2n} = \mathbf{0}) \, \mathbb{P}(S_{2n+2j} \neq \mathbf{0} \text{ for every } j \geq 1 | S_{2n} = \mathbf{0})$$
$$= \mathbb{P}(S_{2n} = \mathbf{0}) \, \mathbb{P}(S_{2j} \neq \mathbf{0} \text{ for every } j \geq 1)$$
$$= \mathbb{P}(S_{2n} = \mathbf{0}) \, \mathbb{P}(T = \infty).$$

Summing over $n \geq 0$, using the fact that the event $\{L < \infty\}$ is equivalent to $\{N < \infty\}$, together with (5.2), we get

$$\mathbb{P}(N < \infty) = \mathbb{E}(N) \, \mathbb{P}(T = \infty). \qquad (5.3)$$

We now distinguish between the two cases where $\mathbb{E}(N) = \infty$ and $0 < \mathbb{E}(N) < \infty$ (note that $\mathbb{E}(N)$ is strictly positive because $\mathbb{P}(S_2 = \mathbf{0}) = \frac{1}{2d}$).

**SAQ 2**  Why is $\mathbb{P}(S_2 = \mathbf{0}) = \frac{1}{2d}$?

In the first case, from Eq. (5.3), it necessarily follows that $\mathbb{P}(T = \infty) = 0$, indicating recurrence. In the second case, $\mathbb{P}(N < \infty) = 1$ and $\mathbb{P}(T = \infty) =$

$\frac{1}{\mathbb{E}(N)} > 0$, indicating transience. Using Eq. (5.2) and the fact that $\mathbb{P}(S_0 = \mathbf{0}) = 1$, the return probability $\mathbb{P}(T < \infty)$ is given by

$$\mathbb{P}(T < \infty) = 1 - \mathbb{P}(T = \infty) = 1 - \frac{1}{\mathbb{E}(N)} = \frac{\mathbb{E}(N) - 1}{\mathbb{E}(N)}$$

$$= \frac{\sum_{n=1}^{\infty} \mathbb{P}(S_{2n} = \mathbf{0})}{1 + \sum_{n=1}^{\infty} \mathbb{P}(S_{2n} = \mathbf{0})}, \tag{5.4}$$

The question of whether the symmetric random walk on $\mathbb{Z}^d$ is recurrent or transient is thus determined by the behavior of the infinite series $\sum_{n=1}^{\infty} \mathbb{P}(S_{2n} = \mathbf{0})$. If this series diverges, the walk is recurrent; otherwise, it is transient. Note that with the convention $\frac{\infty}{1+\infty} := 1$, the above equation remains valid in the case where $\sum_{n=1}^{\infty} \mathbb{P}(S_{2n} = \mathbf{0}) = \infty$.

In the case $d = 1$, using (2.29) and the notation introduced in (1.1), we have

$$\mathbb{P}(S_{2n} = 0) = \frac{\binom{2n}{n}}{2^{2n}} \sim \frac{1}{\sqrt{\pi n}} \quad \text{as } n \to \infty. \tag{5.5}$$

Since $\sum_{n=1}^{\infty} n^{-1/2} = \infty$, we have confirmed the recurrence of the symmetric random walk on $\mathbb{Z}$.

In the case $d = 2$, the event $\{S_{2n} = \mathbf{0}\}$ occurs if and only if the walk moves an equal number of times to the right and left, and an equal number of times up and down, in the first $2n$ steps. Letting $k$ be the number of steps to the right, it follows that

$$\mathbb{P}(S_{2n} = \mathbf{0}) = \frac{1}{4^{2n}} \sum_{k=0}^{n} \binom{2n}{k}\binom{2n-k}{k}\binom{2n-2k}{n-k} \tag{5.6}$$

$$= \frac{1}{4^{2n}} \binom{2n}{n} \sum_{k=0}^{n} \binom{n}{k}\binom{n}{n-k} = \frac{1}{4^{2n}} \binom{2n}{n}^2 \tag{5.7}$$

$$= \left[\frac{1}{2^{2n}}\binom{2n}{n}\right]^2 \sim \frac{1}{\pi n} \quad \text{as } n \to \infty. \tag{5.8}$$

Here, in the penultimate equality, we used the normalization condition for the hypergeometric distribution. Since the harmonic series diverges, the symmetric random walk in the plane is recurrent.

**SAQ 3** Why does the equality in (5.6) hold?

## 5.1 Recurrence and Transience

(b) It suffices to consider the case $d = 3$ (Exercise 5.9). In this case, you return to the origin after exactly $2n$ steps if, for each $j \in \{1, 2, 3\}$, $k_j$ steps are taken in the direction of the $j$-th unit vector $e_j$, and $k_j$ steps in the direction $-e_j$. Here, $k_1, k_2, k_3$ are non-negative integers that sum to $n$. Denoting $C_n := \{k := (k_1, k_2, k_3) \in \mathbb{N}_0^3 : k_1 + k_2 + k_3 = n\}$, it follows that

$$\mathbb{P}(S_{2n} = 0) = \left(\frac{1}{6}\right)^{2n} \sum_{k \in C_n} \frac{(2n)!}{k_1!^2 k_2!^2 k_3!^2} \tag{5.9}$$

$$= \frac{1}{2^{2n}} \binom{2n}{n} \sum_{k \in C_n} \left[\frac{1}{3^n} \cdot \frac{n!}{k_1! k_2! k_3!}\right]^2.$$

**SAQ 4** Why does the equality in (5.9) hold?

Inside the square brackets are the probabilities of a multinomial distribution, whose sum over all $k \in C_n$ equals one. Thus, we obtain the estimate

$$\mathbb{P}(S_{2n} = 0) \leq \frac{1}{2^{2n}} \binom{2n}{n} \max_{k \in C_n} \left(\frac{1}{3^n} \cdot \frac{n!}{k_1! k_2! k_3!}\right).$$

Choosing $n = 3m$ for some $m \in \mathbb{N}$, the maximum is attained (according to Exercise 5.8 b) with $d = 3$, $n = 3m$, and $p_i = \frac{1}{3}$) when the inequalities $m - 1 < k_j \leq m + \frac{2}{3}$ (for $j = 1, 2, 3$) are satisfied, and thus $k_1 = k_2 = k_3 = m$. It follows that

$$\mathbb{P}(S_{6m} = 0) \leq \frac{1}{2^{6m}} \binom{6m}{3m} \frac{1}{3^{3m}} \cdot \frac{(3m)!}{m!^3}, \qquad m \geq 1. \tag{5.10}$$

Using (7.16) and Stirling's formula (7.17), the right-hand side of (5.10) is asymptotically, as $m \to \infty$,

$$\frac{1}{\sqrt{\pi 3m}} \cdot \frac{e^{-3m}(3m)^{3m}\sqrt{2\pi 3m}}{3^{3m}e^{-3m}m^{3m}(2\pi m)^{3/2}} = \frac{1}{2\pi^{3/2}m^{3/2}}.$$

Since $\sum_{m=1}^{\infty} m^{-3/2} < \infty$, it follows that $\sum_{m=1}^{\infty} \mathbb{P}(S_{6m} = 0) < \infty$. Furthermore,

$$\mathbb{P}(S_{6m-2} = 0) \cdot \left(\frac{1}{6}\right)^2 \leq \mathbb{P}(S_{6m} = 0), \tag{5.11}$$

and

$$\mathbb{P}(S_{6m-4} = \mathbf{0}) \cdot \left(\frac{1}{6}\right)^4 \leq \mathbb{P}(S_{6m} = \mathbf{0}). \tag{5.12}$$

Thus, we obtain $\sum_{n=1}^{\infty} \mathbb{P}(S_{2n} = \mathbf{0}) < \infty$, which implies that the symmetric random walk is transient in the three-dimensional case. ∎

**SAQ 5** Why do the inequalities (5.11) and (5.12) hold?

**Determination of $\mathbb{P}(T < \infty)$ in the Case $d \geq 3$**

After establishing that the symmetric random walk on the integer lattice $\mathbb{Z}^d$ is transient for $d \geq 3$, the question arises whether the return probability $\mathbb{P}(T < \infty)$ to the origin of such a walk can be explicitly determined. According to (5.4), we need to evaluate the sum $\sum_{n=1}^{\infty} \mathbb{P}(S_{2n} = \mathbf{0})$ and, in particular, find a manageable expression for the probability $\mathbb{P}(S_{2n} = \mathbf{0})$ as given in (5.9). For brevity, we write

$$P_d(n) := \mathbb{P}(S_{2n} = \mathbf{0})$$

to make the dependence on $d$ clear through the indexing. The recursion formula (Griffin [GR]) holds:

$$P_d(n) = \sum_{j=0}^{n} \binom{2n}{2j} \frac{P_1(j) P_{d-1}(n-j)(d-1)^{2n-2j}}{d^{2n}} \tag{5.13}$$

(for $d \geq 2$, $j \in \{0, 1, \ldots, n\}$). Here, $P_d(0) := 1$. Equation (5.13) follows from the observation that if a walk returns to the origin after $2n$ steps, it must return to 0 in each coordinate individually. The event $\{S_{2n} = \mathbf{0}\}$ is broken down by the number $2j$, where $j \in \{0, \ldots, n\}$, of steps taken along the *first* coordinate axis. The factor $\binom{2n}{2j}$ gives the number of ways to choose exactly $2j$ out of the $2n$ time points $0, 1, \ldots, 2n - 1$ at which the walk will move along the first coordinate axis. Given these time points, the probability that each of these $2j$ steps occur *in either direction* along the first coordinate axis is $(1/d)^{2j}$. Additionally,

$$P_1(j) = \frac{\binom{2j}{j}}{2^{2j}} \tag{5.14}$$

is the probability that the walk returns to 0 after $2j$ steps. For each of the remaining $2n - 2j$ steps, the walk must not move along the first coordinate axis. The probability of this is $((d-1)/d)^{2n-2j}$. Under this condition, the probability of returning to the origin after $2n - 2j$ steps is $P_{d-1}(n-j)$, which leads to Eq. (5.13).

**Table 5.1** Return probabilities of the symmetric random walk on $\mathbb{Z}^d$

| $d$ | 3 | 4 | 5 | 6 | 7 | 8 |
|---|---|---|---|---|---|---|
| $\mathbb{P}(T < \infty)$ | 0.340537 | 0.193202 | 0.135179 | 0.104715 | 0.085845 | 0.072913 |

According to (5.7), $P_2(n) = P_1(n)^2$, so the recursion in (5.13) with the initial condition (5.14) begins with $d = 3$. Griffin [GR] used (5.13) together with an acceleration of convergence method to compute the values of $\mathbb{P}(T < \infty)$ for dimensions $3 \leq d \leq 8$, as shown in Table 5.1.

We conclude this section with a remark on *asymmetric* random walks on $\mathbb{Z}^d$ for $d \geq 2$. An asymmetric random walk is characterized by a non-uniform distribution over the $2d$ neighboring points of a given point. Such a random walk is always transient, unless it occurs on at most two of the coordinate axes and is symmetric on those axes. This result follows directly from the transience of the one-dimensional random walk where $p \neq \frac{1}{2}$. Without loss of generality, consider the steps made in the direction of the first coordinate axis. If the probability that the first coordinate of the wandering particle never returns to 0 is positive, then the probability that the particle does not return to the origin is also positive.

## 5.2 The Number of Visited Sites

How "curious" is a random walk in discovering new, previoulsy unvisited sites? More precisely, in this section, we ask how the number

$$R_n := |\{S_0, S_1, \ldots, S_n\}|$$

of *different* sites visited up to time $n$ by a random walk on $\mathbb{Z}^d$ behaves as $n$ increases. We also consider the case $d = 1$ and the possibility that the independent and identically distributed random vectors $X_1, X_2, \ldots$ do not necessarily follow a uniform distribution over the set $E_d = \{e_1, -e_1, \ldots, e_d, -e_d\}$. In the literature, the somewhat misleading term *range* is commonly used for $R_n$, although this designation would be more appropriate for the *set* $\{S_0, S_1, \ldots, S_n\}$.

An example is the random walk shown in Fig. 5.1. After 21 steps, it has visited exactly 15 different sites. Very little is known about the exact distribution of $R_n$, even in the simplest case of the symmetric one-dimensional random walk. As confirmed by counting paths, the relations $\mathbb{P}(R_3 = 3) = \mathbb{P}(R_3 = 4) = \frac{1}{4}$ and $\mathbb{P}(R_3 = 2) = \frac{1}{2}$ hold in this case. For larger values of $n$, the distribution of $R_n$ can be determined using a computer. Naturally, the general conditions $R_0 = 1$ and $2 \leq R_n \leq n+1$ hold for any $n \geq 1$.

In the following, we first examine the expectation of $R_n$. This value will also grow as $n$ increases, but how quickly? In the case of the *one-dimensional* symmetric random walk, we can immediately provide a closed expression for $\mathbb{E}(R_n)$, as $R_n = 1 + M_n - m_n$, where $M_n$ denotes the maximum and $m_n$ the minimum of the partial sums $S_0, S_1, \ldots, S_n$ (see Sect. 2.7). Since $-m_n$ and $M_n$ have the same distribution and thus also the same expectation due to symmetry, it follows that $\mathbb{E}(R_n) = 1 + 2\mathbb{E}(M_n)$. Using Theorem 2.12 (b), we thus obtain the following result:

**Theorem 5.2 (Expectation of $R_n$ for the Simple Symmetric Random Walk on $\mathbb{Z}$)** *In the case of the simple symmetric random walk on $\mathbb{Z}$,*

$$\mathbb{E}(R_{2n}) = (4n+1)\frac{\binom{2n}{n}}{2^{2n}}, \qquad \mathbb{E}(R_{2n+1}) = (4n+2)\frac{\binom{2n}{n}}{2^{2n}}.$$

Due to Eq. (5.5), the asymptotic equality follows from the result above:

$$\mathbb{E}(R_n) \sim \frac{4}{\sqrt{\pi}} \cdot \sqrt{n} \tag{5.15}$$

as $n \to \infty$. Therefore, the expectation of $R_n$ in the case of the simple symmetric random walk grows approximately proportional to the square root of $n$. We will see that $\mathbb{E}(R_n)$ for the *asymmetric* random walk on $\mathbb{Z}$ grows much faster, namely approximately proportional to $n$. If there is a trend in a specific direction, with $p = \mathbb{P}(X_1 = 1) \neq \frac{1}{2}$, it becomes, intuitively speaking, easier to visit new sites. In the extreme case where $p = 1$, $\mathbb{P}(R_n = n + 1) = 1$, and thus $\mathbb{E}(R_n) = n + 1$.

The representation of $R_n$ as $1 + M_n - m_n$ is naturally restricted to the case $d = 1$. For random walks in higher dimensions, it is initially unclear how to obtain information about $\mathbb{E}(R_n)$. Here, the insight that $R_n$ can be represented as a sum of indicator variables proves helpful. With the definition $I_0 := 1$ and

$$I_k := \mathbf{1}\{S_k \neq S_j \text{ for } j = 0, 1, \ldots, k-1\}$$

for $k = 1, 2, \ldots, n$, we have

$$R_n = \sum_{k=0}^{n} I_k. \tag{5.16}$$

The random variable $I_k$ indicates whether a previously unvisited site is reached at time $k$. Since taking expectations is an additive operation, $\mathbb{E}(R_n) = \sum_{k=0}^{n} \mathbb{E}(I_k)$. For symmetry reasons, and with the return time $T$ to the origin defined in Eq. (5.1), it follows that for every $k \geq 1$

$$\begin{aligned}
\mathbb{E}(I_k) &= \mathbb{P}(S_k \neq S_{k-1}, S_k \neq S_{k-2}, \ldots, S_k \neq S_1, S_k \neq S_0(=0)) \\
&= \mathbb{P}(X_k \neq 0, X_k + X_{k-1} \neq 0, \ldots, X_k + \ldots + X_1 \neq 0) \\
&= \mathbb{P}(X_1 \neq 0, X_1 + X_2 \neq 0, \ldots, X_1 + \ldots + X_k \neq 0) \\
&= \mathbb{P}(T \geq k+1).
\end{aligned}$$

## 5.2 The Number of Visited Sites

Since $\mathbb{P}(T \geq 1) = 1$ and $I_0 = 1$, this equation also holds for $k = 0$, and thus we get

$$\mathbb{E}(R_n) = \sum_{k=0}^{n} \mathbb{P}(T \geq k+1).$$

The following result shows that $R_n$ grows at different rates, depending on whether $\mathbb{P}(T = \infty) = 0$ or $\mathbb{P}(T = \infty) > 0$, that is, whether recurrence or transience is present (see, e.g., [SP], p. 35 ff.).

**Theorem 5.3 (Stochastic Convergence of $R_n/n$)** *For the number $R_n$ of visited states, we have*

$$\lim_{n \to \infty} \mathbb{P}\left( \left| \frac{R_n}{n} - \mathbb{P}(T = \infty) \right| \geq \varepsilon \right) = 0 \quad \text{for every } \varepsilon > 0.$$

*Thus, the sequence $(R_n/n)$ converges in probability to $\mathbb{P}(T = \infty)$.*

**Proof** We set $p_\infty := \mathbb{P}(T = \infty)$ and $\overline{R}_n := R_n/n$. First, we show the convergence

$$\lim_{n \to \infty} \mathbb{E}\left( \overline{R}_n \right) = p_\infty. \tag{5.17}$$

Then we prove

$$\lim_{n \to \infty} \mathbb{V}\left( \overline{R}_n \right) = 0. \tag{5.18}$$

To show (5.17), note that $\mathbb{E}(I_k) = \mathbb{P}(T \geq k+1)$ converges to $p_\infty$ as $k \to \infty$. Therefore, $\mathbb{E}(\overline{R}_n)$ is (except for the irrelevant factor $n/(n+1)$) the arithmetic mean of the first terms of a convergent sequence. Hence, by the Cauchy limit theorem in analysis (see, e.g., [KN], Sec. 2.4), it converges to the same limit, proving (5.17).

To prove (5.18), we first handle the recurrent case $p_\infty = 0$. In this case, we use the inequality $0 \leq 1\{\overline{R}_n \geq \varepsilon\} \leq \overline{R}_n/\varepsilon$ for $\varepsilon > 0$ and the monotonicity of taking expectations to obtain $\mathbb{P}(\overline{R}_n \geq \varepsilon) \leq \mathbb{E}(\overline{R}_n)/\varepsilon$, and the right-hand side converges to 0 according to (5.17). In the remaining case $p_\infty > 0$, starting from the representation (5.16), we use the additivity of expectations and the inequality $\mathbb{E}(I_k)(1 - \mathbb{E}(I_k)) \leq \mathbb{E}(I_k)$ to get

$$\mathbb{V}(R_n) = \mathbb{E}\left( \sum_{k=0}^{n} I_k \right)^2 - \left( \mathbb{E}\left( \sum_{k=0}^{n} I_k \right) \right)^2$$

$$= \mathbb{E}\left( \sum_{j=0}^{n} \sum_{k=0}^{n} I_j I_k \right) - \left( \sum_{k=0}^{n} \mathbb{E}(I_k) \right)^2$$

$$= \sum_{j=0}^{n} \sum_{k=0}^{n} \Big(\mathbb{E}(I_j I_k) - \mathbb{E}(I_j)\mathbb{E}(I_k)\Big)$$

$$\leq 2 \sum_{0 \leq j < k \leq n} \Big(\mathbb{E}(I_j I_k) - \mathbb{E}(I_j)\mathbb{E}(I_k)\Big) + \sum_{k=0}^{n} \mathbb{E}(I_k).$$

To find a suitable upper bound for $\mathbb{V}(R_n)$, we note that for $j < k$, the inequality

$$\mathbb{E}(I_j I_k) \leq \mathbb{E}(I_j)\mathbb{E}(I_{k-j}) \qquad (5.19)$$

is satisfied. For a proof, we define the following events:

$$A := \{S_j \neq S_i \text{ for every } i \in \{0, \ldots, j-1\}\},$$
$$B := \{S_k \neq S_\ell \text{ for every } \ell \in \{0, \ldots, k-1\}\},$$
$$C := \{S_k \neq S_\ell \text{ for every } \ell \in \{j, \ldots, k-1\}\}.$$

Then $B \subseteq C$, and the events $A$ and $C$ are independent because $A$ depends only on $X_1, \ldots, X_j$ and $C$ only on $X_{j+1}, \ldots, X_k$. Due to symmetry, $\mathbb{P}(C) = \mathbb{E}(I_{k-j})$, and hence inequality (5.19) follows from

$$\mathbb{E}(I_j I_k) = \mathbb{P}(A \cap B) \leq \mathbb{P}(A \cap C) = \mathbb{P}(A)\mathbb{P}(C) = \mathbb{E}(I_j)\mathbb{E}(I_{k-j}).$$

**SAQ 6** What are the symmetry reasons for $\mathbb{P}(C) = \mathbb{E}(I_{k-j})$?

Inserting (5.19) into the obtained upper bound for $\mathbb{V}(R_n)$, we get

$$\mathbb{V}(R_n) \leq 2 \sum_{j=0}^{n-1} \mathbb{E}(I_j) \sum_{k=j+1}^{n} \Big(\mathbb{E}(I_{k-j}) - \mathbb{E}(I_k)\Big) + \mathbb{E}(R_n).$$

Since $\mathbb{E}(I_1) \geq \mathbb{E}(I_2) \geq \ldots \geq \mathbb{E}(I_n)$, the sum $\sum_{k=j+1}^{n} \big(\mathbb{E}(I_{k-j}) - \mathbb{E}(I_k)\big)$ is maximal for $j = \lceil \frac{n}{2} \rceil = \min\{\ell \in \mathbb{N} : \ell \geq \frac{n}{2}\}$, which is the smallest integer $\ell$ such that $\ell \geq \frac{n}{2}$. Hence, we obtain

$$\mathbb{V}(R_n) \leq 2 \sum_{j=0}^{n-1} \mathbb{E}(I_j) \left\{\mathbb{E}(R_{n-\lceil n/2 \rceil}) + \mathbb{E}(R_{\lceil n/2 \rceil}) - \mathbb{E}(R_n)\right\} + \mathbb{E}(R_n).$$

## 5.2 The Number of Visited Sites

By omitting the parentheses when taking expectations, we thus get

$$\mathbb{V}(\overline{R}_n) = \frac{\mathbb{V}(R_n)}{n^2}$$

$$\leq 2\frac{\mathbb{E}R_n}{n}\left(\frac{\mathbb{E}R_{n-\lceil n/2 \rceil}}{n-\lceil n/2 \rceil} \cdot \frac{n-\lceil n/2 \rceil}{n} + \frac{\mathbb{E}R_{\lceil n/2 \rceil}}{\lceil n/2 \rceil} \cdot \frac{\lceil n/2 \rceil}{n} - \frac{\mathbb{E}R_n}{n}\right) + \frac{\mathbb{E}R_n}{n^2}.$$

Using (5.17), we now obtain (5.18) as a result of

$$\limsup_{n \to \infty} \mathbb{V}(\overline{R}_n) \leq 2p_\infty\left(\frac{p_\infty}{2} + \frac{p_\infty}{2} - p_\infty\right) + 0.$$

∎

The convergence (5.17) implies the asymptotic equality

$$\mathbb{E}(R_n) \sim n\,\mathbb{P}(T = \infty) \quad \text{as } n \to \infty.$$

In the case of a transient random walk, the expectation of $R_n$ thus grows approximately proportional to $n$ for large $n$. For the symmetric walk on $\mathbb{Z}^d$ with $d \geq 3$, $\mathbb{E}(R_n) \sim (1 - \mathbb{P}(T < \infty))n$, where $\mathbb{P}(T < \infty)$ for $d \in \{3, 4, \ldots, 8\}$ is given in Table 5.1. In the case of the asymmetric random walk on $\mathbb{Z}$, Theorem 4.5 yields $\mathbb{E}(R_n) \sim |2p - 1|n$, in stark contrast to the $\sqrt{n}$-growth (5.15) when $p = \frac{1}{2}$.

Theorem 5.3 tells us that, for the symmetric random walk on the planar lattice $\mathbb{Z}^2$, the ratio $\mathbb{E}(R_n)/n$ converges to zero. This means that the expectation of $R_n$ grows more slowly than $n$, since $\mathbb{P}(T = \infty) = 0$. Here, Dvoretzky and Erdös [DE] showed the asymptotic equality

$$\mathbb{E}(R_n) \sim \frac{\pi n}{\log n} \quad \text{as } n \to \infty. \tag{5.20}$$

Downham and Fotopoulos [DF] derived the inequalities

$$\frac{2\pi n}{1.16\pi - 1 + \log n} < \mathbb{E}(R_{2n-1}) < \frac{2\pi n}{1.066\pi - 1 + \log n}, \quad n \geq 7,$$

from which, due to $R_{2n-1} \leq R_{2n} \leq R_{2n-1} + 1$, (5.20) follows.

Since $R_n$ is a sum of random variables according to (5.16), it is natural to look for a central limit theorem for $R_n$ that would strengthen the law of large numbers given in Theorem 5.3. The technical difficulty here is that the summands in (5.16) are not independent. However, the stochastic dependence is weak enough in the case $d \geq 3$ that a central limit theorem for $R_n$ indeed holds. Jain and Orey [JO] were able to show, among other things, the convergence in distribution

$$\frac{R_n - \mathbb{E}R_n}{\sigma\sqrt{n}} \xrightarrow{\mathcal{D}} \mathrm{N}(0, 1) \quad \text{as } n \to \infty. \tag{5.21}$$

for the walk on $\mathbb{Z}^d$ (which is also allowed to be asymmetric) in the case $d \geq 5$. Here, $\sigma^2 > 0$ is a positive constant that depends, among other things, on the assumed positive return probability $\mathbb{P}(T < \infty)$ to the origin. This result also holds in the case $d = 4$ [JP1]. In the case $d = 3$, (5.21) remains valid if we include the factor $\sqrt{\log n}$ in the denominator of the fraction (see [JP1]). The asymptotic behavior of $R_n$ for the symmetric random walk on $\mathbb{Z}^2$ was long an open problem. In 1985, Le Gall [LG] was able to show that $(\log n)^2(R_n - \mathbb{E}(R_n))/n$ has a limiting distribution as $n \to \infty$, which is not a normal distribution. In the case of the one-dimensional symmetric random walk, the following holds:

$$\frac{R_n}{\sqrt{n}} = \frac{M_n - m_n + 1}{\sqrt{n}} \xrightarrow{\mathcal{D}} \max_{0 \leq t \leq 1} W(t) - \min_{0 \leq t \leq 1} W(t) \quad \text{as } n \to \infty$$

(see [JP2]). Here, $W(t)$ denotes the Brown–Wiener process introduced in Sect. 2.15.

## 5.3 The Discrete Dirichlet Problem

In this section, we will learn about the so-called *discrete Dirichlet[2] problem*, which is a far-reaching generalization of the gambler's ruin problem from Sect. 4.3. This problem also has interesting connections to electrical networks (see Sect. 6.4).

Let $M$ denote any *non-empty, finite* subset of the integer lattice $\mathbb{Z}^d$. For $x =: (\xi_1, \ldots, \xi_d) \in \mathbb{Z}^d$, we set

$$|x| := \max(|\xi_1|, \ldots, |\xi_d|).$$

With this notation, the boundary of $M$ is defined as $\partial M := \{y \in \mathbb{Z}^d \setminus M : \exists x \in M \text{ such that } |y - x| = 1\}$. The set $\overline{M} := M \uplus \partial M$ is called the *closed hull* or simply the *closure* of $M$, and $M$ is called the *inner core* or the *interior* of $\overline{M}$. For each $x \in \mathbb{Z}^d$,

$$\mathcal{N}(x) := \{y \in \mathbb{Z}^d : |y - x| = 1\}$$

is the set of *neighboring points* of $x$. A symmetric random walk on $\mathbb{Z}^d$, located at a point $x$, reaches one of the $2d$ neighboring points of $x$ with equal probability in the next step. Using the canonical unit vectors $e_1, \ldots, e_d$ of $\mathbb{R}^d$, it holds that $\mathcal{N}(x) = \{x + e_j : j = 1, \ldots, d\} \uplus \{x - e_j : j = 1, \ldots, d\}$.

A function $F : \overline{M} \to \mathbb{R}$ has the *mean value property* if, for every $x \in M$, the function value $F(x)$ is equal to the arithmetic mean of the function values of all

---

[2] Peter Gustav Lejeune Dirichlet (1805–1859), studied in Paris under, among others, Laplace, Legendre, Poisson, Fourier and Cauchy. From 1831 to 1855 he was a professor at the University of Breslau, and in 1855 he was appointed to the chair previously held by C.F. Gauss in Göttingen. His main areas of work were number theory, analysis, and mathematical physics.

## 5.3 The Discrete Dirichlet Problem

neighboring points of $x$. That is,

$$F(x) = \frac{1}{2d} \sum_{y \in \mathcal{N}(x)} F(y), \qquad x \in M. \tag{5.22}$$

We have already encountered this property in the proof of Theorem 4.10 in connection with the gambler's ruin problem. In that case, the probability of success per game for each of the two players is equal, i.e., $p = \frac{1}{2}$. There, $d = 1$, $M = \{1, \ldots, r-1\}$, $\overline{M} = \{0, 1, \ldots, r-1, r\}$, and $F(x)$ was the probability that a symmetric random walk starting at $x \in \overline{M}$ is absorbed at the point $r$. The Eq. (4.20) then state $F(x) = \frac{1}{2}(F(x+1) + F(x-1))$ for each $x \in M$, and thus $F$ possesses the mean value property.

If $D \subset \mathbb{R}^d$ is an open set and $g : D \to \mathbb{R}$ is a twice continuously differentiable function, then the *Laplace operator*, named after P.S. Laplace, is defined by

$$\Delta g(x) := \sum_{k=1}^{d} \frac{\partial^2 g(x)}{\partial x_k^2}, \qquad x = (x_1, \ldots, x_d) \in D.$$

The *Laplace equation* for $g$ states that $\Delta g(x) = 0$ for each $x \in D$. In this case, $g$ is called a *harmonic function*. The *Dirichlet problem for the Laplace equation* seeks to find a function $f : \overline{D} \to \mathbb{R}$ such that $f$ satisfies the Laplace equation on $D$ and extends a continuous function $f_1$ given on the sufficiently smooth boundary of a bounded open set $D \subset \mathbb{R}^d$ (see, e.g., [LAW], Section 2.3). In what follows, we discuss a discrete version of this problem.

For a function $F$ defined on the lattice $\mathbb{Z}^d$, differentiation is not possible, but in each of the $d$ coordinates, one can form the differences $(\delta_k F)(x) := F(x + e_k) - F(x)$ and $(\delta_k F)(x - e_k) = F(x) - F(x - e_k)$, for $k \in \{1, \ldots, d\}$. The second differences are defined as $(\delta_k^2 F)(x) := (\delta_k F)(x) - (\delta_k F)(x - e_k)$, which simplifies to

$$(\delta_k^2 F)(x) = F(x + e_k) + F(x - e_k) - 2F(x), \qquad k \in \{1, \ldots, d\}, \ x \in \mathbb{R}^d.$$

Summing over $k$ and dividing by $2d$, we obtain the discrete analog of the Laplace operator, also known as the *Laplace filter*:

$$\mathcal{L}F(x) := \frac{1}{2d} \sum_{y \in \mathcal{N}(x)} (F(y) - F(x)), \qquad x \in \mathbb{R}^d. \tag{5.23}$$

A function $F : \overline{M} \to \mathbb{R}$ is called *(discrete) harmonic* if $\mathcal{L}F(x) = 0$ for each $x \in M$. Comparing (5.23) with (5.22) shows that a function $F : \overline{M} \to \mathbb{R}$ is harmonic if and only if it satisfies the mean value property.

**SAQ 7** Why is this statement true?

We assume tacitly that the set $M \subset \mathbb{Z}^d$ is *connected*, meaning that for any two points $x$ and $y$ in $M$, there exists a finite sequence of points $x_1, \ldots, x_n$ from $M$ such that $|x - x_1| = 1$, $|x_n - y| = 1$ and $|x_j - x_{j+1}| = 1$ for each $j \in \{1, \ldots, n-1\}$.

**Theorem 5.4 (Harmonic Functions and Extreme Values)** *Let $M \subset \mathbb{Z}^d$ be a finite set, and let $F : \overline{M} \to \mathbb{R}$ be a harmonic function. Then both the maximum and the minimum of $F$ are attained on the boundary $\partial M$ of $M$.*

*Proof* Since the set $\overline{M}$ is finite, the maximum of $F$ is attained at some point in $\overline{M}$. Let $x_0 \in M$ be an *interior* point of $\overline{M}$, where $F(x_0) \geq F(y)$ for each $y \in \overline{M}$. Since $F$ satisfies the mean value property, we have

$$F(x_0) = \frac{1}{2d} \sum_{y \in \mathcal{N}(x)} F(y).$$

Because $F(y) \leq F(x_0)$ for every $y \in \overline{M}$, this equation can only be satisfied if $F(y) = F(x_0)$ for every $y \in \mathcal{N}(x_0)$. By repeatedly applying this argument to each neighboring point, it follows that $F(y) = F(x_0)$ for every $y \in \overline{M}$. The same reasoning applies to the minimum of $F$. Therefore, if the maximum or minimum of $F$ is attained in the interior of $\overline{M}$, then $F$ must be constant. ∎

**Corollary 5.5** *Let $M \subset \mathbb{Z}^d$ be a finite set, and let $F : \overline{M} \to \mathbb{R}$ be a harmonic function. If $F$ is constant on the boundary $\partial M$ of $M$, then $F$ is constant throughout $\overline{M}$.*

**SAQ 8** Can you prove this corollary?

**Theorem 5.6 (Harmonic Functions Are Determined by Boundary Values)** *Let $M \subset \mathbb{Z}^d$ be a finite set, and let $F : \overline{M} \to \mathbb{R}$ be a harmonic function. Then $F$ is uniquely determined by its values on the boundary $\partial M$.*

*Proof* Suppose $G : \overline{M} \to \mathbb{R}$ is another harmonic function such that $G(x) = F(x)$ for every $x \in \partial M$. According to Exercise 5.13, the function $H := F - G$ is harmonic, and $H(x) = 0$ for every $x \in \partial M$. By Corollary 5.5, $H(x) = 0$ for every $x \in \overline{M}$, implying $F = G$ on all of $\overline{M}$. ∎

## 5.3 The Discrete Dirichlet Problem

The *discrete Dirichlet problem* is defined as follows: Let $M$ be a non-empty, bounded subset of the integer lattice $\mathbb{Z}^d$, and let $F_0 : \partial M \to \mathbb{R}$ be an arbitrary function. The problem asks: Is there a harmonic function $F : \overline{M} \to \mathbb{R}$ such that $F(x) = F_0(x)$ for every $x \in \partial M$? If such a "harmonic extension" of $F_0$ on $\overline{M}$ exists, is it uniquely determined?

At first glance, it is not immediately clear whether the discrete Dirichlet problem has a solution, as both the set $M$ and the function $F_0$ are *arbitrary*. In the simplest case, where $M =: \{x\}$ is a singleton, the solution is straightforward. Here, $\partial M$ consists of the $2d$ neighboring points $x \pm e_j$, where $j \in \{1, \ldots, d\}$. Since a harmonic function on $\overline{M}$ satisfies the mean value property,

$$F(x) = \frac{1}{2d} \sum_{j=1}^{d} \big(F_0(x + e_j) + F_0(x - e_j)\big).$$

If we define $F(y) := F_0(y)$ for $y \in \partial M$, then $F$ is the unique solution of this simplest form of the discrete Dirichlet problem. The following example suggests that the solution to the discrete Dirichlet problem may be linked to linear algebra.

**Example 5.7** Let $d = 2$ and $M := \{(0, 0), (1, 0)\}$; see Fig. 5.2. The points in $M$ are marked by small black circles, while the points in $\partial M$ are marked by red circles. The given values of the function $F_0$ are assigned to the red points, i.e., $F_0((1, 1)) := f_1$, $F_0((2, 0)) := f_2$, etc. Let $u := F((1, 0))$ and $v := F((0, 0))$ represent the two function values of $F$ on $M$ that we need to determine. By the mean value property, we have

$$u = \frac{1}{4}(f_1 + f_2 + f_3 + v), \quad v = \frac{1}{4}(f_4 + f_5 + f_6 + u).$$

These two equations in the unknowns $u$ and $v$ have the solution

$$u = \frac{\delta + 4\gamma}{15}, \quad v = \frac{\gamma + 4\delta}{15},$$

where $\gamma := f_1 + f_2 + f_3$ and $\delta := f_4 + f_5 + f_6$.

**Fig. 5.2** Example of the discrete Dirichlet problem using a two-element set

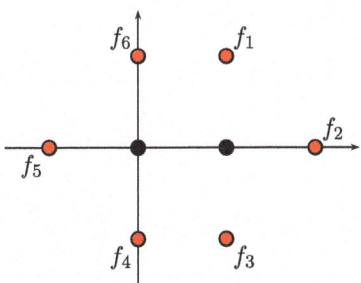

We will now demonstrate how the symmetric random walk on $\mathbb{Z}^d$ leads to a solution of the general discrete Dirichlet problem. Consider a random walk that starts at time 0 from the point $x \in \overline{M}$. Let $S_n := x + \sum_{j=1}^{n} X_j$ for $n \geq 1$, with $S_0 := x$. The random variable

$$T_x := \min \left\{ k \in \mathbb{N}_0 \,\middle|\, S_k \in \partial M \right\} \tag{5.24}$$

denotes the time at which this random walk first reaches a point on the boundary of $M$. This *first entry time* into $\partial M$ is also called a *stopping time*. We will first establish that this stopping time is finite with probability one.

**Theorem 5.8 (Almost Sure Finiteness of the Stopping Time $T_x$)** *Let $M$ be a non-empty, bounded subset of $\mathbb{Z}^d$. Then, for the stopping time $T_x$ defined in (5.24),*

$$\mathbb{P}(T_x < \infty) = 1 \quad \text{for every } x \in \overline{M}.$$

**Proof** Define $F(x) := \mathbb{P}(T_x < \infty | S_0 = x)$ for $x \in \overline{M}$. Since the random variable $T_x$ takes the value zero when $S_0 = x$ with $x \in \partial M$, it follows that $F(x) = 1$ for every $x \in \partial M$. In the case where $x \in M$, we decompose the event $\{T_x = n\}$ for each $n \geq 1$ according to the first step of the random walk starting at $x$. Since each of the $2d$ possible such steps is equally likely, it follows by the law of total probability (7.4) that

$$\mathbb{P}(T_x = n) = \frac{1}{2d} \sum_{y \in \mathcal{N}(x)} \mathbb{P}(T_x = n | S_1 = y) = \frac{1}{2d} \sum_{y \in \mathcal{N}(x)} \mathbb{P}(T_y = n - 1).$$

The second equality holds because, under the condition $S_1 = y$, the event $T_x = n$ occurs if and only if $T_y = n - 1$ holds. Summing over all $n$, we obtain

$$F(x) = \mathbb{P}(T_x < \infty) = \frac{1}{2d} \sum_{y \in \mathcal{N}(x)} \mathbb{P}(T_y < \infty) = \frac{1}{2d} \sum_{y \in \mathcal{N}(x)} F(y).$$

Thus, the function $F : \overline{M} \to \mathbb{R}$ satisfies the mean value property and is therefore harmonic. Since $F$ is constant (equal to 1) on $\partial M$, it follows from Corollary 5.5 that $F(x) = 1$ for every $x \in \overline{M}$. ∎

Given that the random variable $T_x$ takes a finite value with probability one for every $x \in \overline{M}$, we can define a random variable $S_{T_x} : \Omega \to \mathbb{R}$ on the canonical sample space $\Omega$ for random walks on $\mathbb{Z}^d$ (see Sect. 7.2) by

$$S_{T_x}(\omega) := \begin{cases} S_{T_x(\omega)}(\omega), & \text{if } T_x(\omega) < \infty, \\ y^*, & \text{if } T_x(\omega) = \infty, \end{cases} \quad \omega \in \Omega.$$

## 5.3 The Discrete Dirichlet Problem

Here, $y^* \in \partial M$ is an arbitrary point on the boundary of $M$. The random variable $S_{T_x}$ models the random *entry point* into $\partial M$ when a symmetric random walk starting at $x \in M$ *first* leaves the set $M$. The $\mathbb{P}$-almost sure finiteness of $T_x$ for every $x \in \overline{M}$ allows us to apply a function $F_0$ defined on the boundary of $M$ to $S_{T_x}$. Furthermore, the uniform distribution of each individual step of the random walk in all $2d$ directions $\pm e_j$, $j \in \{1, \ldots, d\}$, corresponds to the mean value property. Hence, a symmetric random walk on $\mathbb{Z}^d$ provides the following elegant solution to the discrete Dirichlet problem:

**Theorem 5.9 (Solution of the Discrete Dirichlet Problem)** *Let $M$ be a non-empty, finite subset of $\mathbb{Z}^d$, and let $F_0 : \partial M \to \mathbb{R}$ be an arbitrary function. Then there exists exactly one harmonic function $F : \overline{M} \to \mathbb{R}$ such that $F(x) = F_0(x)$ for every $x \in \partial M$. This function is given by*

$$F(x) := \mathbb{E}[F_0(S_{T_x})] = \sum_{y \in \partial M} F_0(y)\, \mathbb{P}(S_{T_x} = y), \qquad x \in \overline{M}. \tag{5.25}$$

*Proof* By Theorem 5.8, the function $F$ is well-defined. For each $x \in \partial M$, $\mathbb{P}(T_x = 0) = 1$, and therefore $\mathbb{P}(S_{T_x} = x) = 1$. By the definition of $F$, it follows that $F(x) = F_0(x)$. Thus, $F$ is an extension of $F_0$ on $\overline{M}$. Now, we will show that the function $F$ satisfies the mean value property and is thus harmonic. According to Exercise 5.14, we can assume without loss of generality that $F_0$ is of the form $F_0(y_0) = 1$ and $F_0(x) = 0$ for all $x \in \partial M \setminus \{y_0\}$, for some $y_0 \in \partial M$, making it an indicator function. For this special case, the function value $F(x)$, defined in (5.25), is equal to the probability $\mathbb{P}(S_{T_x} = y_0)$, which represents the probability that a symmetric random walk starting at $x$ exits the set $M$ at the point $y_0 \in \partial M$. By decomposing the event $\{S_{T_x} = y_0\}$ according to the first step of the random walk, and applying the law of total probability (7.4), we obtain

$$F(x) = \frac{1}{2d} \sum_{z \in \mathcal{N}(x)} \mathbb{P}(S_{T_x} = y_0 | S_1 = z).$$

Since the random walk is memoryless, the conditional probability $\mathbb{P}(S_{T_x} = y_0 | S_1 = z)$ is equal to the probability that a random walk starting at the neighboring point $z$ of $x$ exits the set $M$ at the point $y_0$. Thus, $\mathbb{P}(S_{T_z} = y_0) = F(z)$. Note that $z = y_0$ is possible, in which case $\mathbb{P}(S_{T_z} = y_0) = 1$. Therefore, the function $F$ satisfies the mean value property. By Theorem 5.6, $F$ is uniquely determined. ∎

**SAQ 9** What is the probability that a random walk starting at $(0, 0)$ in Fig. 5.2 first reaches the boundary of $\{(0, 0), (1, 0)\}$ at the point $(2, 0)$?

It should be emphasized that the solution to the discrete Dirichlet problem is not necessarily unique if the set $M$ is unbounded (see Exercise 5.15). Theorem 5.9 asserts that the function value $F(x)$ of the unique solution $F$ to the discrete Dirichlet problem is equal to the expected value of $F_0$, applied to the first exit point into the set $\partial M$ for a symmetric random walk starting at $x$.

In principle, expected values can be estimated by averaging the results of simulations, where many random walks are started at the point $x$. An exact solution, however, can be obtained by exploiting the mean value property and solving the corresponding linear system of equations. Specifically, for every $x \in M$,

$$2d\, F(x) = \sum_{y \in \mathcal{N}(x)} F(y), \qquad x \in M. \tag{5.26}$$

If $M$ has $r$ elements, this yields a system of $r$ linear equations in the $r$ unknowns $F(x)$ for $x \in M$. Since the discrete Dirichlet problem has a unique solution, this system of equations has exactly one solution.

**Example 5.10** Let $d = 2$ and $M := \{(0,0), (0,1), (0,-1), (1,0), (2,0)\}$. This set is depicted in Fig. 5.3, where small black circles represent the points of $M$, and red circles represent the boundary of $M$. The function $F_0 : \partial M \to \mathbb{R}$ takes only the values 1 and 0. For instance, $F_0(1,1) = 1$ and $F_0(0,2) = 0$. For brevity, we use the notation $F_0(a,b)$ to denote $F_0((a,b))$.

Let $t_1 := F(0,1)$, $t_2 := F(0,0)$, $t_3 := F(0,-1)$, $t_4 := F(1,0)$, and $t_5 := F(2,0)$. Then, the equations from (5.26) take the following form:

$$4t_1 = 1 + t_2,$$

$$4t_2 = t_1 + t_3 + t_4,$$

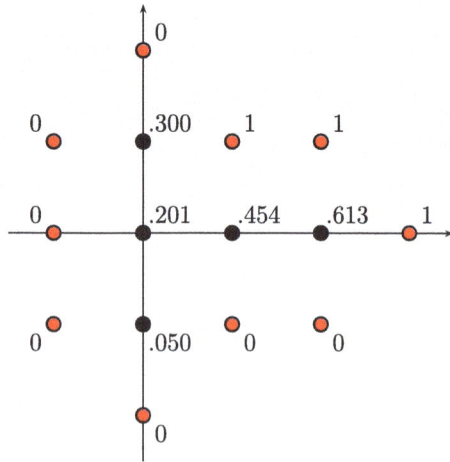

**Fig. 5.3** Probabilities of absorption in $\{(3,0), (1,1), (2,1)\}$

## 5.3 The Discrete Dirichlet Problem

$$4t_3 = t_2,$$
$$4t_4 = 1 + t_2 + t_5,$$
$$4t_5 = 2 + t_4.$$

This system of linear equations has the solution:

$$t_1 = \frac{233}{776}, \quad t_2 = \frac{39}{194}, \quad t_3 = \frac{39}{776}, \quad t_4 = \frac{44}{97}, \quad t_5 = \frac{119}{194}.$$

The numerical values rounded to three decimal places are indicated in Fig. 5.3. They represent, for each point $x \in M$, the respective probability that a symmetric random walk starting at $x$ will first exit the set $M$ at one of the points $(3, 0)$, $(1, 1)$, or $(2, 1)$.

The linear system of equations given by (5.26) can also be solved numerically using the *relaxation method* (see, e.g., [DS], pp. 23–27): First, the function $F_0$ defined on the boundary $\partial M$ of $M$ is extended with *arbitrary* values $F_0(x)$ for $x \in M$. For example, one can set $F_0(x) := 0$ for $x \in M$. Thus, $F_0$ now represents a function defined on the entire set $\overline{M}$. Then, for each $n \geq 1$, we define $F_n(x) := F_0(x)$ if $x \in \partial M$, and for $x \in M$:

$$F_{n+1}(x) := \frac{1}{2d} \sum_{y \in \mathcal{N}(x)} F_n(y). \tag{5.27}$$

At each iteration step, while maintaining the boundary condition $F_n(x) = F_0(x)$ for all $n$ and $x \in \partial M$, the new function values $F_{n+1}(x)$ for $x \in M$ are computed as the arithmetic mean of the function values $F_n(y)$ at the $2d$ neighboring points $y$ of $x$.

**Theorem 5.11 (Convergence of the Relaxation Method)** *For any choice of initial values $F_0(x)$ for $x \in M$, the sequence of functions $(F_n)_{n \geq 1}$, recursively defined by (5.27), converges as $n \to \infty$ to the solution $F$ of the discrete Dirichlet problem, with boundary values given by $F_0(x)$ for $x \in \partial M$.*

**Proof** Let the set $M$ have $r$ elements, which we label as $x_1, \ldots, x_r$. Denote the corresponding function values $F(x_j)$ of the solution $F$ to the discrete Dirichlet problem by $t_j := F(x_j)$, for $j = 1, \ldots, r$. Then the linear system of equations (5.26) can be written as

$$\begin{pmatrix} t_1 \\ t_2 \\ \vdots \\ t_r \end{pmatrix} = \frac{1}{2d} B \begin{pmatrix} t_1 \\ t_2 \\ \vdots \\ t_r \end{pmatrix} + \begin{pmatrix} c_1 \\ c_2 \\ \vdots \\ c_r \end{pmatrix}. \tag{5.28}$$

Here, the $(r \times r)$-matrix $B = (b_{i,j})$ is defined such that by $b_{i,j} := 1$ if $x_j \in \mathcal{N}(x_i)$ (i.e., $x_i$ and $x_j$ are neighbors) and $b_{i,j} := 0$ otherwise. The matrix $B$ is symmetric and satisfies $b_{i,i} = 0$ for all $i \in \{1, \ldots, r\}$. Since each point $x_i$ has up to $2d$ neighboring points, the matrix $B$ contains at most $2d$ non-zero entries per row and column. The components of the vector $(c_1, \ldots, c_r)^\top$ are determined by the values of $F_0$ on the boundary of $M$. Specifically, $c_i = 0$ if all neighboring points of $x_i$ belong to $M$. We define

$$A := \frac{1}{2d} B \qquad (5.29)$$

and let $\mathbf{t} := (t_1, \ldots, t_r)^\top$ and $\mathbf{c} := (c_1, \ldots, c_r)^\top$. Then the system of equations (5.28) can be written compactly as

$$\mathbf{t} = A\mathbf{t} + \mathbf{c}.$$

Thus, (5.27) becomes the *iterative scheme*

$$\mathbf{t}_{n+1} := A\mathbf{t}_n + \mathbf{c}, \qquad n \geq 1. \qquad (5.30)$$

We now show that for the matrix $A$ defined in (5.29),

$$\lim_{n \to \infty} A^n = 0_{r \times r}. \qquad (5.31)$$

Here, $0_{r \times r}$ denotes the zero matrix of order $r$. This will quickly conclude the proof, since if $\lambda$ is any eigenvalue of the (symmetric) matrix $A$ with a corresponding non-zero eigenvector $z$, it follows from (5.31) that $A^n z = \lambda^n z \to 0$. Thus, $|\lambda| < 1$. Consequently, the *spectral norm* of $A$, denoted $|||A|||_2$ and defined as the positive square root of the largest eigenvalue of $A^\top A = A^2$ (see, e.g., [KSC], p. 52), is less than 1.

Let $\|\cdot\|_2$ denote the Euclidean norm in $\mathbb{R}^d$. It follows that $\|Ax\|_2 \leq |||A|||_2 \cdot \|x\|_2$ for all $x \in \mathbb{R}^d$ (see e.g., [KSC], pp. 51–52). This means that the mapping $T : \mathbb{R}^d \to \mathbb{R}^d$, defined by $T(x) := Ax + \mathbf{c}$, $x \in \mathbb{R}^d$, is a *contraction*, since

$$\|T(x) - T(y)\|_2 = \|A(x-y)\|_2 \leq |||A|||_2 \cdot \|x - y\|_2, \qquad x, y \in \mathbb{R}^d,$$

where $|||A|||_2 < 1$. By the Banach fixed-point theorem (see, e.g., [KSC], Theorem 4.4), $T$ has exactly one fixed point, which is the limit of the sequence $(\mathbf{t}_n)$ defined by $\mathbf{t}_{n+1} := T(\mathbf{t}_n)$, $n \geq 0$, for any initial value $\mathbf{t}_0 \in \mathbb{R}^d$. Thus, it remains to prove (5.31).

By the definition of $B$, the entry $b_{i,j}^{(n)}$ in the $i$-th row and $j$-th column of the $n$-th power of $B$ equals the number of paths of length $n$ within the set $M$, starting at $x_i$ and ending at $x_j$. This corresponds to the number of tuples $(x_i, y_1, \ldots, y_{n-1}, x_j)$, where $y_1, \ldots, y_{n-1} \in M$, and each pair of consecutive points in this tuple are neighbors.

## 5.3 The Discrete Dirichlet Problem

An upper bound for $b_{i,j}^{(n)}$ can be obtained by relaxing the condition that $y_1, \ldots, y_{n-1}$ belong to $M$, and considering free random walks of length $n$ that start at $x_i$ and end at $x_j$, as discussed at the beginning of this chapter.

Now, we distinguish between the cases where $n = 2\ell$ and $n = 2\ell + 1$, with $\ell \in \mathbb{N}$. According to Exercise 5.17 b), an upper bound for $b_{i,j}^{(2\ell)}$ is largest when $i = j$. Due to the spatial homogeneity of the symmetric random walk on $\mathbb{Z}^d$, we can assume without loss of generality that $x_i = x_1 = \mathbf{0}$. Using the return probability $\mathbb{P}(S_{2\ell} = \mathbf{0})$ to $\mathbf{0}$ after $2\ell$ steps (examined in Sect. 5.1), and with $A^n =: (a_{i,j}^{(n)})_{r \times r}$, it follows from Exercise 5.17 b) that

$$a_{1,1}^{(2\ell)} = \frac{1}{(2d)^{2\ell}} b_{1,1}^{(2\ell)} \leq \mathbb{P}(S_{2\ell} = \mathbf{0}).$$

In the case $d \geq 3$, due to the transience of the symmetric random walk (cf. the proof of Theorem 5.1), $\sum_{\ell=1}^{\infty} \mathbb{P}(S_{2\ell} = \mathbf{0}) < \infty$ and therefore $\mathbb{P}(S_{2\ell} = \mathbf{0}) \to 0$. This convergence also holds in the cases where $d = 2$ (cf. (5.8)) and $d = 1$ (cf. (5.5)). Hence, $A^{2\ell}$ converges to the zero matrix of order $r$ as $\ell \to \infty$. Since $b_{i,j}^{(2\ell+1)} = \sum_{m=1}^{r} b_{i,m}^{(2\ell)} b_{m,j}$, in the case $n = 2\ell + 1$, we have

$$\max_{i,j=1,\ldots,r} a_{i,j}^{(2\ell+1)} \leq \frac{1}{(2d)^{2\ell+1}} \cdot 2d \cdot \max_{i,j=1,\ldots,r} b_{i,j}^{(2\ell)} \leq \frac{1}{(2d)^{2\ell}} b_{1,1}^{(2\ell)} \leq \mathbb{P}(S_{2\ell} = \mathbf{0}).$$

Thus, we conclude that $\lim_{n \to \infty} A^n = 0_{r \times r}$. ∎

**Example 5.12 (Continuation of Example 5.10)** In the situation of Example 5.10, with $r = 5, d = 2$, the matrix $B$ in (5.28) is

$$B = \begin{pmatrix} 0 & 1 & 0 & 0 & 0 \\ 1 & 0 & 1 & 1 & 0 \\ 0 & 1 & 0 & 0 & 0 \\ 0 & 1 & 0 & 0 & 1 \\ 0 & 0 & 0 & 1 & 0 \end{pmatrix}, \quad \begin{pmatrix} c_1 \\ c_2 \\ c_3 \\ c_4 \\ c_5 \end{pmatrix} = \begin{pmatrix} 0.25 \\ 0 \\ 0 \\ 0.25 \\ 0.5 \end{pmatrix}.$$

Starting the iterative scheme (5.30) with $t_0 := (0, 0, 0, 0, 0)^\top$, Table 5.2 shows that the values in Fig. 5.3 are reached after 10 iterations.

### Answers to the Self-Assessment Questions

**Answer 1** Let $X_1 =: (X_{1,1}, \ldots, X_{1,d})$. Since $X_1$ has a uniform distribution over the set $\{e_1, -e_1, \ldots, e_d, -e_d\}$, for each $j \in \{1, \ldots, d\}$, the random variable $X_{1,j}$ takes the values 1 and $-1$ with probability $\frac{1}{2d}$, and the value 0 with probability $1 - \frac{1}{d}$. It follows that $\mathbb{E}(X_{1,j}) = 0$ for $j \in \{1, \ldots, d\}$, and therefore $\mathbb{E}(X_1) = \mathbf{0}$.

**Table 5.2** Numerical values obtained from the iterative scheme (5.30) for Example 5.10

| $n$ | $t_{n,1}$ | $t_{n,2}$ | $t_{n,3}$ | $t_{n,4}$ | $t_{n,5}$ |
|---|---|---|---|---|---|
| 0 | 0.000 | 0.000 | 0.000 | 0.000 | 0.000 |
| 1 | 0.250 | 0.125 | 0.000 | 0.375 | 0.563 |
| 2 | 0.281 | 0.156 | 0.031 | 0.422 | 0.594 |
| 3 | 0.289 | 0.184 | 0.039 | 0.438 | 0.605 |
| 4 | 0.296 | 0.191 | 0.046 | 0.447 | 0.609 |
| 5 | 0.298 | 0.197 | 0.048 | 0.450 | 0.612 |
| 6 | 0.299 | 0.199 | 0.049 | 0.452 | 0.613 |
| 7 | 0.300 | 0.200 | 0.050 | 0.453 | 0.613 |
| 8 | 0.300 | 0.201 | 0.050 | 0.453 | 0.613 |
| 9 | 0.300 | 0.201 | 0.050 | 0.453 | 0.613 |
| 10 | 0.300 | 0.201 | 0.050 | 0.454 | 0.613 |

**Answer 2** Regardless of the first step of the random walk, for the second step, only one of $2d$ equally likely possibilities will return to the origin. A formal proof uses the law of total probability (7.4).

**Answer 3** In total, there are $4^{2n}$ $2n$-tuples with entries $\leftarrow, \rightarrow, \uparrow$ and $\downarrow$, which determine the respective step directions. The binomial coefficient $\binom{2n}{k}$ represents the number of ways to choose $k$ components as $\rightarrow$ from the tuple. For each of these choices, there are $\binom{2n-k}{k}$ ways to choose $k$ from the remaining components as $\leftarrow$. Finally, there are $\binom{2n-2k}{n-k}$ ways to choose $n-k$ positions for the upward steps $\uparrow$. The remaining $n-k$ steps will be downward steps $\downarrow$.

**Answer 4** The argument is analogous to that of the previous question. In total, there are $6^{2n}$ $2n$-tuples with entries corresponding to the six possible step directions (forward/backward in the first coordinate, up/down in the second coordinate, and up/down in the third coordinate). The multinomial coefficient counts the number of ways to first choose $k_1$ positions for forward steps, then (from the remaining $2n - k_1$ positions) $k_1$ for backward steps, then (from the remaining $2n - 2k_1$ positions) $k_2$ for upward steps in the second coordinate, continuing similarly for the other directions. The product of the respective binomial coefficients is the multinomial coefficient.

**Answer 5** These inequalities follow from the subset relations

$$\{S_{6m-2} = \mathbf{0}\} \cap \{X_{6m-1} + X_{6m} = \mathbf{0}\} \subset \{S_{6m} = \mathbf{0}\},$$

$$\{S_{6m-4} = \mathbf{0}\} \cap \{X_{6m-3} + X_{6m-2} = \mathbf{0}\} \cap \{X_{6m-1} + X_{6m} = \mathbf{0}\} \subset \{S_{6m} = \mathbf{0}\},$$

along with the independence of the sets involved in the respective intersections, and the fact that $\mathbb{P}(X_1 + X_2 = \mathbf{0}) = \left(\frac{1}{6}\right)^2$.

## 5.3 The Discrete Dirichlet Problem

**Answer 6** We have $\mathbb{P}(C) = \mathbb{P}(X_k \neq \mathbf{0}, X_k+X_{k-1} \neq \mathbf{0}, \ldots, X_k+\ldots+X_{j+1} \neq \mathbf{0})$. Furthermore, $\mathbb{E}(I_{k-j}) = \mathbb{P}(X_{k-j} \neq \mathbf{0}, X_{k-j}+X_{k-j-1} \neq \mathbf{0}, \ldots, X_{k-j}+\ldots+X_1 \neq \mathbf{0})$. These probabilities are equal, since $X_1, \ldots, X_k$ are i.i.d.

**Answer 7** For every $x \in M$,

$$0 = \mathcal{L}F(x) = \frac{1}{2d} \sum_{y \in \mathcal{N}(x)} \big(F(y) - F(x)\big) = \frac{1}{2d} \sum_{y \in \mathcal{N}(x)} F(y) - F(x),$$

which is equivalent to

$$F(x) = \frac{1}{2d} \sum_{y \in \mathcal{N}(x)} F(y).$$

**Answer 8** Let $F(y) = c$ for every $y \in \partial M$. If $F(x) > c$ or $F(x) < c$ for any $x \in M$, then the maximum or minimum of $F$ on $\overline{M}$ would be attained in the interior of $\overline{M}$. By Theorem 5.4, this only occurs if $F$ is constant on $\overline{M}$. Hence, $F(x) = c$ for every $x \in \overline{M}$.

**Answer 9** This probability results if we set $f_2 := 1$ and $f_3 = \ldots = f_6 = f_1 := 0$. Then $\gamma = 1$ and $\delta = 0$. Therefore, the desired probability $v$ is $\frac{1}{15}$.

### Exercises

**Exercise 5.1** A particle performs a two-dimensional, symmetric random walk starting at the origin. Show that if $(U_n, V_n)$ denotes the position of the particle after $n$ steps, then

$$\mathbb{E}(U_n^2 + V_n^2) = n.$$

**Exercise 5.2** A particle performs a two-dimensional symmetric random walk starting at the origin. Show that for every $n \geq 1$, the probability that after $2n$ steps an even number of steps have been taken in each of the two coordinates is equal to $\frac{1}{2}$.
**Hint:** 0 is an even number.

**Exercise 5.3** Consider a symmetric random walk starting at $\mathbf{0}$ on $\mathbb{Z}^d$. The random vectors $X_1, \ldots, X_n$ are represented as column vectors. Show:

(a) $\mathbb{E}(X_1^\top X_2) = 0$. Here, $X_1^\top X_2$ denotes the dot product of $X_1$ and $X_2$.
(b) $\mathbb{E}\|S_n\|^2 = n$.

**Exercise 5.4** Let $(S_n)_{n\geq 1}$ with $S_n = X_1 + \ldots + X_n$ be a symmetric random walk on $\mathbb{Z}^d$ starting at $\mathbf{0}$. Show that for any positive integers $k$ and $n$, and any choice of $x, y \in \mathbb{Z}^d$:

$$\mathbb{P}(S_{n+k} = y | S_n = x) = \mathbb{P}(S_{n+k} = x | S_n = y).$$

**Exercise 5.5** Let $(S_n)_{n\geq 1}$ with $S_n = X_1 + \ldots + X_n$ be a symmetric random walk on $\mathbb{Z}^d$ starting at $\mathbf{0}$. Show the following temporal homogeneity property: For any positive integers $k, m$ and $n$ and any choice of $x, y \in \mathbb{Z}^d$:

$$\mathbb{P}(S_{n+k} = y | S_n = x) = \mathbb{P}(S_{m+k} = y | S_m = x).$$

**Exercise 5.6** Let $e_j := (0, \ldots, 0, 1, 0, \ldots, 0)$ (with the 1 in the $j$-th position) be the $j$-th canonical unit vector in $\mathbb{R}^d$, where $d \geq 2$. Let $X_1, X_2, \ldots$ be a sequence of i.i.d. random vectors with $\mathbb{P}(X_1 = e_j) = p_j$ and $\mathbb{P}(X_1 = -e_j) = q_j$ for $j = 1, \ldots, d$, where $p_j > 0$ and $q_j > 0$ for $j = 1, \ldots, d$ and $\sum_{j=1}^{d}(p_j + q_j) = 1$. We define a random walk on the integer lattice $\mathbb{Z}^d$ by $S_0 := \mathbf{0}$ and $S_n := X_1 + \ldots + X_n$ for $n \geq 1$. In the special case where $p_j = q_j = \frac{1}{2d}$ for $j = 1, \ldots, d$, this results in the symmetric random walk. Show:

(a) $\mathbb{P}(S_2 = \mathbf{0}) = 2\sum_{j=1}^{d} p_j q_j$.

(b) With probability one, this random walk will eventually revisit a previously visited site, i.e.,

$$\mathbb{P}\left(\bigcup_{n=1}^{\infty} \{S_n \in \{S_0, S_1, \ldots, S_{n-1}\}\}\right) = 1.$$

**Exercise 5.7** According to Eq. (5.8), the probability that a symmetric random walk on $\mathbb{Z}^2$, starting at $\mathbf{0}$, returns to $\mathbf{0}$ after $2n$ steps is equal to the probability that two particles, starting simultaneously at 0 and independently taking symmetric random walks on $\mathbb{Z}$, meet at 0 after each has taken $2n$ steps. Why is this the case?

**Exercise 5.8** Let $p_1, \ldots, p_d \in (0, 1)$ with $p_1 + \ldots + p_d = 1$ and $n \in \mathbb{N}$. Furthermore, let $k_1, \ldots, k_d \in \mathbb{N}_0$ with $k_1 + \ldots + k_d = n$.

(a) Show that for the maximum term of the multinomial probabilities

$$\frac{n!}{k_1! \cdot \ldots \cdot k_d!} p_1^{k_1} \cdot \ldots \cdot p_d^{k_d}, \quad (k_1, \ldots, k_d) \in \mathbb{N}_0^d \text{ with } \sum_{j=1}^{d} k_j = n,$$

the inequalities $p_i k_j \leq p_j(k_i + 1)$ hold for each pair $(i, j)$.

## 5.3 The Discrete Dirichlet Problem

(b) From this, deduce

$$np_i - 1 < k_i \leq (n+d-1)p_i, \qquad i \in \{1, \ldots, d\}.$$

**Hint for (b):** Sum the inequalities in part (a) over all $j$ as well as over all $i$ with $i \neq j$.

**Exercise 5.9** Show that if the symmetric random walk on $\mathbb{Z}^d$ is transient for $d = 3$, then it is transient for any $d \geq 4$.

**Exercise 5.10** Let $R_3 := |\{S_0, S_1, S_2, S_3\}|$ be the number of distinct sites that a two-dimensional symmetric random walk, starting at $\mathbf{0}$, has visited after 3 steps. What is the distribution of $R_3$?

**Exercise 5.11** Let $R_2 := |\{S_0, S_1, S_2\}|$ be the number of distinct sites visited by a $d$-dimensional symmetric random walk, starting in $\mathbf{0}$, after two steps. Show that:

$$\mathbb{P}(R_2 = 2) = \frac{1}{2d}, \qquad \mathbb{P}(R_2 = 3) = \frac{2d-1}{2d}.$$

**Exercise 5.12** What does the set $N := \{\mathcal{N}(y) : y \in \mathcal{N}(\mathbf{0})\}$ look like in the case $d = 2$?

**Exercise 5.13** Let $M \subset \mathbb{Z}^d$ be a non-empty, finite set. Show that the set $V := \{F : \overline{M} \to \mathbb{R} : F \text{ is harmonic}\}$, consisting of all harmonic functions defined on $\overline{M}$, forms a vector space over $\mathbb{R}$.

**Exercise 5.14** Show that if the discrete Dirichlet problem for an indicator function of the form $F_0(x) = 1$ if $x = y_0$, and $F_0(x) := 0$, if $x \in \partial M \setminus \{y_0\}$, is solvable for each $y_0 \in \partial M$, then it is solvable for any function $F_0 : \partial M \to \mathbb{R}$.
**Hint:** Consider Exercise 5.13.

**Exercise 5.15** Let $d = 1$ and $M := \mathbb{N} = \{1, 2, 3, \ldots\}$. Then $\partial M = \{0\}$ and $\overline{M} = \mathbb{N}_0$. Show that the discrete Dirichlet problem, defined by $F_0 : \partial M \to \mathbb{R}$ with $F_0(0) := 0$, does not have a unique solution $F : \overline{M} \to \mathbb{R}$. What is the underlying reason for this?

**Exercise 5.16** In the scenario depicted in Fig. 5.2, a player bets one euro and starts a symmetric random walk at the point $(0, 0)$. The game ends when the player first reaches one of the boundary points. If the walk ends at one of the points $(-1, 0)$, $(0, 1)$, or $(0, -1)$, the player loses 2 euros each time (i.e., $f_4 = f_5 = f_6 = -2$). If it ends at one of the points $(1, 1)$ or $(1, -1)$, the player wins 3 euros (i.e., $f_1 = f_3 = 3$). If the player reaches the point $(2, 0)$, he wins $g$ euros.

a) How can the game be played using two fair dice?

b) What value of $g$ makes the game fair?

**Exercise 5.17** Let $X_1, X_2, \ldots$ be as at the beginning of this chapter, and let $z \in \mathbb{Z}^d$. Define $S_0^z := z$ and $S_m^z := z + X_1 + \ldots + X_m$ for $m \geq 1$. Then $S_\ell^z$, $\ell \geq 0$, is a symmetric random walk on $\mathbb{Z}^d$ starting at $z$. For $\ell \geq 1$ and $z \in \mathbb{Z}^d$, let $M(z, \ell)$ denote the (finite) range of $S_\ell^z$, i.e., the set of all sites that the random walk starting at $z$ can reach after $\ell$ steps. Show that:

(a) $\mathbb{P}(S_{2n}^0 = x) = \sum_{y \in M(0,n)} \mathbb{P}(S_n^0 = y) \mathbb{P}(S_n^x = y), \quad x \in \mathbb{Z}^d,$

(b) $\mathbb{P}(S_{2n}^0 = x) \leq \mathbb{P}(S_{2n}^0 = 0), \quad x \in \mathbb{Z}^d.$

**Hint for (b):** Apply the Cauchy–Schwarz inequality appropriately to the sum in part (a).

# Outlook 6

Since this book is intended for second-year mathematics students who have only taken an introductory course in stochastics, we have focused on topics that can be explored using relatively simple mathematical tools. These include geometric and combinatorial methods, generating functions, and concepts from an introductory analysis course. I hope this book has sparked your curiosity and, to further inspire your interest, I would like to briefly mention a few additional topics for your consideration.

## 6.1 Self-Avoiding Walks

Figure 6.1 shows the first 30 steps of a so-called *self-avoiding walk* on $\mathbb{Z}^2$. Such a walk is characterized by the condition that, regardless of its previous course, it moves from any point with equal probability to any *unvisited* neighboring point. The key restriction is that each point can only be visited once. In the case $d \geq 2$, such a walk may terminate after a finite number of steps, as all neighboring points could eventually be visited. For the walk shown in Fig. 6.1, this would occur if the walk moved left twice from the last position and then up twice.

Self-avoiding random walks were proposed by P. Flory[1] as models for unbranched polymers (see, e.g., [MS], Section 2.2). Many open mathematical problems are associated with these walks. The challenge of finding a general formula for the number of all $n$-step self-avoiding random walks on $\mathbb{Z}^d$, denoted by $c_n(d)$, remains ambitious, and this number is only known for relatively small values of $n$. For example, in the case $d = 2$, there are exactly 16,741,957,935,348 self-avoiding

---

[1] Paul John Flory (1910–1985), American chemist and Nobel laureate (1975).

**Fig. 6.1** The first 30 steps of a self-avoiding random walk on $\mathbb{Z}^2$

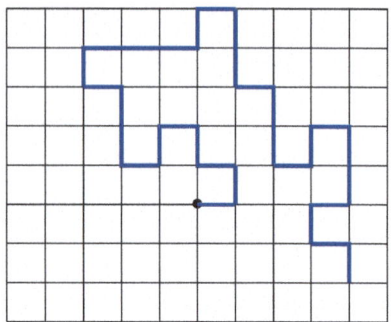

random walks of length 30 (see [MS], p. 394). In general, the following inequalities hold:

$$d^n \leq c_n(d) \leq 2d(2d-1)^{n-1} \tag{6.1}$$

(see [MS], p. 3). It can be shown that for each $d$, the limit known as the *connective constant*

$$\mu(d) := \lim_{n \to \infty} c_n(d)^{1/n}$$

exists (see, e.g., [MS], Section 1.2, or [HU], p. 424). Due to (6.1), $d \leq \mu(d) \leq 2d - 1$, but an exact value for the connective constant is known only for random walks on a *planar hexagonal lattice*, where it is $\sqrt{2 + \sqrt{2}}$ (see [DUS]). This lattice, in which each point has 6 neighbors, is spanned by all integer linear combinations of the vectors $\frac{1}{2}(\sqrt{3}, 1)$ and $\frac{1}{2}(-\sqrt{3}, 1)$.

## 6.2 Walks with Arbitrary Distributions for $X_1$

A common theme in Chaps. 2, 4, and 5 is that the position $S_n$ of a random walk after $n$ steps can be expressed as $S_n = X_1 + \ldots + X_n$, where $X_1, X_2, \ldots$ are *independent* and *identically distributed* random variables or random vectors. In Chaps. 2 and 4, the $X_j$ are random variables taking values of $+1$ and $-1$. In Chap. 5, they are $d$-dimensional random vectors with their distribution concentrated on the canonical unit vectors and their inverses. Many generalizations can be made in this context, and several advanced monographs explore these topics at a significantly higher technical level than this book.

A direct generalization of the framework introduced in Chap. 5 arises if, while maintaining the independence and identical distribution of $X_1, X_2, \ldots$, we assume that $\mathbb{P}(X_1 = e_j) = \mathbb{P}(X_1 = -e_j) =: p_j$ for $j \in \{1, \ldots, d\}$, and $\mathbb{P}(X_j = \mathbf{0}) = 1 - 2(p_1 + \ldots + p_d)$ for positive numbers $p_1, \ldots, p_d$ such that $p_1 + \ldots + p_d \leq \frac{1}{2}$. This generalized symmetry assumption, compared to the uniform distribution

## 6.2 Walks with Arbitrary Distributions for $X_1$

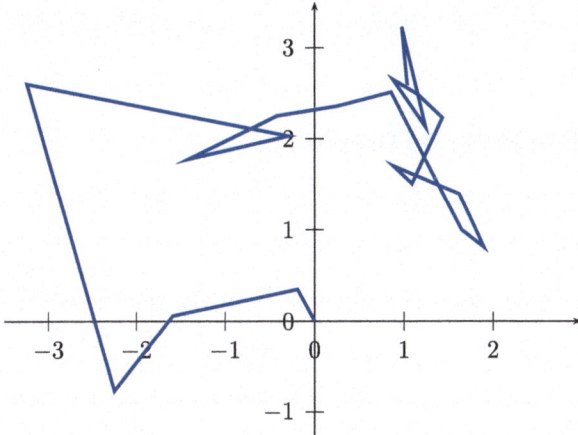

**Fig. 6.2** The first 20 steps of a random walk with Exp(1)-distributed step lengths and uniformly distributed directions

on $\{e_1, -e_1, \ldots, e_d, -e_d\}$, forms the basis of the monograph [LL]. Meanwhile, the monograph [CO] is focused, among other things, on random walks on $\mathbb{N}_0$ and $\mathbb{N}_0 \times \mathbb{N}_0$ with *reflecting* or *absorbing boundaries*, and on general distributions on $\mathbb{N}_0$ or $\mathbb{N}_0 \times \mathbb{N}_0$ for the increments $S_{n+1} - S_n$.

Additionally, the restriction that $X_1$ takes only integer lattice points can be relaxed to allow random walks on $\mathbb{R}^d$, where $X_1$ follows an *arbitrary non-degenerate d-dimensional distribution*. For instance, Fig. 6.2 shows the first 20 steps of a random walk for $d = 2$, where $X_1$ has the same distribution as $R(\cos \Psi, \sin \Psi)$. Here, $R$ and $\Psi$ are independent random variables, with $R$ having an exponential distribution with a mean of 1, $\Psi$ being uniformly distributed over the interval $[0, 2\pi]$. Consequently, the step lengths of this random walk follow an Exp(1)-distribution, while the directions are uniformly distributed.

The book [KS] provides an introduction to the theory and applications of random walks, allowing for a general distribution of $X_1$, with the possibility if random time intervals between changes in position. An important technical tool in this context is the use of *characteristic functions*. The monographs [BO] and [BOB] focus on the asymptotic behavior of probabilities related to certain rare events, often referred to as *large deviations*, in such random walks. The book [LA] addresses, among other things, the probability that a random walk does not hit a given set, or that two independent random walks do not intersect. It also inludes a chapter on self-avoiding random walks, as does the book [HU]. With an eye on applications, the book [GU] primarily examines random walks with non-negative increments that are stopped at random times, which arise, for example, in renewal theory. The monograph [GS] explores random walks on $\mathbb{Z}^d$, where $X_1, X_2, \ldots$ are independent, but *not identically distributed*. Here, while $\mathbb{P}(X_i \in \{e_j, -e_j\}) = \frac{1}{d}$ for each $i \geq 1$

and $j \in \{1, \ldots, d\}$, the probability $\mathbb{P}(X_i = e_j)$ depends on $i$ and $j$ through a *dynamic system*.

## 6.3 Random Walks on Graphs

Aside from Chap. 3, the position of a random walk after $n$ steps has always been the result of adding $n$ independent and identically distributed random variables or $d$-dimensional random vectors. However, there are also extensive studies on random walks on general graphs (see, e.g., [TE], or [WO]; the latter also considers random walks on groups).

A *graph* is a pair $G = (V, E)$. Here, $V$ is a non-empty set of *vertices* (also called *nodes* or *points*), and $E$ is a collection of two-element subsets of $V$. These two-element subsets, called *edges*, are written as $\{v, w\}$, where $v$ and $w$ are distinct vertices in $V$. If $\{v, w\} \in E$, we say that the vertices $v$ and $w$ are *adjacent*. A sequence of vertices $(v_1, \ldots, v_n)$ is called a *path* from $v_1$ to $v_n$, if every consecutive pair of vertices is connected by an edge, that is, $\{v_j, v_{j+1}\} \in E$ for each $j$ from 1 to $n - 1$. A graph $G$ is called *connected* if there exists a path between any two vertices $v$ and $w$. The *degree* of a vertex $v$, denoted by $\deg(v)$, is the number of vertices adjacent to $v$, i.e., the number of edges connected to $v$.

A *simple random walk* on a graph $G = (V, E)$ is a process where a particle starts at some vertex $v_0 \in V$ and then moves to one of its adjacent (or *neighboring*) vertices, chosen at random with equal probability. Specifically, the particle moves from $v_0$ to a neighboring vertex $w$ with probability $\frac{1}{\deg(v_0)}$. Upon arriving at $w$, the particle again moves to one of its neighbors with probability $\frac{1}{\deg(w)}$, and this process repeats.

The random walk on the graph $G$, starting at vertex $v_0$, can be represented by a sequence $Z_0, Z_1, Z_2, \ldots$ of $V$-valued random variables, where $Z_n$ denotes the position of the particle after $n$ steps. Initially, $\mathbb{P}(Z_0 = v_0) = 1$, and, for each $n \geq 0$ and $v \in V$, the transition probabilities are given by

$$\mathbb{P}(Z_{n+1} = w | Z_n = v) = \frac{1}{\deg(v)} \quad \text{for each } w \in V \text{ such that } \{v, w\} \in E \tag{6.2}$$

These transition probabilities depend only on the current vertex $v$, and not on the step number $n$ or the entire history of the walk. Consequently, the random variables $Z_0, Z_1, Z_2, \ldots$ form a *Markov chain*. The theory of Markov chains, especially concerning the long-term behavior of the walk as $n \to \infty$, is applicable here (see, e.g., [LA1], pp. 9–63).

The following fact is easy to understand: If you start a simple random walk on $G$ by randomly choosing the initial vertex according to

$$\mathbb{P}(Z_0 = v) := \frac{\deg(v)}{2|E|}, \quad v \in V, \tag{6.3}$$

## 6.3 Random Walks on Graphs

then using (6.2) and the law of total probability (7.4), we obtain for each vertex $w \in V$:

$$\mathbb{P}(Z_1 = w) = \sum_{v \in V} \mathbb{P}(Z_1 = w | Z_0 = v) \mathbb{P}(Z_0 = v)$$

$$= \sum_{v \in V} \frac{\mathbf{1}\{\{v, w\} \in E\}}{\deg(v)} \cdot \frac{\deg(v)}{2|E|}$$

$$= \frac{1}{2|E|} \sum_{v \in V} \mathbf{1}\{\{v, w\} \in E\} = \frac{\deg(w)}{2|E|}$$

$$= \mathbb{P}(Z_0 = w).$$

By induction, $\mathbb{P}(Z_n = w) = \mathbb{P}(Z_0 = w)$ for every $n \geq 0$ and $w \in V$. The distribution given by (6.3), which assigns to each vertex a probability proportional to its degree, is thus a so-called *invariant distribution*.

I would like to highlight one aspect of random walks on finite graphs, namely the *cover times*, which were first discussed in a textbook by Blom et al. [BHS]. If $G_n$ is any connected graph with $n$ vertices, the *cover time* of a simple random walk starting at the vertex $v$ of $G_n$ is the expected number of steps, denoted by $\mathbb{E}_v(G_n)$, until every vertex has been visited for the first time. The order of magnitude of $\mathbb{E}_v(G_n)$ can vary significantly, depending on the structure of the graph $G_n$. When considering cover times, the following inequalities hold (see [FG1], [FG2]):

$$(1 + o(1)) n \log n \leq \mathbb{E}_v(G_n) \leq \frac{4}{27} n^3 (1 + o(1)). \tag{6.4}$$

Here, $o(1)$ is a term that converges to zero as $n \to \infty$.

For certain special graphs, the cover time can be explicitly determined. One example is the *complete graph* with $n$ vertices, denoted by $K_n$, in which every pair of vertices is connected by an edge. Figure 6.3 (left) shows the graph $K_5$. If you start a simple random walk on $K_n$ at any vertex, and you have already visited $j$ vertices, where $j < n$, the probability of visiting a new vertex in the next step is $\frac{n-j}{n-1}$, regardless of which $j$ vertices have already been visited.

**Fig. 6.3** Complete graph $K_5$ (left) and cyclic graph $C_5$ (right)

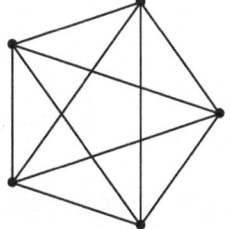

**Fig. 6.4** Lollipop graph with 9 vertices

The distribution of the number of steps until all vertices are visited follows the distribution of the sum of independent, geometrically distributed random variables. Therefore, the cover time is

$$\mathbb{E}(K_n) = 1 + \frac{n-1}{n-2} + \frac{n-1}{n-3} + \ldots + \frac{n-1}{1} = (n-1) \sum_{\ell=1}^{n-1} \frac{1}{\ell}.$$

Since $\sum_{\ell=1}^{n} \frac{1}{\ell} \sim \log n$ as $n \to \infty$, the asymptotic lower bound in (6.4) applies to $K_n$. In contrast, for a *cyclic graph* with $n$ vertices, denoted by $C_n$, the cover time grows quadratically with $n$, since $\mathbb{E}(C_n) = \frac{n(n-1)}{2}$ (see [BHS], p. 151). In the graph $C_n$—where the vertices are numbered from 1 to $n$—vertices 1 and $n$, as well as each $j \in \{1, \ldots, n-1\}$ and $j+1$, are connected by an edge. Figure 6.3 (right) shows the graph $C_5$.

The asymptotic upper bound in (6.4) is achieved when the complete graph $K_n$ is coupled with a linear graph with $n$ vertices to form a so-called *Lollipop graph* (see Fig. 6.4). In this case, the random walk starts at the vertex where the complete graph and the linear graph are connected.

## 6.4 Random Walks and Electrical Networks

As the following explanations illustrate, there is a close relationship between random walks and electrical networks (see, e.g., [DS], Chapter 1). Figure 6.5 shows an electrical circuit in which $n$ equal resistors are connected in series. A voltage of 1 Volt is applied across the terminal nodes 0 and $n$. The node 0 is grounded.

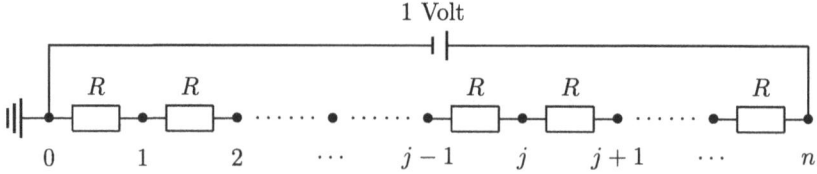

**Fig. 6.5** Electrical circuit (series connection of equal resistors)

## 6.4 Random Walks and Electrical Networks

Let $U(j)$ denote the voltage at node $j$, $j \in \{0, 1, \ldots, n\}$, so $U(n) = 1$ and $U(0) = 0$. According to Kirchhoff's[2] first law, the sum of incoming and outgoing current in each node $j \in \{1, \ldots, n-1\}$ is zero. By Ohm's[3] law, the current flowing from node $j$ to node $j - 1$ is equal to the voltage difference $U(j) - U(j - 1)$, divided by the resistance $R$. Thus,

$$\frac{U(j-1) - U(j)}{R} + \frac{U(j+1) - U(j)}{R} = 0, \qquad j \in \{1, \ldots, n-1\}.$$

Multiplying these equations by $R$ gives

$$U(j) = \frac{U(j-1) + U(j+1)}{2}, \qquad j \in \{1, \ldots, n-1\}. \tag{6.5}$$

The function $U : \{0, 1 \ldots, n\} \to \mathbb{R}$ thus has the mean value property and is therefore harmonic. The boundary conditions $U(0) = 0$ and $U(n) = 1$, along with the mean value equations (6.5), are the same as (4.19) and (4.20) (with $p = \frac{1}{2}$) for the ruin probability of player A in the gambler's ruin problem in Sect. 4.3 (with $r = n$ and $P_j = U(j)$, $j \in \{0, 1 \ldots, n\}$). Due to the uniqueness of the solution to the discrete Dirichlet problem (Theorem 5.9) and Theorem 4.10, $U(j) = \frac{j}{k}$ for $j \in \{0, \ldots, n\}$. Consequently, the voltages at the nodes can be interpreted as ruin probabilities.

In Example 5.10 we solved a discrete Dirichlet problem for the set

$$M = \{(0, 0), (0, 1), (0, -1), (1, 0), (2, 0)\}$$

(see Fig. 5.3). Figure 6.6 shows the corresponding electrical circuit. A voltage of 1 Volt is applied to the boundary points $(1, 1)$, $(2, 1)$, $(3, 0)$ of $M$, while the other boundary points are grounded. What is the voltage at the point $(i, j) \in M$?

To answer this question, we refer again to Kirchhoff's first law, which states that the sum of currents flowing into $(i, j)$ and out of $(i, j)$ is zero. If each of the marked resistors has a resistance of $R$ (Ohms), then

$$0 = \frac{U(i+1, j) - U(i, j)}{R} + \frac{U(i-1, j) - U(i, j)}{R}$$
$$+ \frac{U(i, j+1) - U(i, j)}{R} + \frac{U(i, j-1) - U(i, j)}{R},$$

---

[2] Gustav Robert Kirchhoff (1824–1887), a German physicist, worked at the universities of Breslau, Heidelberg, and Berlin. Kirchhoff received numerous honors. His so-called *Kirchhoff's laws* are fundamental for the construction and analysis of electrical circuits.

[3] Georg Simon Ohm (1789–1854), a German physicist, became a professor at the Royal Polytechnic School in Nuremberg in 1833 and a professor at the University of Munich in 1849. The law named after him, which describes the proportionality between electric current and voltage in a conductor, was discovered by Ohm in 1826.

**Fig. 6.6** Electrical circuit corresponding to Fig. 5.3

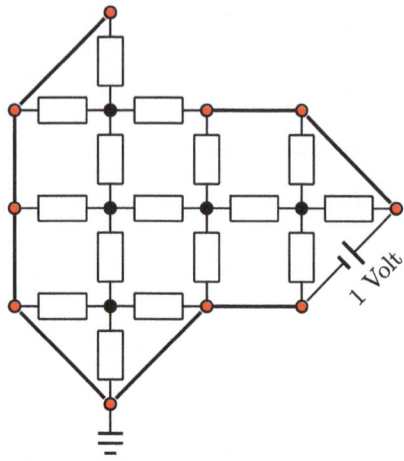

which simplifies to

$$U(i,j) = \frac{U(i+1,j) + U(i,j+1) + U(i-1,j) + U(i,j-1)}{4}.$$

Therefore, the function $U : \overline{M} \to \mathbb{R}$ satisfies the mean value property and is thus harmonic. Since the boundary points $(0, 2)$, $(-1, 1)$, $(-1, 0)$, $(-1, -1)$, $(0, -2)$, $(1, -1)$, $(2, -1)$ of $M$ have a voltage of 0 due to grounding, the boundary conditions for $U$ as are the same as for the function $F$ in Example 5.10. By the uniqueness of the solution to the discrete Dirichlet problem (Theorem 5.9), the voltages at the points in $M$ are as shown in Fig. 5.3; for instance, $U(1, 0) \approx 0.454$ (to three decimal places).

Conversely, the voltage $U(i, j)$ at point $(i, j)$ can be interpreted as the probability that a symmetric random walk starting at $(i, j)$ will leave the set $M$ for the first time at a boundary point where the voltage is 1. The book [DS] contains a wealth of connections between random walks and electrical networks.

# 7 Tools from Stochastics, Combinatorics, and Analysis

This chapter consolidates various terms and results referenced in earlier chapters. Many of these can be found in sources such as [BT], [CA], [FE1], [GSN], [RO], or [ST].

## 7.1 Probability Spaces

According to *Kolmogorov's axiomatic system*, a *probability space* is a triple $(\Omega, \mathcal{A}, \mathbb{P})$. Here,

- $\Omega$ is any non-empty set,
- $\mathcal{A}$ is a collection of subsets of $\Omega$ that satisfy:
  - $\Omega \in \mathcal{A}$,
  - if $A \in \mathcal{A}$, then the complement $A^c := \Omega \setminus A \in \mathcal{A}$,
  - if $A_n \in \mathcal{A}$ for each $n \geq 1$, then $\bigcup_{n=1}^{\infty} A_n \in \mathcal{A}$.
  
  Such a collection of sets, that contains $\Omega$, includes the complement of each set, and is closed under countable unions is called a $\sigma$-*algebra* on $\Omega$.
- $\mathbb{P} : \mathcal{A} \to \mathbb{R}$ is a function with the following properties:
  - $\mathbb{P}(A) \geq 0$ for all $A \in \mathcal{A}$      (*non-negativity*),
  - $\mathbb{P}(\Omega) = 1$      (*normalization*),
  - $\mathbb{P}\left(\biguplus_{n=1}^{\infty} A_n\right) = \sum_{n=1}^{\infty} \mathbb{P}(A_n)$ for any sequence $A_1, A_2, \ldots$ of pairwise disjoint sets in $\mathcal{A}$      ($\sigma$-*additivity*).

The sets in $\mathcal{A}$ are called *events*, and the function $\mathbb{P}$ is called a *probability measure* on $\mathcal{A}$. For each $A \in \mathcal{A}$, $\mathbb{P}(A)$ represents the *probability of* event $A$.

If $\Omega$ is a countable set, then $\mathcal{A} := \mathcal{P}(\Omega)$ is the power set of $\Omega$. If $\Omega = \mathbb{R}^d$ for an integer $d$, we use the $\sigma$-*algebra of Borel sets*, denoted by $\mathcal{B}^d$ (also known as the

*Borel $\sigma$-algebra*). This is the smallest $\sigma$-algebra on $\mathbb{R}^d$ that contains all open subsets of $\mathbb{R}^d$.

Based on Kolmogorov's axioms, we derive the following conclusions for arbitrary events $A$, $B$, and sequences of events $A_1, A_2, \ldots$:

- $\mathbb{P}(A^c) = 1 - \mathbb{P}(A)$      (*rule of complementary events*),
- if $A \subset B$, then $\mathbb{P}(A) \leq \mathbb{P}(B)$      (*monotonicity*),
- $\mathbb{P}\left(\bigcup_{n=1}^{\infty} A_n\right) \leq \sum_{n=1}^{\infty} \mathbb{P}(A_n)$      ($\sigma$-*subadditivity*).

Additionally, we require the following *continuity properties* of $\mathbb{P}$.

- If $(A_n)_{n \in \mathbb{N}}$ is an *increasing* sequence of events, i.e., $A_n \subset A_{n+1}$ for each $n \geq 1$, then

$$\mathbb{P}\left(\bigcup_{n=1}^{\infty} A_n\right) = \lim_{n \to \infty} \mathbb{P}(A_n) \quad (continuity\ from\ below). \tag{7.1}$$

- If $(A_n)_{n \in \mathbb{N}}$ is a *decreasing* sequence of events, i.e., $A_n \supset A_{n+1}$ for each $n \geq 1$, then

$$\mathbb{P}\left(\bigcap_{n=1}^{\infty} A_n\right) = \lim_{n \to \infty} \mathbb{P}(A_n) \quad (continuity\ from\ above). \tag{7.2}$$

The proof of (7.1) utilizes the $\sigma$-additivity of $\mathbb{P}$. By defining $B_1 := A_1$ and $B_j := A_j \cap A_1^c \cap \ldots \cap A_{j-1}^c$ for each $j \geq 2$, the sets $B_1, B_2, \ldots$ are pairwise disjoint. Setting $A := \bigcup_{j=1}^{\infty} A_j$, we obtain

$$A = \biguplus_{j=1}^{\infty} B_j, \quad A_1 \cup \ldots \cup A_n = B_1 \uplus \ldots \uplus B_n, \quad n \geq 1.$$

Since $A_n = A_1 \cup \ldots \cup A_n$ for each $n \geq 1$,

$$\mathbb{P}(A) = \mathbb{P}\left(\biguplus_{j=1}^{\infty} B_j\right) = \sum_{j=1}^{\infty} \mathbb{P}(B_j) = \lim_{n \to \infty} \sum_{j=1}^{n} \mathbb{P}(B_j)$$

$$= \lim_{n \to \infty} \mathbb{P}\left(\biguplus_{j=1}^{n} B_j\right) = \lim_{n \to \infty} \mathbb{P}\left(\bigcup_{j=1}^{n} A_j\right) = \lim_{n \to \infty} \mathbb{P}(A_n).$$

Continuity from above follows by applying (7.1) to the increasing sequence $(A_n^c)_{n \geq 1}$ and using the rule of complementary events.

## 7.2 A Canonical Probability Space

If $A$ and $B$ are events with $\mathbb{P}(A) > 0$, the *conditional probability of $B$ given $A$* is defined as

$$\mathbb{P}(B|A) := \frac{\mathbb{P}(A \cap B)}{\mathbb{P}(A)}. \tag{7.3}$$

If $A_1, \ldots, A_n$ (for $n \geq 2$) are events such that $\Omega = A_1 \uplus \ldots \uplus A_n$ and $\mathbb{P}(A_j) > 0$ for each $j \in \{1, \ldots, n\}$, then for any event $B$:

$$\mathbb{P}(B) = \sum_{j=1}^{n} \mathbb{P}(A_j) \cdot \mathbb{P}(B|A_j) \quad \textit{(law of total probability)}. \tag{7.4}$$

The conditional probabilities mentioned here are typically not calculated according to (7.3), but are provided as model components.

Events $A_1, \ldots, A_n$ (for $n \geq 2$) are *(mutually) independent* if the multiplication rule

$$\mathbb{P}\left(\bigcap_{j \in T} A_j\right) = \prod_{j \in T} \mathbb{P}(A_j)$$

holds for every set $T \subset \{1, \ldots, n\}$ with at least two elements. In total, these are $2^n - n - 1$ equations. Infinitely many events are *independent* if any finite number of them is independent.

Random vectors $X_j : \Omega \to \mathbb{R}^d$, $j = 1, \ldots, n$, are *independent* if for any Borel sets $B_1, \ldots, B_n \in \mathcal{B}^d$, the events $X_j^{-1}(B_j) := \{\omega \in \Omega : X_j(\omega) \in B_j\}$, $j = 1, \ldots, n$, are independent. Here, the dimension $d$ may depend on $j$. Infinitely many random vectors are *independent* if any finite number of them is independent. Note that the term *random vector* implies that for each $j \in \{1, \ldots, n\}$, the *preimage* (or *inverse image*) $X_j^{-1}(B_j)$ of $B_j$ under $X_j$ belongs to $\mathcal{A}$.

### 7.2 A Canonical Probability Space

For much of this book, we use the model of a discrete uniform distribution over a finite set $\Omega$. In this model, the probability of an event $A$ is the number of elements in $A$, divided by the number of elements in $\Omega$. However, at various points, we require a more complex probability space $(\Omega, \mathcal{A}, \mathbb{P})$ that supports an infinite sequence $X_1, X_2, \ldots$ of independent $\mathbb{R}^d$-valued random vectors with specified distributions. According to general theorems from measure theory (see, e.g., [BI2], p. 27 ff.), we can use the sample space

$$\Omega := (\mathbb{R}^d)^{\mathbb{N}} := \{\omega := (\omega_j)_{j \geq 1} : \omega_j \in \mathbb{R}^d \text{ for } j \geq 1\},$$

where sequences have terms belonging to $\mathbb{R}^d$. For $X_j$, we take the $j$-th coordinate mapping, i.e., $X_j(\omega) := \omega_j$ for $\omega \in \Omega$ and $j \geq 1$. The $\sigma$-algebra $\mathcal{A}$ on $\Omega$ is the smallest $\sigma$-algebra on $\Omega$ such that for each $j \geq 1$ and any Borel set $B \in \mathcal{B}^d$, the preimage $X_j^{-1}(B)$ belongs to $\mathcal{A}$.

Let $Q_1, Q_2, \ldots$ be any sequence of probability measures on $\mathcal{B}^d$. If we choose the probability measure $\mathbb{P}$ on $\mathcal{A}$ as the product of $Q_1, Q_2, \ldots$, i.e., $\mathbb{P} := \otimes_{j \geq 1} Q_j$, then $X_1, X_2, \ldots$ are independent random vectors, and $X_j$ has the distribution $Q_j$ for $j \geq 1$. The probability measure $\mathbb{P}$ is uniquely determined by the equations

$$\mathbb{P}\left(\{\omega = (\omega_j)_{j \geq 1} \in \Omega : \omega_j \in B_j \text{ for } j = 1, \ldots, n\}\right) = \prod_{j=1}^{n} Q_j(B_j), \qquad (7.5)$$

where $n \geq 1$ and $B_1, \ldots, B_n \in \mathcal{B}^d$ are any Borel sets. Note that the left-hand side of (7.5) equals $\mathbb{P}(X_1 \in B_1, \ldots, X_n \in B_n)$.

As a special case, we can model an infinite random walk on the $d$-dimensional integer lattice $\mathbb{Z}^d$. If the random vector $X_j$ represents the $j$-th step of such a walk, it has some distribution on the set

$$E_d := \{e_1, -e_1, e_2, -e_2, \ldots, e_d, -e_d\},$$

which consists of the canonical unit vectors $e_1, \ldots, e_d$ in $\mathbb{R}^d$ and their reflections $-e_1, \ldots, -e_d$ at the origin. As a sample space, we may also choose the subset $E_d^{\mathbb{N}}$ of $(\mathbb{R}^d)^{\mathbb{N}}$. In this space, a sequence $\omega = (\omega_j)_{j \geq 1} \in \Omega$ represents an infinitely long path, with the $j$-th step given by $\omega_j$. In the case of a simple symmetric random walk on $\mathbb{Z}^d$, the distribution of $X_j$ is uniform on $E_d$. The probabilities of events like

$$A := \{(\omega_j)_{j \geq 1} \in \Omega : (\omega_1, \ldots, \omega_n) \in A_n\},$$

where $A_n \subset E_d^n$, are defined by a finite initial segment of an infinitely long path and are given by the ratio

$$\mathbb{P}(A) = \frac{|A_n|}{(2d)^n},$$

where $|A_n|$ is the number of initial segments that belong to $A_n$.

## 7.3 Convergence in Distribution

Let $Z, Z_1, Z_2, \ldots$ be random variables with distribution functions $F(z) := \mathbb{P}(Z \leq z)$ and $F_n(z) := \mathbb{P}(Z_n \leq z)$ for $z \in \mathbb{R}$ and $n \geq 1$. We say $Z_n$ *converges in distribution* to $Z$, denoted by

$$Z_n \xrightarrow{\mathcal{D}} Z \quad \text{as } n \to \infty, \qquad (7.6)$$

if $\lim_{n \to \infty} F_n(z) = F(z)$ for each continuity point $z$ of $F$.

## 7.3 Convergence in Distribution

If $F$ is continuous, then $Z_n \xrightarrow{\mathcal{D}} Z$ as $n \to \infty$ entails even uniform convergence of $F_n$ to $F$, that is,

$$\lim_{n \to \infty} \sup_{z \in \mathbb{R}} |F_n(z) - F(z)| = 0. \tag{7.7}$$

To prove this, fix any $\varepsilon > 0$. Since $F$ is continuous there exists an integer $k$ and points $z_1, \ldots, z_k \in \mathbb{R}$ such that $z_1 < \ldots < z_k$ and $F(z_1) \leq \varepsilon$, $F(z_{j+1}) - F(z_j) \leq \varepsilon$ for each $j = 1, \ldots, k-1$, and $F(z_k) \geq 1 - \varepsilon$. We can thus choose $n_0$ sufficiently large such that

$$\max_{j=1,\ldots,k} |F_n(x_j) - F(x_j)| \leq \varepsilon$$

for every $n \geq n_0$. Now fix any real number $z$. We consider three cases: $z \leq z_1$, $z_j < z \leq z_{j+1}$ for some $j \in \{1, \ldots, k-1\}$, or $z > z_k$. If $z_j < z \leq z_{j+1}$, the monotonicity of $F_n$ and $F$, combined with the previous inequality, implies

$$F_n(z) - F(z) \leq F_n(z_{j+1}) - F(z_j) \leq F(z_{j+1}) + \varepsilon - F(z_j) \leq 2\varepsilon,$$
$$F_n(z) - F(z) \geq F_n(z_j) - F(z_{j+1}) \geq F(z_j) - \varepsilon - F(z_{j+1}) \geq -2\varepsilon$$

for each $n \geq n_0$. Thus, $|F_n(z) - F(z)| \leq 2\varepsilon$ for $n \geq n_0$. Using $0 \leq F_n(z), F(z) \leq 1$ the cases $z \leq z_1$ and $z > z_k$ can be handled similarly, leading to the conclusion that (7.7) holds.

If $z$ is a continuity point of $F$ and $(z_n)_{n \geq 1}$ is a sequence that converges to $z$, then $Z_n \xrightarrow{\mathcal{D}} Z$ implies

$$\lim_{n \to \infty} F_n(z_n) = F(z). \tag{7.8}$$

Additionally,

$$\lim_{n \to \infty} \mathbb{P}(Z_n < z) = F(z). \tag{7.9}$$

To prove (7.8), fix any $\varepsilon > 0$. Since $F$ has at most countably many discontinuity points, there exists some $\delta > 0$ such that $F(z + \delta) \leq F(z) + \varepsilon$ and $z + \delta$ is a continuity point of $F$. Since $z_n \to z$, we have $z_n \leq z + \delta$ for sufficiently large $n$, and thus $F_n(z_n) \leq F_n(z + \delta)$ for such $n$ (because $F_n$ is an increasing function). It follows that

$$\limsup_{n \to \infty} F_n(z_n) \leq \lim_{n \to \infty} F_n(z + \delta) = F(z + \delta) \leq F(z) + \varepsilon.$$

Thus, $\limsup_{n\to\infty} F_n(z_n) \leq F(z)$ since $\varepsilon > 0$ was arbitrary. Similarly,

$$\liminf_{n\to\infty} F_n(z_n) \geq F(z),$$

which proves (7.8). Moreover, because $F_n(z - \delta) \leq \mathbb{P}(Z_n < z) \leq F_n(z)$ for every $\delta > 0$, (7.6) yields

$$F(z - \delta) \leq \liminf_{n\to\infty} \mathbb{P}(Z_n < z) \leq \limsup_{n\to\infty} \mathbb{P}(Z_n < z) \leq F(z),$$

if $z - \delta$ is a continuity point of $F$. Letting $\delta$ approach zero under this condition, we obtain (7.9).

## 7.4 Central Limit Theorems

Suppose $Y_1, Y_2, \ldots$ are i.i.d. random variables with $\mathbb{E}(Y_1^2) < \infty$ and positive variance $\sigma^2 := \mathbb{V}(Y_1)$. Let $\mu := \mathbb{E}(Y_1)$. The *Lindeberg–Lévy central limit theorem* states that for any $a, b$ with $-\infty \leq a < b \leq \infty$:

$$\lim_{n\to\infty} \mathbb{P}\left(a \leq \frac{\sum_{j=1}^{n} Y_j - n\mu}{\sigma\sqrt{n}} \leq b\right) = \int_a^b \frac{1}{\sqrt{2\pi}} \exp\left(-\frac{t^2}{2}\right) dt.$$

For those who are familiar with characteristic functions, the proof of this result is straightforward (see, e.g., [DU], p. 124). Alternatively, the Stein-Chen method also provides a proof (see, e.g., [CGS], p. 16).

In the special case where $Y_j$ follows a Bernoulli distribution Bernoulli($p$), with $0 < p < 1$, we have the celebrated *De Moivre-Laplace central limit theorem* for the binomial distribution, which states that

$$\lim_{n\to\infty} \mathbb{P}\left(a \leq \frac{\sum_{j=1}^{n} Y_j - np}{\sqrt{np(1-p)}} \leq b\right) = \int_a^b \frac{1}{\sqrt{2\pi}} \exp\left(-\frac{t^2}{2}\right) dt,$$

where $-\infty < a < b < \infty$. This theorem can be proved by fairly elementary methods, such as Stirling's formula (7.17) (see, e.g., [FE1], p. 182 ff.).

## 7.5 Inequalities for the Natural Logarithm

We frequently utilize the following inequalities for the natural logarithm:

$$\log t \leq t - 1, \tag{7.10}$$

$$\log t \geq 1 - \frac{1}{t}, \tag{7.11}$$

**Fig. 7.1** Illustrating the inequalities (7.10) and (7.11)

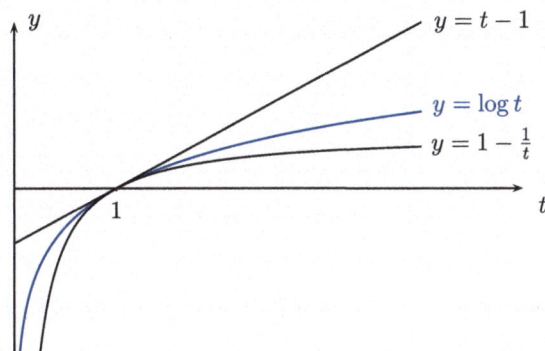

(refer to Fig. 7.1). The first inequality holds because the natural logarithm is a concave function, as its second derivative is strictly negative. The second inequality follows from the first by substituting $t$ with $\frac{1}{t}$. In both inequalities, equality holds only when $t = 1$.

## 7.6 Expectation and Variance of Non-negative Integer-valued Random Variables

The expectation $\mathbb{E}(Y)$ of an $\mathbb{N}_0$-valued random variable $Y$ exists if and only if the series $\sum_{n=1}^{\infty} \mathbb{P}(Y \geq n)$ converges. In this case,

$$\mathbb{E}(Y) = \sum_{n=1}^{\infty} \mathbb{P}(Y \geq n). \tag{7.12}$$

To prove this, note that

$$\sum_{k=1}^{\infty} k\mathbb{P}(Y=k) = \sum_{k=1}^{\infty} \left(\sum_{n=1}^{k} 1\right) \mathbb{P}(Y=k) = \sum_{n=1}^{\infty} \sum_{k=n}^{\infty} \mathbb{P}(Y=k) = \sum_{n=1}^{\infty} \mathbb{P}(Y \geq n).$$

Moreover, these equalities remain valid even if $\sum_{k=1}^{\infty} k\mathbb{P}(Y=k) = \infty$. Thus, the assertion follows. Similarly, $\mathbb{E}(Y^2)$ exists if and only if the series $\sum_{n=1}^{\infty}(2n-1)\mathbb{P}(Y \geq n)$ converges. In this case,

$$\mathbb{E}(Y^2) = \sum_{n=1}^{\infty}(2n-1)\mathbb{P}(Y \geq n). \tag{7.13}$$

For a proof, note that

$$\sum_{n=1}^{\infty}(2n-1)\mathbb{P}(Y\geq n)=\sum_{n=1}^{\infty}(2n-1)\sum_{k=n}^{\infty}\mathbb{P}(Y=k)=\sum_{k=1}^{\infty}\left(\sum_{n=1}^{k}(2n-1)\right)\mathbb{P}(Y=k)$$

$$=\sum_{k=1}^{\infty}k^2\mathbb{P}(Y=k)=\mathbb{E}(Y^2).$$

Since $\mathbb{V}(Y) = \mathbb{E}(Y^2) - (\mathbb{E}Y)^2$, Eqs. (7.12) and (7.13) can be used to calculate the variance of $Y$.

At certain points, we need the following result, which is a special case of the monotone convergence theorem from integration theory (see, e.g., [BI2], p. 208): If $Y_1, Y_2, \ldots$ is a sequence of non-negative integer-valued random variables that may also take the value $\infty$, then

$$\mathbb{E}\left(\sum_{j=1}^{\infty} Y_j\right) = \sum_{j=1}^{\infty} \mathbb{E}(Y_j). \tag{7.14}$$

Here, it is possible for both sides of this equation to be $\infty$. This may occur if $\mathbb{E}(Y_j) = \infty$ for some $j$, or if the right-hand side is a series of real numbers that does not converge.

## 7.7 The Wallis Product Representation for $\pi$

The famous product representation for $\pi$, due to John Wallis,[1] states that

$$\pi = \lim_{n\to\infty} \frac{2^2 \cdot 4^2 \cdot \ldots \cdot (2n)^2}{1^2 \cdot 3^2 \cdot \ldots \cdot (2n-1)^2} \cdot \frac{1}{n}. \tag{7.15}$$

To prove this, define $\sin^n x := (\sin x)^n$ and $\cos^n x := (\cos x)^n$ for each $n \geq 0$ and $x \in \mathbb{R}$, and let

$$I_n := \int_0^{\pi/2} \sin^n x \, dx, \quad n \in \mathbb{N}_0.$$

---

[1] John Wallis (1616–1703), British mathematician. Wallis influenced, among other things, the development of analysis before Newton. He is credited with the infinity symbol $\infty$.

## 7.7 The Wallis Product Representation for $\pi$

We have $I_0 = \frac{\pi}{2}$ and $I_1 = 1$, and for $k \geq 2$, partial integration gives

$$I_k = \int_0^{\pi/2} \sin^{k-1} x \sin x \, dx = \left[ -\sin^{k-1} x \cos x \right]_0^{\pi/2}$$
$$+ (k-1) \int_0^{\pi/2} \sin^{k-2} x \cos^2 x \, dx.$$

Here, the first term vanishes, and with $\cos^2 x = 1 - \sin^2 x$, it follows that $I_k = (k-1)I_{k-2} - (k-1)I_k$. Thus,

$$I_k = \frac{k-1}{k} \cdot I_{k-2}, \quad k \geq 2.$$

Because $I_0 = \frac{\pi}{2}$ and $I_1 = 1$, we obtain

$$I_{2n} = \frac{2n-1}{2n} \cdot \frac{2n-3}{2n-2} \cdot \ldots \cdot \frac{3}{4} \cdot \frac{1}{2} \cdot \frac{\pi}{2},$$

$$I_{2n+1} = \frac{2n}{2n+1} \cdot \frac{2n-2}{2n-1} \cdot \ldots \cdot \frac{4}{5} \cdot \frac{2}{3} \cdot 1.$$

Noting that $I_{2n+1} \leq I_{2n} \leq I_{2n-1}$, it follows that

$$\frac{2 \cdot 4 \cdot \ldots \cdot (2n)}{3 \cdot 5 \cdot \ldots \cdot (2n+1)} \leq \frac{1 \cdot 3 \cdot (2n-1)}{2 \cdot 4 \cdot \ldots \cdot (2n)} \cdot \frac{\pi}{2} \leq \frac{2 \cdot 4 \cdot \ldots \cdot (2n-2)}{3 \cdot 5 \cdot \ldots \cdot (2n-1)}.$$

Defining

$$a_n := \frac{2^2 \cdot 4^2 \cdot \ldots \cdot (2n)^2}{1^2 \cdot 3^2 \cdot \ldots \cdot (2n-1)^2} \cdot \frac{1}{2n+1},$$

$$b_n := \frac{2^2 \cdot 4^2 \cdot \ldots \cdot (2n)^2}{1^2 \cdot 3^2 \cdot \ldots \cdot (2n-1)^2} \cdot \frac{1}{2n},$$

and $c_n := a_n(2n+1)$, we have $a_n \leq \frac{\pi}{2} \leq b_n$, as well as

$$0 \leq b_n - \frac{\pi}{2} \leq b_n - a_n = c_n \left( \frac{1}{2n} - \frac{1}{2n+1} \right) = \frac{c_n}{2n(2n+1)} = \frac{a_n}{2n}.$$

Since $\lim_{n \to \infty}(b_n - a_n) = 0$ and $a_n \leq \frac{\pi}{2}$, it follows that $\lim_{n \to \infty} b_n = \frac{\pi}{2}$ which, after multiplication by 2, yields (7.15).

If we take the root in (7.15) and switch to the reciprocal, we get

$$\frac{1}{\sqrt{\pi}} = \lim_{n \to \infty} \frac{1 \cdot 3 \cdot \ldots \cdot (2n-1)}{2 \cdot 4 \cdot \ldots \cdot (2n)} \cdot \sqrt{n}.$$

Expanding the fraction by $2^n n!$, it follows that

$$\lim_{n\to\infty} \frac{\binom{2n}{n}}{2^{2n}} \sqrt{\pi n} = 1. \tag{7.16}$$

## 7.8 Stirling's Formula

Stirling's[2] formula provides an important approximation for the factorial of large numbers. It states that

$$\lim_{n\to\infty} \frac{n!\, e^n}{n^n \sqrt{2\pi n}} = 1. \tag{7.17}$$

This can be equivalently written as $n! \sim n^n e^{-n} \sqrt{2\pi n}$, indicating that Stirling's formula approximates $n!$ for large $n$.

The following elementary proof of (7.17) is adapted from [COL]. Define

$$q_n := \frac{n!\, e^n}{n^n \sqrt{n}},$$

which makes (7.17) equivalent to $q_n \to \sqrt{2\pi}$. We start with

$$\frac{q_{2n}}{q_n^2} = \frac{(2n)!\, e^{2n}\, n^{2n}\, n}{(2n)^{2n} \sqrt{2n}\, n!^2\, e^{2n}} = \binom{2n}{n} \frac{1}{2^{2n}} \sqrt{n} \cdot \frac{1}{\sqrt{2}}.$$

Using (7.16), it follows that $q_{2n}/q_n^2 \to 1/\sqrt{2\pi}$. Assuming $q_n \to q$ for some $q > 0$, we have

$$\frac{q_{2n}}{q_n^2} \to \frac{q}{q^2} = \frac{1}{q} = \frac{1}{\sqrt{2\pi}},$$

which gives $q = \sqrt{2\pi}$. Thus, we need to show $(q_n)$ converges to *some* $q > 0$. Consider the logarithm of $q_n$:

$$\log q_n = \log n! - n \log n + \frac{1}{2} \log n.$$

Noting that $\log n! = \sum_{k=1}^{n} \log k$, we look at the graph of $\log x$ over the interval $[k, k+1]$. Since $\log x$ is concave, the area of the trapezoid in Fig. 7.2 (left), denoted

---

[2] James Stirling (1692–1770) became a member of the London Royal Society in 1726 and served as the managing director of the Scottish Mining company in Leadhills from 1735. His main areas of work were algebraic curves, difference calculation, and asymptotic developments.

## 7.8 Stirling's Formula

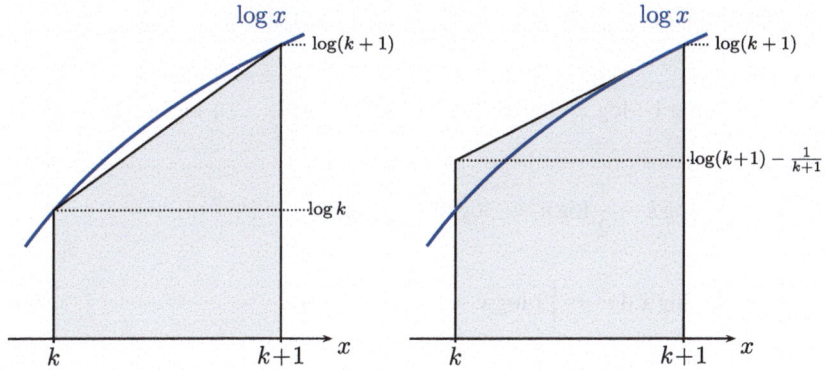

**Fig. 7.2** Trapezoidal areas as bounds for $\int_k^{k+1} \log x \, dx$

by $s_k$, is smaller than the integral of $\log x$ over $[k, k+1]$, i.e.,

$$s_k := \frac{\log k + \log(k+1)}{2} < \int_k^{k+1} \log x \, dx. \tag{7.18}$$

The tangent to $\log x$ at $(k+1, \log(k+1))$ provides the upper bound

$$\int_k^{k+1} \log x \, dx < \log(k+1) - \frac{1}{2(k+1)}. \tag{7.19}$$

Similarly, considering the tangent at $(k, \log k)$, we get

$$\int_k^{k+1} \log x \, dx < \log k + \frac{1}{2k}. \tag{7.20}$$

The mean of the upper bounds in (7.19) and (7.20) is also an upper bound for $\int_k^{k+1} \log x \, dx$. In view of (7.18), we obtain

$$s_k < \int_k^{k+1} \log x \, dx < s_k + \frac{1}{4}\left(\frac{1}{k} - \frac{1}{k+1}\right), \quad k \in \mathbb{N}. \tag{7.21}$$

By the definition of $s_k$, and using $\log 1 = 0$ and (7.18), it follows that

$$\sum_{k=1}^{n-1} s_k = \frac{1}{2}\big(\log 1 + \log 2 + \log 2 + \log 3 + \log 3 + \log 4 + \ldots + \log(n-1) + \log n\big)$$

$$= \sum_{k=1}^{n} \log k - \frac{1}{2}\log n = \log n! - \frac{1}{2}\log n$$

$$< \int_1^n \log x \, dx = \big[x \log x - x\big]_1^n \qquad (7.22)$$

$$= n \log n - n + 1.$$

If we consider the second inequality in (7.21) and sum over $k$ from 1 to $n-1$, a telescoping effect gives

$$\sum_{k=1}^{n-1} s_k < n \log n - n + 1 \ \left(= \int_1^n \log x \, dx\right)$$

$$< \log n! - \frac{1}{2}\log n + \frac{1}{4} - \frac{1}{4n}. \qquad (7.23)$$

Define the sequence $(b_n)$ by

$$b_n := \log n! - \frac{1}{2}\log n - (n \log n - n + 1) = \log n! - \left(n + \frac{1}{2}\right)\log n + n - 1, \quad n \geq 1.$$

Due to the strict inequality in (7.22), $b_n < 0$. Moreover, the sequence $(b_n)$ is strictly decreasing, because $-b_n$ is the area between the $x$-axis and the graph of $\log x$ over $[1,n]$, reduced by a smaller sum of trapezoidal areas. Thus, the sequence $(-b_n)$ is increasing, and therefore $(b_n)$ is decreasing. Due to (7.23) and (7.22), we also have $b_n > -\frac{1}{4}$, so the decreasing sequence $(b_n)$ is bounded from below. Therefore, $\lim_{n\to\infty} b_n = b$ for some $b < 0$. Consequently, $e^{b_n} \to e^b$. Since $e^{b_n} = q_n/e$, it follows that $q_n \to e^{b+1} =: q$, completing the proof.

## 7.9  Generating Functions

Power series are fundamental in introductory analysis courses. For a sequence of real numbers $a := (a_k)_{k \geq 0}$, the power series

$$g(t) := \sum_{k=0}^{\infty} a_k t^k$$

## 7.9 Generating Functions

is called the *generating function of a*. We assume that this series converges absolutely for some $t \neq 0$, implying it has a non-zero radius of convergence. The function $g$ is defined on an interval $(-t_0, t_0)$ with $0 < t_0 < \infty$, and on this interval, $g$ can be differentiated infinitely often. Moreover, differentiation can be performed term by term. For any $r \geq 1$,

$$g^{(r)}(t) = \frac{d^r}{dt^r} g(t) = \sum_{k=r}^{\infty} k(k-1)\ldots(k-r+1)\, a_k\, t^{k-r}, \qquad |t| < t_0,$$

implying that $g^{(r)}(0) = r!\, a_r$ (see, e.g., [TA], Proposition 4.2.6). Together with $a_0 = g(0)$, it follows that the sequence $a$ is uniquely determined by its generating function.

A random variable $N$ taking values in $\mathbb{N}_0 \cup \{\infty\}$ defines a special sequence $(a_k)$, where $a_k := \mathbb{P}(N = k)$ for $k \geq 0$. In this case, the generating function of the sequence $(a_k)$ is called the *generating function of $N$*, denoted as

$$g_N(t) := \sum_{k=0}^{\infty} \mathbb{P}(N = k)\, t^k. \tag{7.24}$$

Since this series converges for $t = 1$, the distribution of $N$ is uniquely determined by the function $g_N$. Note that $g_N(1) = \mathbb{P}(N < \infty)$. We assume $g_N(1) = 1$, meaning $\mathbb{P}(N = \infty) = 0$, which is necessary for $N$ to have a finite expectation.

If $\sum_{k=0}^{n} \mathbb{P}(N = k) = 1$ for some integer $n$ with $\mathbb{P}(N = n) > 0$, then $g_N$ is a polynomial of degree $n$. This holds, for example, for a binomially distributed random variable $N$ with parameters $n$ and $p$. In this case,

$$g_N(t) = \sum_{k=0}^{n} \binom{n}{k} p^k (1-p)^{n-k} t^k = \sum_{k=0}^{n} \binom{n}{k} (pt)^k (1-p)^{n-k} = (1 - p + pt)^n.$$

If $M$ and $N$ are independent $\mathbb{N}_0$-valued random variables with generating functions $g_M(t)$ and $g_N(t)$, respectively, the discrete convolution formula

$$\mathbb{P}(M + N = k) = \sum_{j=0}^{k} \mathbb{P}(M = j)\mathbb{P}(N = k - j)$$

for each $k \in \mathbb{N}_0$ and $t \in [-1, 1]$ implies

$$g_M(t)\, g_N(t) = \left( \sum_{m=0}^{\infty} \mathbb{P}(M = m)\, t^m \right) \left( \sum_{n=0}^{\infty} \mathbb{P}(N = n)\, t^n \right)$$

$$= \sum_{m=0}^{\infty} \sum_{n=0}^{\infty} \mathbb{P}(M = m)\, \mathbb{P}(N = n)\, t^{m+n}$$

$$= \sum_{k=0}^{\infty} \left( \sum_{j=0}^{k} \mathbb{P}(M=j)\,\mathbb{P}(N=k-j) \right) t^k$$

$$= \sum_{k=0}^{\infty} \mathbb{P}(M+N=k)\, t^k$$

$$= g_{M+N}(t).$$

Thus, the generating function of the sum of independent random variables is the product of the generating functions of the summands.

Differentiating the generating function $g_N$ in (7.24) twice yields

$$g'_N(t) = \sum_{k=1}^{\infty} k\,\mathbb{P}(N=k)\, t^{k-1}, \qquad |t|<1,$$

$$g''_N(t) = \sum_{k=2}^{\infty} k(k-1)\,\mathbb{P}(N=k)\, t^{k-2}, \qquad |t|<1.$$

If these derivatives remain bounded as $t \uparrow 1$, i.e., if the left-sided first and second derivatives of $g_N$ at the point 1, denoted by $g'_N(1)$ and $g''_N(1)$, exist, then the expectation of $N$ and the *second factorial moment* $\mathbb{E}\big(N(N-1)\big)$ of $N$ are finite. Specifically,

$$\mathbb{E}(N) = g'_N(1) = \sum_{k=1}^{\infty} k\,\mathbb{P}(N=k),$$

$$\mathbb{E}(N(N-1)) = g''_N(1) = \sum_{k=2}^{\infty} k(k-1)\,\mathbb{P}(N=k).$$

The variance of $N$ is then given by

$$\mathbb{V}(N) = g''_N(1) + g'_N(1) - \big(g'_N(1)\big)^2.$$

## 7.10 Some Identities involving Binomial Coefficients

We need the following results, which hold for each $n \geq 1$ (see [RI], p. 34):

$$\sum_{k=1}^{n} k \binom{2n}{n+k} = n \binom{2n-1}{n-1}, \tag{7.25}$$

## 7.10 Some Identities involving Binomial Coefficients

$$\sum_{k=1}^{n} k \binom{2n+1}{n+k+1} = \frac{n+1}{2}\binom{2n+1}{n} - 2^{2n-1}, \qquad (7.26)$$

$$\sum_{k=1}^{n} k^2 \binom{2n}{n+k} = n\, 2^{2n-2}, \qquad (7.27)$$

$$\sum_{k=1}^{n} k^2 \binom{2n+1}{n+k+1} = (n+1)2^{2n-1} - \frac{2n+1}{2}\binom{2n}{n}. \qquad (7.28)$$

We will first prove the first two equations and denote $a_n$ as the left-hand side of (7.25) and $b_n$ as the left-hand side of (7.26). Using the recursion formula

$$\binom{m+1}{\ell} = \binom{m}{\ell-1} + \binom{m}{\ell}, \qquad m \geq 1,\ \ell = 1,\ldots,m,$$

for the binomial coefficients, a direct calculation yields

$$a_n = 2b_{n-1} + 2^{2n-2}, \quad n \geq 2, \qquad (7.29)$$

$$b_n = 2a_n - \frac{1}{2}\left(2^{2n} - \binom{2n}{n}\right), \quad n \geq 1. \qquad (7.30)$$

Define

$$A(x) := \sum_{n=1}^{\infty} a_n x^n, \qquad B(x) := \sum_{n=1}^{\infty} b_n x^n,$$

as the generating functions of the sequences $(a_n)$ and $(b_n)$, respectively. Since $\binom{2n}{n+k} \leq \binom{2n}{n}$ and $\binom{2n+1}{n+k+1} \leq \binom{2n+2}{n+1}$, as well as $\binom{2n}{n} = u_{2n}2^{2n}$, with $u_{2n}$ as in (2.15), (2.29) implies that the above infinite series converge for $|x| < \frac{1}{4}$. This condition is tacitly assumed in the following. From (7.29) and the geometric series formula $\sum_{k=0}^{\infty} y^k = \frac{1}{1-y}$ for $|y| < 1$, we get

$$A(x) = 2x B(x) + \frac{x}{1-4x}. \qquad (7.31)$$

Now, (7.33) and (7.34) imply

$$\sum_{n=1}^{\infty} \binom{2n}{n} x^n = \sum_{n=1}^{\infty} \binom{-\frac{1}{2}}{n}(-4x)^n = (1-4x)^{-1/2} - 1.$$

Using (7.30), it follows that

$$B(x) = 2A(x) - \frac{1}{2} \cdot \frac{1}{1-4x} + \frac{1}{2} \cdot \frac{1}{\sqrt{1-4x}}. \tag{7.32}$$

From (7.31) and (7.32), we thus obtain

$$A(x) = \frac{x}{(1-4x)^{3/2}}, \qquad B(x) = \frac{1}{2(1-4x)^{3/2}} - \frac{1}{2(1-4x)}.$$

Using the binomial expansion formula (7.34) for $\alpha = -\frac{3}{2}$, we get

$$A(x) = \sum_{n=1}^{\infty} n \binom{2n-1}{n-1} x^n,$$

$$B(x) = \sum_{n=1}^{\infty} \left( \frac{n+1}{2} \binom{2n+1}{n} - 2^{2n-1} \right) x^n,$$

from which (7.25) and (7.26) follow. The proofs of (7.27) and (7.28) are analogous. Let $u_n$ and $v_n$ denote the left-hand sides of (7.27) and (7.28), respectively. We get

$$u_n = 2v_{n-1} + n\binom{2n-1}{n-1}, \qquad n \geq 2,$$

$$v_n = 2u_n - \frac{2n+1}{2}\binom{2n}{n} + 2^{2n-1}, \qquad n \geq 1.$$

Define $U(x) := \sum_{n=1}^{\infty} u_n x^n$ and $V(x) := \sum_{n=1}^{\infty} v_n x^n$ for $|x| < \frac{1}{4}$. The above equations imply

$$U(x) = 2xV(x) + \frac{x}{(1-4x)^{3/2}},$$

$$V(x) = 2U(x) + \frac{1}{2(1-4x)} - \frac{1}{2(1-4x)^{3/2}}.$$

Thus, we have

$$U(x) = \frac{x}{(1-4x)^2}, \qquad V(x) = \frac{1}{2(1-4x)^2} - \frac{1}{2(1-4x)^{3/2}}.$$

By expanding these functions into power series, we derive (7.27) and (7.28).

## 7.11 The Binomial Series

For any real number $\alpha$ and any non-negative integer $n$, the binomial coefficient $\binom{\alpha}{n}$ is defined by $\binom{\alpha}{0} := 1$ and for $n \geq 1$,

$$\binom{\alpha}{n} := \frac{\alpha \cdot (\alpha - 1) \cdot \ldots \cdot (\alpha - n + 1)}{n!}.$$

In particular,

$$\binom{1/2}{n} = \frac{1}{n!} \cdot \frac{1}{2} \cdot \left(\frac{1}{2} - 1\right) \cdot \left(\frac{1}{2} - 2\right) \cdot \ldots \cdot \left(\frac{1}{2} - n + 1\right)$$

$$= \frac{(-1)^{n-1}}{2^n \cdot n!} \cdot 1 \cdot 3 \cdot 5 \cdot \ldots \cdot (2n - 3) = \frac{(-1)^{n-1}(2n - 2)!}{2^{2n-1}(n - 1)!n!}$$

$$= (-1)^{n-1} \frac{2}{n} \frac{\binom{2n-2}{n-1}}{2^{2n}}.$$

Similarly,

$$\binom{-1/2}{n} = (-1)^n 2^{-2n} \binom{2n}{n}. \tag{7.33}$$

With the definition of $\binom{\alpha}{n}$, for any $x$ with $|x| < 1$, the following holds:

$$(1 + x)^\alpha = \sum_{n=0}^{\infty} \binom{\alpha}{n} x^n. \tag{7.34}$$

The series that appears here is called the *binomial series* (see, e.g., [AP], p. 244).

We also need the binomial series

$$(1 + z)^{-1/2} = \sum_{n=0}^{\infty} (-1)^n \frac{\binom{2n}{n}}{2^{2n}} z^n, \qquad z \in \mathbb{C}, \ |z| < 1, \tag{7.35}$$

for a complex-valued variable. If we multiply the right-hand side of this equation by itself and note that the sum of the probabilities appearing in part (a) of Theorem 2.4

is equal to 1, it follows that

$$\sum_{k,\ell=0}^{\infty} (-1)^{k+\ell} \frac{\binom{2k}{k}\binom{2\ell}{\ell}}{2^{2(k+\ell)}} z^{k+\ell} = \sum_{n=0}^{\infty} \sum_{k=0}^{n} \frac{\binom{2k}{k}\binom{2(n-k)}{n-k}}{2^{2n}} (-z)^n$$

$$= \sum_{n=0}^{\infty} (-z)^n = \frac{1}{1+z}.$$

Consequently, Eq. (7.35) holds for any complex number $z$ with $|z| < 1$.

## 7.12 Legendre Polynomials

If two sides of a triangle, one of length 1 and the other of length $x < 1$, enclose the angle $\vartheta$, then according to the cosine rule, the reciprocal of the length of the third side is given by $1/\sqrt{1 - 2x \cos \vartheta + x^2}$. Considering this expression as a function of $x$ and expanding it into a power series around 0, the resulting coefficient of $x^n$ is a polynomial of degree $n$ in $\cos \vartheta$. This polynomial is called the *Legendre*[3] *polynomial of order n*, denoted by $P_n$. By definition,

$$\frac{1}{\sqrt{1 - 2xt + x^2}} =: \sum_{n=0}^{\infty} P_n(t)\, x^n, \qquad t \in \mathbb{R}, \tag{7.36}$$

for every $x$ whose absolute value is sufficiently small. To provide a closed expression for $P_n(t)$, we set $y := x^2 - 2xt$. If $|y| < 1$, then, using the binomial series (7.34), the binomial theorem, as well as (7.33), we obtain

$$\frac{1}{\sqrt{1 - 2xt + x^2}} = (1+y)^{-1/2} = \sum_{r=0}^{\infty} \binom{-1/2}{r} y^r$$

$$= \sum_{r=0}^{\infty} \frac{\binom{2r}{r}}{2^{2r}} (-1)^r \sum_{k=0}^{r} \binom{r}{k} x^{2k} (-2tx)^{r-k}$$

$$= \sum_{r=0}^{\infty} \sum_{k=0}^{r} (-1)^k \frac{\binom{2r}{r}}{2^{2r+k}} \binom{r}{k} t^{r-k} x^{r+k}.$$

---

[3] Adrien-Marie Legendre (1752–1833), a French mathematician, was a member of the "Bureau de Longitudes" from 1813 to 1833, succeeding J.L. Lagrange. His main areas of work included celestial mechanics, the calculus of variations, the least squares method, number theory, and the foundations of geometry.

## 7.12 Legendre Polynomials

Introducing the summation index $n := r + k$ and noting that for a fixed $n$, the index $k$ runs from 0 to $\lfloor \frac{n}{2} \rfloor$, it follows that

$$\frac{1}{\sqrt{1 - 2xt + x^2}} = \sum_{n=0}^{\infty} \left( \frac{1}{2^n} \sum_{k=0}^{\lfloor n/2 \rfloor} (-1)^k \frac{(2n - 2k)!}{k!(n - k)!(n - 2k)!} t^{n-2k} \right) x^n.$$

This results in the closed-form expression

$$P_n(t) = \frac{1}{2^n} \sum_{k=0}^{\lfloor n/2 \rfloor} (-1)^k \frac{(2n - 2k)!}{k!(n - k)!(n - 2k)!} \cdot t^{n-2k} \tag{7.37}$$

for the $n$-th Legendre polynomial. The first four Legendre polynomials are:

$$P_0(t) = 1, \quad P_1(t) = t, \quad P_2(t) = \frac{1}{2}(3t^2 - 1), \quad P_3(t) = \frac{1}{2}(5t^3 - 3t).$$

Another representation of the Legendre polynomials is

$$P_n(t) = \frac{1}{2^n n!} \cdot \frac{d^n}{dt^n} (t^2 - 1)^n. \tag{7.38}$$

The equality of (7.37) and (7.38) becomes apparent when one considers

$$\frac{d^n}{dt^n}(t^2 - 1)^n = \sum_{k=0}^{n} (-1)^{n-k} \binom{n}{k} \frac{d^n}{dt^n} t^{2k}.$$

The derivative on the right-hand side vanishes for $k < \frac{n}{2}$; otherwise, it equals $\frac{t^{2k-n}(2k)!}{(2k-n)!}$. Using $t^2 - 1 = (t - 1)(t + 1)$ and Leibniz's rule for the differentiation of products, (7.38) yields the further representation

$$P_n(t) = \frac{1}{2^n} \sum_{k=0}^{n} \binom{n}{k}^2 (t - 1)^k (t + 1)^{n-k}.$$

From this, using direct computation, the relationship

$$P_n'(1) = \frac{n(n + 1)}{2}, \tag{7.39}$$

required in Sect. 2.3, follows.

Due to the orthogonality property $\int_{-1}^{1} P_n(t) P_m(t)\, dt = 0$ for $m \neq n$, the Legendre polynomials play an important role in approximation theory and numerical quadrature (see, e.g., [BEL], Chap. 3).

## 7.13 The Borel–Cantelli Lemma

If $A_1, A_2, \ldots$ is a sequence of events in a probability space $(\Omega, \mathcal{A}, \mathbb{P})$, the *limit superior* of this sequence is defined by

$$\limsup_{n \to \infty} A_n := \bigcap_{n=1}^{\infty} \bigcup_{k=n}^{\infty} A_k.$$

Note that $\omega \in \Omega$ belongs to $\limsup_{n \to \infty} A_n$ if and only if it belongs to infinitely many of the sets $A_1, A_2, \ldots$.

The famous *Borel–Cantelli lemma* states the following:

(a) If $\sum_{n=1}^{\infty} \mathbb{P}(A_n) < \infty$, then $\mathbb{P}\left(\limsup_{n \to \infty} A_n\right) = 0$.

(b) If $A_1, A_2, \ldots$ are independent events, then:

$$\sum_{n=1}^{\infty} \mathbb{P}(A_n) = \infty \implies \mathbb{P}\left(\limsup_{n \to \infty} A_n\right) = 1.$$

**Proof**

(a) Let $B_k := \bigcup_{n=k}^{\infty} A_k$ for $k \geq 1$. The sequence of sets $(B_k)$ is decreasing, and $\limsup_{n \to \infty} A_n = \bigcap_{k=1}^{\infty} B_k$. Since a probability measure is continuous from above, it follows that $\mathbb{P}\left(\limsup_{n \to \infty} A_n\right) = \lim_{k \to \infty} \mathbb{P}(B_k)$. The $\sigma$-subadditivity of $\mathbb{P}$ yields $\mathbb{P}(B_k) \leq \sum_{n=k}^{\infty} \mathbb{P}(A_n)$, and the latter expression, being the tail of a convergent series, converges to zero as $k \to \infty$, which is what we wanted to show.

(b) Note that $1 - x \leq e^{-x}$ for any real number $x$. Setting $x = \mathbb{P}(A_k)$, we have

$$1 - \exp\left(-\sum_{k=n}^{m} \mathbb{P}(A_k)\right) \leq 1 - \prod_{k=n}^{m}(1 - \mathbb{P}(A_k)) \leq 1$$

for any integers $n$ and $m$ with $1 \leq n < m$. Taking the limit as $m \to \infty$ yields

$$\lim_{m \to \infty} \prod_{k=n}^{m}(1 - \mathbb{P}(A_k)) = 0. \tag{7.40}$$

## 7.13 The Borel–Cantelli Lemma

Since $A_n, A_{n+1}, \ldots, A_m$ are independent events, we obtain

$$\begin{aligned}
1 - \mathbb{P}\left(\limsup_{n\to\infty} A_n\right) &= 1 - \lim_{n\to\infty} \mathbb{P}\left(\bigcup_{k=n}^{\infty} A_k\right) \\
&= \lim_{n\to\infty} \mathbb{P}\left(\bigcap_{k=n}^{\infty} A_k^c\right) \\
&= \lim_{n\to\infty}\left[\lim_{m\to\infty} \mathbb{P}\left(\bigcap_{k=n}^{m} A_k^c\right)\right] \\
&= \lim_{n\to\infty}\left[\lim_{m\to\infty} \prod_{k=n}^{m}(1 - \mathbb{P}(A_k))\right] \\
&= 0,
\end{aligned}$$

which was to be shown. The last equality holds due to (7.40).

∎

For further reading on the Borel–Cantelli lemma, see [CH].

# Solutions to the Exercises

**Solution 2.1**

(a) Using (2.8), it follows that

$$\mathbb{E}|S_{2n}| = \sum_{k=-n}^{n} |2k|\, \mathbb{P}(S_{2n}=2k) = \sum_{k=-n}^{n} |2k| \frac{\binom{2n}{n+k}}{2^{2n}}$$

$$= \frac{2}{2^{2n}} \sum_{k=-1}^{-n} |k| \binom{2n}{n+k} + \frac{2}{2^{2n}} \sum_{k=1}^{n} k \binom{2n}{n+k}.$$

According to (7.25), the second sum is equal to $n\binom{2n-1}{n-1}$, and because $\binom{2n}{n+k} = \binom{2n}{n-k}$, the first sum is the same as the second one. Hence,

$$\mathbb{E}|S_{2n}| = \frac{4}{2^{2n}} \cdot n \cdot \binom{2n-1}{n-1} = 2n \frac{\binom{2n}{n}}{2^{2n}}.$$

(b) Setting

$$a_n := \frac{n!\, e^n}{n^n \sqrt{2\pi n}}, \qquad n \in \mathbb{N},$$

we have

$$\frac{a_{2n}}{a_n^2} = \frac{(2n)!\, e^{2n}\, n^{2n}\, 2\pi n}{(2n)^{2n}\, 2\sqrt{\pi n}\, n!^2\, e^{2n}} = \frac{\binom{2n}{n}}{2^{2n}} \sqrt{\pi n}.$$

In view of (7.17), $a_n \to 1$. Thus, $a_{2n}/a_n^2 \to 1$, from which the asymptotic equality (2.9) follows.

**Solution 2.2**

(a) There are $\binom{m+n}{m}$ paths, since out of a total of $m + n$ steps, $m$ must be selected for the rightward steps.

(b) There are $\binom{a+b}{a}$ paths leading from $(0, 0)$ to $(a, b)$. Each of these paths can be continued in $\binom{m-a+n-b}{m-a}$ ways to form a path from $(a, b)$ to $(m, n)$. According to the multiplication rule of combinatorics, there are thus

$$\binom{a+b}{a}\binom{m-a+n-b}{m-a}$$

paths from $(0, 0)$ to $(m, n)$ that visit the point $(a, b)$.

(c) Each path leading from $(0, 0)$ to $(m, n)$ visits exactly one of the points $(m-k, k)$ (located on a descending grid diagonal ending at the point $(m, 0)$), where $k \in \{0, \ldots, n\}$. According to part (b), there are

$$\binom{m-k+k}{m-k}\binom{m-(m-k)+n-k}{m-(m-k)} = \binom{m}{m-k}\binom{n}{k}$$

paths from $(0, 0)$ to $(m, n)$ that visit the point $(m-k, k)$. Summing over $k$ yields the assertion.

**Solution 2.3** The event $E^c$, complementary to $E$, states that A does not always have at least as many votes as B during the entire counting process. Let $D$ be the event that the first ballot is for B. Then $E^c = D \uplus D^c \cap E^c$, and thus $\mathbb{P}(E) = 1 - \mathbb{P}(D) - \mathbb{P}(D^c \cap E^c)$. According to the reasoning before Sect. 2.2, $\mathbb{P}(D) = \binom{a+b-1}{a} / \binom{a+b}{a} = \frac{b}{a+b}$. The paths corresponding to $D^c \cap E^c$ are those which start with an upward step and then, at some point, reach the height $-1$ for the first time. If one reflects the path segment from $(0, 0)$ to this first hitting point of height $-1$ across the horizontal line $y = -1$, one gets a path starting at $(0, -2)$, which initially takes a downward step and ends at the point $(a+b, a-b)$. According to (2.11), there are $\binom{a+b-1}{a+1}$ paths leading from $(1, -3)$ to $(a+b, a-b)$. By the reflection principle, it follows that $\mathbb{P}(D^c \cap E^c) = \binom{a+b-1}{a+1} / \binom{a+b}{a} = \frac{b(b-1)}{(a+1)(a+b)}$. Thus,

$$\mathbb{P}(E) = 1 - \mathbb{P}(E^c) = 1 - \mathbb{P}(D) - \mathbb{P}(D^c \cap E^c)$$
$$= \frac{a}{a+b} - \frac{b(b-1)}{(a+b)(a+1)} = \frac{a-b+1}{a+1}.$$

**Solution 2.4** Let

$$a_n := \sum_{j=1}^{n} \frac{\binom{2j}{j}}{2^{2j}}, \quad b_n := (2n+1)\frac{\binom{2n}{n}}{2^{2n}} - 1, \quad n \geq 1.$$

Solutions to the Exercises

According to (2.35), it is to be shown that $a_n = b_n$ for every $n \in \mathbb{N}$. We have $a_1 = b_1 = \frac{1}{2}$, which provides the base case of a proof by induction. For the induction step $n \to n+1$, we use

$$a_{n+1} = a_n + \frac{\binom{2(n+1)}{n+1}}{2^{2(n+1)}} = b_n + \frac{\binom{2(n+1)}{n+1}}{2^{2(n+1)}} = (2n+2)\frac{\binom{2(n+1)}{n+1}}{2^{2(n+1)}} - 1 + \frac{\binom{2(n+1)}{n+1}}{2^{2(n+1)}} = b_{n+1},$$

and the claim follows.

**Solution 2.5** Summing the terms on the left-hand side of Eq. (2.36) over $j$ from 0 to $n-2$, and using the index shift $k := j+2$, the result is

$$(4n^2 + 6n + 2)\sum_{k=2}^{n} p_k - (4n+3)\sum_{k=2}^{n} k p_k + \sum_{k=2}^{n} k^2 p_k.$$

The corresponding sum of the terms on the right-hand side of (2.36) is

$$4n(n-1)\sum_{j=0}^{n-2} p_j - 4(2n-1)\sum_{j=0}^{n-2} j p_j + 4\sum_{j=0}^{n-2} j^2 p_j.$$

Now, $\sum_{k=2}^{n} p_k = 1 - p_0 - p_1 = 1 - \binom{2n}{n}2^{-2n} - \binom{2n-1}{n}2^{-(2n-1)}$. Furthermore,

$$\sum_{k=2}^{n} k p_k = \mathbb{E}(N_{2n}) - p_1 = \mathbb{E}(N_{2n}) - \frac{\binom{2n-1}{n}}{2^{2n-1}}, \quad \sum_{k=2}^{n} k^2 p_k$$

$$= \mathbb{E}(N_{2n}^2) - p_1 = \mathbb{E}(N_{2n}^2) - \frac{\binom{2n-1}{n}}{2^{2n-1}}.$$

Similarly,

$$\sum_{j=0}^{n-2} p_j = 1 - p_n - p_{n-1} = 1 - 1/2^n - (n+1)/2^{n-1},$$

$$\sum_{j=0}^{n-2} j p_j = \mathbb{E}(N_{2n}) - \frac{n}{2^n} - \frac{n^2-1}{2^{n+1}},$$

$$\sum_{j=0}^{n-2} j^2 p_j = \mathbb{E}(N_{2n}^2) - \frac{n^2}{2^n} - \frac{(n-1)^2(n+1)}{2^{n+1}}.$$

Thus, the sum of the terms on the right-hand side of (2.36) is

$$4\mathbb{E}(N_{2n}^2) - 4(2n-1)\big((2n+1)p_0 - 1\big).$$

The sum of the terms on the left-hand side of (2.36) equals

$$\mathbb{E}(N_{2n}^2) + 4n^2 + 10n + 5 - (2n+1)(8n+5)p_0.$$

Equating both sums yields the assertion.

**Solution 2.6** If we set $u_{2n} = \binom{2n}{n}/2^{2n}$ and use Theorem 2.6 (d), it follows that

$$\frac{\mathbb{V}(N_{2n})}{2n} = \frac{2(n+1)}{2n} - \frac{2n+1}{2n}u_{2n} - \frac{(2n+1)^2}{2n}u_{2n}^2.$$

The first term on the right-hand side converges to one, and the second converges to zero because $u_{2n}\sqrt{\pi n} \to 1$ (see (2.29)). Since $u_{2n}^2 \pi n \to 1$, the last term converges to $\frac{2}{\pi}$. This proves the assertion.

**Solution 2.7**

(a) According to (2.45) with $n+1$ instead of $k$, $\mathbb{P}(W \le 2n) = 1 - u_{2n}$. Using Theorem 2.9(a), it follows that

$$\mathbb{P}(W = 2k|W \le 2n) = \frac{\mathbb{P}(W = 2k)}{\mathbb{P}(W \le 2n)} = \frac{1}{1 - u_{2n}} \cdot \frac{u_{2(k-1)}}{2k}, \quad k = 1, \ldots, n,$$

and thus

$$\mathbb{E}[W|W \le 2n] = \sum_{k=1}^{n} 2k\mathbb{P}(W = 2k|W \le 2n) = \frac{1}{1 - u_{2n}} \sum_{k=1}^{n} u_{2(k-1)}.$$

Since $u_0 = 1$, Theorem 2.6(b) and (2.35) (both with $n-1$ instead of $n$) yield

$$\mathbb{E}[W|W \le 2n] = \frac{1}{1 - u_{2n}}\left(1 + (2(n-1)+1)u_{2(n-1)} - 1\right) = \frac{(2n-1)u_{2(n-1)}}{1 - u_{2n}}.$$

(b) The assertion follows because $u_{2(n-1)}\sqrt{\pi(n-1)} \to 1$ (see (2.29)).

**Solution 2.8** Under the condition $X_1 = 1$ (which holds with probability $\frac{1}{2}$) and the condition $S_0 = k$, the random walk starts at time 1 at the height $k+1$. Since the event of having at least one visit to zero refers to an infinite time horizon, we can also start the random walk at time 0 at the height $k+1$, and the probability that this random walk visits zero at least once is, by definition, equal to $p(k+1)$. Similarly, under the condition $X_1 = -1$ (which also holds with probability $\frac{1}{2}$), the random walk starts at height $k-1$ at time 0, leading to the probability $p(k-1)$.

## Solutions to the Exercises

**Solution 2.9** Using Eq. (2.49), it follows that

$$\sum_{n=1}^{k} u_{2n} = \sum_{n=1}^{k}\sum_{r=1}^{n} f_{2r}\, u_{2n-2r} = \sum_{r=1}^{k} f_{2r} \sum_{n=r}^{k} u_{2n-2r}.$$

Because $u_0 = 1$ the last sum can be bounded from above by $1 + \sum_{n=1}^{n} u_{2n-2r}$, which proves the assertion.

**Solution 2.10** Setting $T := \min(W, 2n)$ yields

$$\mathbb{E}(T) = \sum_{k=1}^{n-1} 2k\mathbb{P}(W = 2k) + 2n\mathbb{P}(T \geq 2n).$$

Using Theorem 2.9(a), Eq. (2.45), and the abbreviation $u_{2m} = \binom{2m}{m}/2^{2m}$ for $m \in \mathbb{N}_0$, it follows that

$$\mathbb{E}(T) = 1 + \sum_{j=1}^{n-2} u_{2j} + 2nu_{2(n-1)}.$$

According to (2.35) and Theorem 2.6(b), the sum is equal to $(2n-1)u_{2(n-1)} - 1$. A direct calculation then yields $\mathbb{E}(T) = 4nu_{2n}$. With Solution 2.1(a), it now follows that $\mathbb{E}(T) = 2\mathbb{E}|S_{2n}|$. The second equality to be shown holds because $\mathbb{P}(S_{2n} = 0) = u_{2n}$.

**Solution 2.11** In view of (2.67),

$$\mathbb{E}(M_{2n}) = \frac{4}{2^{2n}} \sum_{k=1}^{n} k\binom{2n}{n+k} - \frac{1}{2^{2n}} \sum_{k=1}^{n} \binom{2n}{n+k}.$$

According to Eq. (2.68), the last sum is equal to $(2^{2n} - \binom{2n}{n})/2$, and using Eq. (7.25), the first sum equals $n\binom{2n-1}{n-1}$. The last expression is equal to $n\binom{2n}{n}/2$, and with $u_{2n} = \binom{2n}{n}/2^{2n}$, it follows that

$$\mathbb{E}(M_{2n}) = \frac{4}{2^{2n}}\binom{2n}{n}\frac{n}{2} - \frac{1}{2^{2n}}\frac{1}{2}\left(2^{2n} - \binom{2n}{n}\right) = 2nu_{2n} - \frac{1}{2} + \frac{u_{2n}}{2} = \left(2n + \frac{1}{2}\right)u_{2n} - \frac{1}{2},$$

which was to be shown.

**Solution 2.12** If one inserts the expressions for $\mathbb{P}(M_{2n} = 2k)$ and $\mathbb{P}(M_{2n} = 2k+1)$ from part (a) of Theorem 2.12 into Eq. (2.69), it follows that

$$\mathbb{E}(M_{2n}^2) = \frac{4}{2^{2n}} \sum_{k=1}^{n} k^2 \binom{2n}{n+k} + \frac{1}{2^{2n}} \sum_{k=0}^{n-1} (2k+1)^2 \binom{2n}{n+k+1}.$$

According to Eq. (7.27), the first sum is equal to $n2^{2n-2}$, and the second sum, with the index shift $j = k+1$ and using $(2k+1)^2 k^2 + 4k + 1$, becomes

$$4 \sum_{j=1}^{n} j^2 \binom{2n}{n+j} - 4 \sum_{j=1}^{n} j \binom{2n}{n+j} + \sum_{j=1}^{n} \binom{2n}{n+j}.$$

In view of (7.27) and (7.25), the first sum is equal to $n2^{2n-2}$, and the second equals $n\binom{2n-1}{n-1}$. The third sum is given by $(2^{2n} - \binom{2n}{n})/2$. Using $\binom{2n-1}{n-1} = \binom{2n}{n}/2$, the assertion follows after substitution and summarizing.

**Solution 2.13** According to the reflection principle, $\mathbb{P}(M_{2n-1} \geq k, S_{2n} = 0) = \mathbb{P}(S_{2n} = 2k)$. Indeed: Suppose there is a $2n$-path leading to the point $(2n, 0)$ whose maximum is greater than or equal to $k$. If we reflect the path segment starting from the first time the height $k$ is reached across the line $y = k$, we get a path starting at $(0, 0)$ with $S_{2n} = 2k$. This mapping is not only injective but also surjective, since every path leading from $(0, 0)$ to $(2n, 2k)$ can be reflected across the straight line $y = k$ from the first time the height $k$ is reached, thus leading to a path with $M_{2n-1} \geq k$ and $S_{2n} = 0$. It follows that

$$\mathbb{P}(M_{2n-1} = k, S_{2n} = 0) = \mathbb{P}(M_{2n-1} \geq k, S_{2n} = 0) - \mathbb{P}(M_{2n-1} \geq k+1, S_{2n} = 0)$$
$$= \mathbb{P}(S_{2n} = k) - \mathbb{P}(S_{2n} = 2(k+1)).$$

**Solution 2.14** With $u_{2n} = \binom{2n}{n}/2^{2n}$, Theorem 2.6(b) yields $\mathbb{E}(N_{2n}) = (2n+1)u_{2n} - 1$. According to Theorem 2.12(b), $\mathbb{E}(M_{2n}) = (2n + \frac{1}{2})u_{2n} - \frac{1}{2}$. Hence

$$\left| \mathbb{E}(N_{2n}) - \mathbb{E}(M_{2n}) \right| = \frac{1}{2} |u_{2n} - 1|.$$

Since $u_2 = \frac{1}{2}$ and $u_{2(n+1)}/u_{2n} = (2n+1)/(2n+2) < 1$ for every $n \geq 1$, it follows that $|u_{2n} - 1| \leq \frac{1}{2}$, and thus the assertion.

**Solution 2.15** Since $a < v < b$, we have $b - a \geq 2$. Thus, both $v + 2k(b - a)$ and $2b - v + 2k(b - a)$ differ by at least four for different values of $k$. Since $X$ only takes the values 1 and $-1$, at most one term can be different from zero in each of the sums on the right-hand side of (2.128). We will consider the cases $v = 1$, $v = -1$, and $v \notin \{-1, 1\}$ separately.

If $v = 1$, the term with $k = 0$ in the first sum is equal to $\frac{1}{2}$. In the second sum, a term can only be non-zero if $2b - 1 + 2k(b - a) \in \{1, -1\}$ for an integer $k$, which is equivalent to $b + k(b - a) \in \{0, 1\}$. The equation $b + k(b - a) = 0$ only has the solution $k = -1$ in the case $a = 0$. The resulting term, to be subtracted, is $\mathbb{P}(X = -1) = \frac{1}{2}$, meaning that the right-hand side of (2.128) vanishes. For $a = 0$, however, the left-hand side of (2.128) is also zero, because the condition $a < 0 = \min(0, X)$ is violated if $v = 1$. Since the equation $b + k(b - a) = 1$ has no integer solution $k$, (2.128) holds for $v = 1$ and all permitted values of $a$ and $b$, because in the case $a \leq -1$ both sides of (2.128) are equal to $\frac{1}{2}$. The case $v = -1$ follows from completely analogous considerations.

If $v \notin \{-1, 1\}$, the left-hand side of (2.128) vanishes. We consider the case that $v \geq 3$. The case $v \leq -3$ follows similarly. Because $a \leq 0$ and $v < b$, it follows that $b - a > v$. There is no integer $k$ with $v + 2k(b - a) \in \{-1, 1\}$, because if $k \geq 0$ then the inequalities $v + 2k(b - a) \geq v \geq 3$ hold, and for $k \leq -1$ we have $v + 2k(b - a) \leq v - 2(b - a) < -v \leq -3$, since $b - a > v$. Thus, the first sum on the right-hand side of (2.128) vanishes. Similarly, it follows that there is also no integer $k$ with $2b - v + 2k(b - a) \in \{-1, 1\}$. Thus, the second sum on the right-hand side of (2.128) is also zero, which completes the proof.

**Solution 2.16** Setting $u_{2k} = \binom{2k}{k}/2^{2k}$, $k \in \mathbb{N}$, Theorem 2.17 (d) gives $\mathbb{E}(Q_{2n}) = 2 - 2u_{2(n+1)}$. In view of (2.77) and $\mathbb{V}(Q_{2n}) = \mathbb{E}(Q_{2n}^2) = (\mathbb{E}(Q_{2n}))^2$, it follows that

$$\mathbb{V}(Q_{2n}) = 6 - 12u_{2(n+2)} + \left(\frac{2}{n+2} - 2\right)u_{2(n+1)} - 4 + 8u_{2(n+1)} - 4u_{2(n+1)}^2.$$

Because

$$u_{2(n+2)} = \frac{2n+3}{2(n+2)}u_{2(n+1)},$$

the assertion follows from direct calculation.

**Solution 2.17** According to part (a) of Theorem 2.21, $\mathbb{P}(V_1 = 2n + 1) = u_{2n}/(2(n+1))$, where $u_{2n} = \binom{2n}{n}/2^{2n}$. In view of (2.29), $u_{2n}\sqrt{\pi n} \to 1$. Hence,

$$\mathbb{P}(V_1 = 2n+1) \sim \frac{C}{n^{3/2}}, \quad \text{where } C = \frac{1}{2\sqrt{\pi}}.$$

**Solution 2.18** According to Theorem 2.24,

$$\mathbb{P}(V_k = n) = \frac{k}{n2^n} \cdot \binom{n}{\frac{n+k}{2}}, \quad n \geq k,$$

where $k$ and $n$ have the same parity. From this, we obtain the initial condition $\mathbb{P}(V_k = k) = 2^{-k}$. Furthermore,

$$\frac{\mathbb{P}(V_k = n+2)}{\mathbb{P}(V_k = n)} = \frac{k}{n+2}\binom{n+2}{\frac{n+2+k}{2}}\frac{1}{2^{n+2}} \cdot \frac{n}{k} \cdot \frac{2^n}{\binom{n}{\frac{n+k}{2}}}$$

$$= \frac{n}{4(n+2)} \cdot \frac{(n+2)!\left(\frac{n+k}{2}\right)!\left(n-\frac{n+k}{2}\right)!}{\left(\frac{n+k}{2}+1\right)!\left(n+2-\frac{n+k}{2}-1\right)!n!}$$

$$= \frac{n(n+1)}{4\left(\frac{n+k}{2}+1\right)\left(n+1-\frac{n+k}{2}\right)} = \frac{n(n+1)}{(n+k+2)(n+2-k)}.$$

Thus, the ratio of the probabilities is

$$\frac{\mathbb{P}(V_k = n+2)}{\mathbb{P}(V_k = n)} = \frac{n(n+1)}{(n+k+2)(n+2-k)}.$$

**Solution 2.19** From (2.90), we first get

$$2\left(1 - \sqrt{1-t}\right) - t = 4\sum_{n=1}^{\infty} \frac{\binom{2n-2}{n-1}}{n 2^{2n}} t^n - t = 4\sum_{n=2}^{\infty} \frac{\binom{2n-2}{n-1}}{n 2^{2n}} t^n.$$

Thus,

$$\frac{2\left(1 - \sqrt{1-t}\right) - t}{t} = 4\sum_{n=2}^{\infty} \frac{\binom{2n-2}{n-1}}{n 2^{2n}} t^{n-1}.$$

If we set $k := n - 1$, the right-hand side becomes

$$\sum_{k=1}^{\infty} \frac{\binom{2k}{k}}{(k+1)2^k} t^k,$$

which was to be shown.

**Solution 2.20** We start from Eq. (2.96), which directly follows from (7.28). Using $\mathbb{V}(C_{2n+1}) = \mathbb{E}(C_{2n-1}^2) - (\mathbb{E}(C_{2n+1}))^2$ and part (b) of Theorem 2.33, together with

$$\frac{\binom{2n}{n}}{2^{2n}} = \frac{2(n+1)}{2n+1} \cdot \frac{\binom{2n+1}{n}}{2^{2n+1}},$$

it follows that

$$\mathbb{V}(C_{2n+1}) = \frac{n+1}{2} - \frac{2n+1}{2} \cdot \frac{2(n+1)}{2n+1} \frac{\binom{2n+1}{n}}{2^{2n+1}} - \left((n+1)\frac{\binom{2n+1}{n}}{2^{2n+1}}\right)^2$$

$$-\frac{1}{4} + (n+1)\frac{\binom{2n+1}{n}}{2^{2n+1}}$$

$$= \frac{n+1}{2} - \frac{1}{4} - \left((n+1)\frac{\binom{2n+1}{n}}{2^{2n+1}}\right)^2.$$

**Solution 2.21** In the dual walk $\{(\ell, S_\ell^*) : \ell = 0, \ldots, 2n\}$ defined by (2.116) and (2.117), the event of interest that the original walk reaches the final height $S_{2n}$ for the first time at time $2k$ is equivalent to the event $\{S_{2n-2k}^* = 0\} \cap \{S_j^* \neq 0 \text{ for } j = 2n - 2k+1, \ldots, 2n\}$. Thus, the dual walk visits zero at time $2n - 2k$ and is subsequently zero-avoiding. The probability of this event is (see the discussion after (2.16)):

$$u_{2k} u_{2(n-k)} = \mathbb{P}(S_{2k} = 0)\mathbb{P}(S_{2n-2k} = 0).$$

**Solution 3.1** Let $(a_1, \ldots, a_{2n}) \in \{-1, 1\}^{2n}$ with $a_1 + \ldots + a_{2n} = 0$. Thus, $n$ of the entries of this tuple are equal to 1 and $n$ are equal to $-1$. Given that $X_1, \ldots, X_{2n}$ are independent with $\mathbb{P}(X_j = 1) = p$ and $\mathbb{P}(X_j = -1) = q := 1 - p$ for $j = 1, \ldots, 2n$, we get $\mathbb{P}(X_1 + \ldots + X_{2n} = 0) = \binom{2n}{n} p^n q^n$. Since $X_1 = a_1, \ldots, X_{2n} = a_{2n}$ implies $X_1 + \ldots + X_{2n} = 0$, we obtain

$$\mathbb{P}(X_1 = a_1, \ldots, X_{2n} = a_{2n} | X_1 + \ldots + X_{2n} = 0) = \frac{\mathbb{P}(X_1 = a_1, \ldots, X_{2n} = a_{2n})}{\mathbb{P}(X_1 + \ldots + X_{2n} = 0)}$$

$$= \frac{p^n q^n}{\binom{2n}{n} p^n q^n} = \frac{1}{\binom{2n}{n}}.$$

**Solution 3.2** Generalizing Solution 3.1, let $a_1 + \ldots + a_{2n} = 2k$. This implies that $n + k$ of the entries in the tuple $(a_1, \ldots, a_{2n}) \in \{-1, 1\}^{2n}$ are equal to 1 and $n - k$ are equal to $-1$. Therefore, $\mathbb{P}(X_1 + \ldots + X_{2n} = 0) = \binom{2n}{n+k} p^{n+k} q^{n-k}$, and thus

$$\mathbb{P}(X_1 = a_1, \ldots, X_{2n} = a_{2n} | X_1 + \ldots + X_{2n} = 0) = \frac{\mathbb{P}(X_1 = a_1, \ldots, X_{2n} = a_{2n})}{\mathbb{P}(X_1 + \ldots + X_{2n} = 0)}$$

$$= \frac{p^{n+k} q^{n-k}}{\binom{2n}{n+k} p^{n+k} q^{n-k}} = \frac{1}{\binom{2n}{n+k}}.$$

**Solution 3.3** Summing both sides of the equations in (3.4) over $k$ from 0 to $n-2$, and using suitable index shifts, we get

$$2\sum_{k=0}^{n-2} kq_k - \sum_{k=1}^{n-1} kq_k = 2(n-1)\sum_{k=0}^{n-2} q_k - 2n\sum_{k=1}^{n-1} q_k.$$

Since $q_0 = 1$, this results in

$$\sum_{k=1}^{n-1} kq_k = 2\left(n-1-\sum_{k=1}^{n-1} q_k\right). \qquad (*)$$

Using (7.12) and (7.13), $\mathbb{E}(N_{2n}^\circ) = \sum_{k=1}^{n-1} q_k$ and $\mathbb{E}(N_{2n}^{\circ\,2}) = 2\sum_{k=1}^{n-1} kq_k - \mathbb{E}(N_{2n}^\circ)$. In view of $(*)$, we then obtain (3.5). The formula for the variance now follows from $\mathbb{V}(N_{2n}^\circ) = \mathbb{E}(N_{2n}^{\circ\,2}) - (\mathbb{E}N_{2n}^\circ)^2$ and part (b) of Theorem 3.1.

**Solution 3.4** For every $x > 0$,

$$1 - \Phi(x) = \frac{1}{\sqrt{2\pi}}\int_x^\infty \exp\left(-\frac{t^2}{2}\right) dt < \frac{1}{\sqrt{2\pi}} \int_x^\infty \frac{t}{x} \cdot \exp\left(-\frac{t^2}{2}\right) dt.$$

Since the function $t\exp(-t^2/2)$ has the antiderivative $-\exp(-t^2/2)$, the assertion follows.

**Solution 3.5** Since $\Phi(0) = \frac{1}{2}$, it follows that $h(0) = 0$. Moreover,

$$h'(x) = 2\varphi(x) - x\exp\left(-\frac{x^2}{2}\right) = \left(\sqrt{\frac{2}{\pi}} - x\right) \exp\left(-\frac{x^2}{2}\right), \quad x > 0.$$

This implies $h'(x) > 0$ if $0 < x < \sqrt{2/\pi}$, which shows $h(x) > 0$ for each $x$ with $0 < x \le \sqrt{2/\pi}$. If $\varphi$ denotes the derivative of $\Phi$ and thus the density of the standard normal distribution, a simple transformation shows that $h(x) > 0$ is equivalent to $(1-\Phi(x))/\varphi(x) < \sqrt{\pi/2}$. According to Exercise 3.4, $(1-\Phi(x))/\varphi(x) < \frac{1}{x}$ for each $x > 0$, and in particular $(1-\Phi(x))/\varphi(x) < \sqrt{\pi/2}$ for each $x > \sqrt{2/\pi}$, which was still to be shown.

**Solution 3.6** Note that

$$\prod_{j=1}^{k-1}\left(1-\frac{j}{n}\right) = \frac{(n-1)\cdot\ldots\cdot(n-k+1)}{n^{k-1}} = \frac{(n-1)!}{(n-k)!n^{k-1}},$$

$$\prod_{j=1}^{k}\left(1-\frac{j}{2n}\right) = \frac{(2n-1)\cdot\ldots\cdot(2n-k)}{(2n)^k} \cdot \frac{2n}{2n} = \frac{(2n)!}{(2n-k-1)!(2n)^{k+1}}.$$

Thus, the right-hand side of (3.13) becomes

$$\frac{k}{2n} \cdot \frac{(n-1)!(2n-k-1)!(2n)^{k+1}}{(n-k)!n^{k-1}(2n)!} = \frac{2^k k \binom{2n-k-1}{n-1}}{n \binom{2n}{n}}.$$

According to Theorem 3.1(a), the expression on the right-hand side of the equality sign is equal to $\mathbb{P}(N_{2n}^\circ = k-1)$, which was to be shown.

**Solution 3.7** According to part (a) of Theorem 2.6, $\mathbb{P}(N_{2n} = k+1) = \binom{2n-k-1}{n}/2^{2n-k-1}$. Furthermore, $\mathbb{P}(S_{2n} = 0, N_{2n-2} \geq k) = \mathbb{P}(S_{2n} = 0)\mathbb{P}(N_{2n-2} \geq k | S_{2n} = 0)$. The first factor on the right-hand side equals $\binom{2n}{n}/2^{2n}$. According to Exercise 3.1, under the condition $S_{2n} = 0$, the underlying random walk is a $2n$-bridge. Therefore, the second factor is equal to the probability that such a bridge visits zero at least $k$ times. According to Eq. (3.3), we get

$$\mathbb{P}(N_{2n-2} \geq k | S_{2n} = 0) = 2^{k+1}\binom{2n-k-1}{n}\bigg/\binom{2n}{n},$$

and thus the assertion.

**Solution 3.8** Using (3.15) and the definitions of $c_{n,k}$ and $x_{n,k}$ as well as $K_n^*$ and $Z_{n,k}^*$, it follows that

$$\mathbb{P}(V_n > K_n | L_n = k) = \mathbb{P}(n + Z_{n,k} > K_n) = \mathbb{P}(Z_{n,k} > K_n - n)$$

$$= \mathbb{P}\left(\frac{Z_{n,k}}{\sqrt{n/2}} > \frac{K_n - n}{\sqrt{n/2}}\right)$$

$$= \mathbb{P}\left(\frac{Z_{n,k} - k/2}{\sqrt{n/2}} + \frac{k/2}{\sqrt{n/2}} > K_n^*\right)$$

$$= \mathbb{P}\left(Z_{n,k}^* \cdot \frac{\sqrt{k/4}}{\sqrt{n/2}} + \frac{k/2}{\sqrt{n/2}} > K_n^*\right)$$

$$= \mathbb{P}(Z_{n,k}^* \cdot c_{n,k} + x_{n,k} > K_n^*).$$

**Solution 3.9** For symmetry reasons, as discussed before Theorem 3.7, $Y$ has the same distribution as $X$. If $X_j := \mathbf{1}\{A_j\}$ for $j = 1, \ldots, 50$, the random vector $(X_1, \ldots, X_{50})$ has a uniform distribution over all $\binom{50}{20}$ 50-tuples with exactly 20 ones and 30 zeros. Thus, in particular, $(X_1, X_2, \ldots, X_{50})$ and $(X_{50}, X_{49}, \ldots, X_1)$ have the same distribution.

**Solution 3.10** Using (3.20), part (b) of Theorem 3.8, and (7.25), it follows that

$$\mathbb{V}(M_{2n}^\circ) = \mathbb{E}[(M_{2n}^\circ)^2] - (\mathbb{E}M_{2n}^\circ)^2$$

$$= \frac{2n}{\binom{2n}{n}}\binom{2n-1}{n-1} - \frac{2^{2n-1}}{\binom{2n}{n}} + \frac{1}{2} - \left(\frac{2^{2n-1}}{\binom{2n}{n}} - \frac{1}{2}\right)^2$$

$$= n - \frac{2^{2n-1}}{\binom{2n}{n}} + \frac{1}{2} - \frac{2^{4n-2}}{\binom{2n}{n}^2} + \frac{2^{2n-1}}{\binom{2n}{n}} - \frac{1}{4} = n + \frac{1}{4} - \frac{2^{4n-2}}{\binom{2n}{n}^2}.$$

**Solution 3.11**

(a) This follows from Lemma 2.1.
(b) The maximum height is $\frac{n+a}{2}$. This is attained when the path first takes $\frac{n+a}{2}$ upward steps and then $\frac{n-a}{2}$ downward steps to reach the point $(n, a)$.
(c) Every path leading from $(0, 0)$ to $(n, a)$ with a maximum height greater than or equal to $k$ corresponds in a one-to-one manner to a path from $(0, 0)$ to $(n, 2k - a)$. This path is obtained by reflecting the original path leading from the first time it reaches the height $k$ across the straight line $y = k$. The final height of this path is $a + 2(k - a) = 2k - a$. According to Lemma 2.1, the number of all paths from $(0, 0)$ to $(n, 2k - a)$ equals the binomial coefficient

$$\binom{n}{\frac{n-a+2k}{2}}.$$

**Solution 3.12** We prove the validity of

$$\frac{1}{\binom{2n}{n}} \sum_{k=1}^{n-1} (2k-1) \cdot 2 \cdot \binom{2n-1}{n+k} = n + 1 - \frac{2^{2n}}{\binom{2n}{n}}. \qquad (*)$$

According to (3.26), the left-hand side of $(*)$ is equal to the second moment of $C_{2n}^\circ$. Part (b) of Theorem 3.11 would then immediately follow from part (b) of this theorem, because $\mathbb{V}(C_{2n}^\circ) = \mathbb{E}(C_{2n}^\circ)^2 - (\mathbb{E}C_{2n}^\circ)^2$. The left-hand side of $(*)$ is equal to

$$\frac{4}{\binom{2n}{n}} \sum_{k=1}^{n-1} k\binom{2n-1}{n+k} - \frac{2}{\binom{2n}{n}} \sum_{k=1}^{n-1} \binom{2n-1}{n+k}.$$

Using (7.26) with $n$ replaced by $n - 1$, the first sum equals $\frac{n}{2}\binom{2n-1}{n-1} - 2^{2n-3}$, and the second sum is equal to $2^{2n-2} - \binom{2n-1}{n-1}$ according to (3.27). Inserting these expressions and noting that $2\binom{2n-1}{n-1}/\binom{2n}{n} = 1$, $(*)$ follows.

**Solution 4.1** If $V_1 = \inf\{k \in \mathbb{N} : S_k = 1\}$ denotes the first-passage time to height $k$, then $\max_{j \in \mathbb{N}_0} S_j = 0$ if and only if $V_1 = \infty$. According to part (b) of Theorem 4.1, it follows that $\mathbb{P}(V_1 = \infty) = 1 - \frac{p}{q}$, and thus $\mathbb{P}(V_1 = \infty) = 2/3$ in the case where $p = \frac{1}{4}$.

**Solution 4.2** If both sides of Eq. (4.4) are multiplied by $t^n$ and summed over $n \geq 2$, the result is $g(t) - pt$ on the left-hand side, taking $a_0 = 0$ and $a_1 = p$ into account. The right-hand side becomes

$$\sum_{n=2}^{\infty}\left(q\sum_{r=2}^{n-1}a_{r-1}a_{n-r}\right)t^n = q\sum_{r=2}^{\infty}a_{r-1}\left(\sum_{n=r+1}^{\infty}a_{n-r}t^{n-r}\right)t^r.$$

Since $a_0 = 0$, the expression within the brackets on the right-hand side equals $g(t)$ (index shift $k := n - r$), and thus the right-hand side becomes

$$q\,g(t)\,t\sum_{r=2}^{\infty}a_{r-1}t^{r-1}.$$

Because $a_0 = 0$, the above series equals $g(t)$, which was to be shown.

**Solution 4.3** Let $x := -4pqt^2$. Since $|t| \leq 1$, it follows that $|x| < 1$. Using the binomial series for $(1+x)^{1/2}$, we get

$$\sqrt{1 - 4pqt^2} = \sum_{n=0}^{\infty}\binom{\frac{1}{2}}{n}x^n = 1 + \sum_{n=1}^{\infty}(-1)^n\binom{\frac{1}{2}}{n}2^{2n}(pq)^n t^{2n}.$$

Note that for each $n \geq 1$,

$$\binom{\frac{1}{2}}{n} = \frac{1}{n!}\cdot\frac{1}{2}\cdot\left(-\frac{1}{2}\right)\cdot\left(-\frac{3}{2}\right)\cdot\ldots\cdot\left(-\frac{2n-3}{2}\right) = \frac{(-1)^{n-1}}{n!2^n}\prod_{j=1}^{n-1}(2j-1)$$

$$= \frac{(-1)^{n-1}}{n!2^n}\prod_{j=1}^{n-1}(2j-1)\cdot\frac{2\cdot 4\cdot\ldots\cdot(2n-2)}{2^{n-1}(n-1)!} = \frac{(-1)^{n-1}}{2^{2n-1}n}\binom{2n-2}{n-1}.$$

Thus,

$$1 - \sqrt{4pqt^2} = 2\sum_{n=1}^{\infty}\frac{1}{n}\binom{2n-2}{n-1}(pq)^n t^{2n},$$

from which the assertion follows.

**Solution 4.4** Using (4.8) and $\sqrt{1-4pq} = p-q$ as well as $1 = p+q$,

$$\mathbb{E}[V_1(V_1-1)] = g''(1) = \frac{8p^2q}{(p-q)^3} + 2 - \frac{2p}{p-q} = \frac{8p^2q}{(p-q)^3} - \frac{2q}{p-q}.$$

Since $\mathbb{E}(V_1) = 1/(p-q)$ and $\mathbb{V}(V_1) = \mathbb{E}(V_1^2) - (\mathbb{E}V_1)^2$, it follows that

$$\mathbb{V}(V_1) = \frac{8p^2q}{(p-q)^3} - \frac{2q}{p-q} + \frac{1}{p-q} - \frac{1}{(p-q)^2} = \frac{8p^2q}{(p-q)^3} + 1 - \frac{1}{(p-q)^2}$$

$$= \frac{8p^2q}{(p-q)^3} - \frac{4pq}{(p-q)^2},$$

and thus the assertion.

**Solution 4.5** Let $t$ be any non-zero real number with $|t| \leq 1$. Using $g(t)$ as given in (4.5) and the last equation in Solution 4.3, it follows that

$$g_2(t) = g(t)^2 = \frac{1}{4q^2t^2}\left(1 - 2\sqrt{1-4pqt^2} + 1 - 4pqt^2\right)$$

$$= \frac{1}{2q^2t^2}\left(1 - \sqrt{1-4pqt^2} - 2pqt^2\right)$$

$$= \frac{1}{q^2t^2}\left(\sum_{n=1}^{\infty} \frac{1}{n}\binom{2n-2}{n-1}(pq)^n t^{2n} - pqt^2\right)$$

$$= \frac{1}{q^2t^2}\sum_{n=2}^{\infty} \frac{1}{n}\binom{2n-2}{n-1}(pq)^n t^{2n}$$

$$= \sum_{n=2}^{\infty} \frac{1}{n}\binom{2n-2}{n-1}p^n q^{n-2} t^{2(n-1)} = \sum_{j=1}^{\infty} \frac{1}{j+1}\binom{2j}{j}p^{j+1}q^{j-1}t^{2j}.$$

**Solution 4.6** We prove (4.40) by induction on $k$. The base case $k = 1$ follows from Theorem 4.1 (b). Using the hint, the induction step $k \mapsto k+1$ follows from

$$\mathbb{P}(V_{k+1} < \infty) = \sum_{\ell=k+1}^{\infty} \mathbb{P}(V_{k+1} = \ell) = \sum_{\ell=k+1}^{\infty} \sum_{j=k}^{\ell-1} \mathbb{P}(V_{k+1} = \ell, V_k = j)$$

$$= \sum_{\ell=k+1}^{\infty} \sum_{j=k}^{\ell-1} \mathbb{P}(V_k = j, V_{k+1} - V_k = \ell - j)$$

# Solutions to the Exercises

$$= \sum_{j=k}^{\infty} \mathbb{P}(V_k = j) \sum_{\ell=j+1}^{\infty} \mathbb{P}(V_1 = \ell - j)$$

$$= \sum_{j=k}^{\infty} \mathbb{P}(V_k = j) \mathbb{P}(V_1 < \infty)$$

$$= \mathbb{P}(V_k < \infty) \cdot \frac{p}{q} = \left(\frac{p}{q}\right)^k \cdot \frac{p}{q} = \left(\frac{p}{q}\right)^{k+1}.$$

Note that the induction hypothesis was used in the penultimate equality.

**Solution 4.7** Let $M := \max_{j \geq 0} S_j$. Note that $M \geq k$ if and only if $V_k < \infty$, where $V_k = \inf\{\ell \geq k : S_\ell = k\}$ denotes the first-passage time to the height $k$. Following Exercise 4.6, $\mathbb{P}(M \geq k) = \mathbb{P}(V_k < \infty) = (p/q)^k$. Thus,

$$\mathbb{P}(M = k) = \mathbb{P}(M \geq k) - \mathbb{P}(M \geq k+1) = \left(\frac{p}{q}\right)^k - \left(\frac{p}{q}\right)^{k+1} = \left(\frac{p}{q}\right)^k \left(1 - \frac{p}{q}\right).$$

Using (7.12), we obtain

$$\mathbb{E}(M) = \sum_{k=1}^{\infty} \mathbb{P}(M \geq k) = \sum_{k=1}^{\infty} \left(\frac{p}{q}\right)^k = \frac{\frac{p}{q}}{1 - \frac{p}{q}} = \frac{p}{q - p}.$$

**Solution 4.8** For $p = \frac{1}{2}$, Eq. (4.10) takes the form

$$\mathbb{P}(V_2 = 2k) = \frac{1}{k+1} \binom{2k}{k} \left(\frac{1}{2}\right)^{2k}, \quad k \geq 1.$$

Substituting the value 2 for $k$ and $2k$ for $n$ in Theorem 2.24, we get

$$\mathbb{P}(V_2 = 2k) = \frac{2}{2k 2^{2k}} \binom{2k}{k+1} = \frac{1}{k 2^{2k}} \cdot \frac{(2k)!}{(k+1)!(k-1)!}$$

$$= \frac{1}{k+1} \binom{2k}{k} \left(\frac{1}{2}\right)^{2k}, \quad k \geq 1.$$

**Solution 4.9** For each $m \in \mathbb{Z}$ with $m \neq 0$, let $V_m := \inf\{j \geq 1 : S_j = m\}$ be the first-passage time of the height $m$. We have $\mathbb{P}(T = 2k+1) = p\mathbb{P}(T = 2k+1|X_1 = 1) + q\mathbb{P}(T = 2k+1|X_1 = -1)$. Under the condition $X_1 = 1$, the random walk must descend from height 1 to height $-1$ in $2k$ steps, so we get $\mathbb{P}(T = 2k+1|X_1 =$

$1) = \mathbb{P}(V_{-2} = 2k)$. Similarly, $\mathbb{P}(T = 2k+1|X_1 = -1) = \mathbb{P}(V_2 = k)$. Using (4.10) twice (once by swapping $p$ and $q$), we obtain

$$\mathbb{P}(T = 2k+1) = \frac{p}{k+1}\binom{2k}{k}q^{k+1}p^{k-1} + \frac{q}{k+1}\binom{2k}{k}p^{k+1}q^{k-1}$$

$$= \frac{1}{k+1}\binom{2k}{k}(pq)^{k-1}(pq^2 + qp^2) = \frac{1}{k+1}\binom{2k}{k}(pq)^k.$$

**Solution 4.10** Let $a_{2n-1} := \mathbb{P}(V_1 = 2n-1)$ and $b_{2n-1} := \mathbb{P}(V_{-1} = 2n-1)$ for $n \geq 1$. We have (compare the proof of Theorem 4.6) $\mathbb{P}(W = 2n) = pb_{2n-1} + qa_{2n-1}$. According to (4.5), $V_1$ has the generating function

$$g_{V_1}(t) = \frac{1 - \sqrt{1 - 4pqt^2}}{2qt}, \qquad |t| \leq 1, \ t \neq 0.$$

By swapping the roles of $p$ and $q$, we get

$$g_{V_{-1}}(t) = \frac{1 - \sqrt{1 - 4pqt^2}}{2pt}, \qquad |t| \leq 1, \ t \neq 0,$$

and thus obtain

$$g_W(t) = \sum_{n=1}^{\infty} \mathbb{P}(W = 2n)t^{2n} = pt\sum_{n=1}^{\infty} b_{2n-1}t^{2n-1} + qt\sum_{n=1}^{\infty} a_{2n-1}t^{2n-1}$$

$$= ptg_{V_{-1}}(t) + qtg_{V_1}(t) = 1 - \sqrt{1 - 4pqt^2}, \qquad |t| \leq 1.$$

**Solution 4.11**

(a) Using (7.16),

$$\mathbb{P}(S_{2n} = 0) = \binom{2n}{n}(pq)^n = \frac{\binom{2n}{n}}{2^{2n}}(4pq)^n \sim \frac{1}{\sqrt{\pi n}} \cdot \varrho^n,$$

where $\varrho := 4pq$. Since $p \neq \frac{1}{2}$, it follows that $0 < \varrho < 1$.

(b) Let $N := \sum_{n=1}^{\infty} \mathbf{1}\{S_{2n} = 0\}$ be the number of visits to zero of the random walk. Using part (a), $\mathbb{E}(N) = \sum_{n=1}^{\infty} \mathbb{P}(S_{2n} = 0) < \infty$, and thus $\mathbb{P}(N < \infty) = 1$.

**Solution 4.12** If A plays with a stake of 2, 5, or 10 euros, the same ruin probabilities for B arise as if A were to play with a stake of one euro and B with a stake of 50, 20, or 10 euros, respectively.

(a) In the case where $p = 0.5$, A should play with a 10-euro stake. According to Theorem 4.10, A then ruins B with a probability of $\frac{1}{11} \approx 0.091$.
(b) In the case where $p = 0.4$, a stake of 10 euros is also advisable. Then B goes bankrupt with a probability of 0.00585.
(c) If $p = 0.8$, A should choose a one-euro stake. A then wins with a probability of 0.750.

**Solution 4.13** Let $x := \frac{q}{p}$. We need to show

$$m_a = \frac{1}{p-q} \cdot \frac{b(1-x^{-a}) - a(x^b - 1)}{x^b - x^{-a}}.$$

Note that $m_a = m_1 + d_1 + \ldots + d_{a-1}$, where

$$m_1 = \frac{1}{p-q}\left(\frac{(a+b)(1-x)}{1-x^{a+b}} - 1\right), \quad d_j = x^j m_1 - \frac{1-x^j}{p-q}$$

(cf. (4.25) and (4.26)). This results in

$$m_a = m_1 + \sum_{j=1}^{a-1}\left(m_1 x^j - \frac{1-x^j}{p-q}\right) = m_1 \sum_{j=0}^{a-1} x^j - \frac{1}{p-q}\left(a - \sum_{j=0}^{a-1} x^j\right)$$

$$= m_1 \frac{x^a - 1}{x-1} - \frac{a}{p-q} + \frac{1}{p-q} \frac{x^a - 1}{x-1}$$

$$= \frac{1}{p-q} \frac{(a+b)(1-x)}{1-x^{a+b}} \frac{x^a - 1}{x-1} - \frac{a}{p-q}$$

$$= \frac{1}{p-q}\left(\frac{(a+b)(1-x^a) - a(1-x^{a+b})}{1-x^{a+b}}\right).$$

If we multiply both the numerator and the denominator by $-x^{-a}$, the assertion follows.

**Solution 4.14**

(a) $\mathbb{E}(T)$ takes the smallest value for $a = 1$ and the largest value for $a = 95$, which are 2.5 and 233.89, respectively.
(b) $\mathbb{E}(T)$ takes the smallest value for $a = 99$ and the largest value for $a = 4$, which are 1.67 and 159.35, respectively.

**Solution 4.15** Setting $q := 1 - p$, the answer is

$$\binom{2k+a-1}{k} p^k q^{a+k}.$$

The last game goes to player B, and of the $2k + a - 1$ games before that, exactly $k$ go to A and $k + a - 1$ go to B. The binomial coefficient represents the number of ways to choose the $k$ games that A wins from these $2k + a - 1$ games.

**Solution 4.16** The situation is equivalent to a gambler's ruin problem with $k$ euros for player A and $r - k$ euros for player B. However, A cannot be ruined because if he loses his last euro, it is given back to him by B, and the game starts anew with one euro for player A and $r - 1$ euros for player B. According to Theorem 4.10(a), A will be ruined with a probability of $1 - \frac{k}{r}$. The random walk therefore reaches the height 0 with this probability before it is absorbed. With a probability of $\frac{k}{r}$, it ends before reaching the height 0, which means $Y = 0$, proving part (a).

If A starts with 1 euro, he will be ruined according to Theorem 4.10(a) with a probability of $1 - \frac{1}{r}$. For the random walk to reach the height 0 $\ell$ times from the height $k$ before it ends at the height $r$, A must first go bankrupt and then go bankrupt $\ell - 1$ more times, starting with 1 euro each time. Afterwards, he must win with 1 euro against B, who starts with $r - 1$ euros. Since the random walk, having reached any height, restarts from there without memory, part (b) follows.

**Solution 4.17**

(a) The event $\{R_1 = k\}$ occurs if and only if the sequence $(X_j)$ begins with $k$ hits, followed by a miss, or vice versa. This leads to part (a).
(b) Using the derivative of the geometric series, we get

$$\mathbb{E}(R_1) = \sum_{k=1}^{\infty} k \left(p^k q + q^k p\right) = \frac{qp}{(1-p)^2} + \frac{pq}{(1-q)^2} = \frac{p}{q} + \frac{q}{p} = \frac{1}{pq} - 2.$$

(c) For the event $\{R_2 = k\}$ to occur, there are two possibilities, namely $X_1 = 1$ or $X_1 = -1$. In the first case, which occurs with probability $p$, the second run automatically starts with a miss, and then $k - 1$ more misses and a hit must follow. For the second case, which occurs with probability $q$, the sequence $(X_j)$ starts with a miss, and the second run automatically starts with a hit. Then $k - 1$ more hits and a miss must follow for $R_2$ to take the value $k$. These considerations clarify the occurrence of the exponents $k - 1$ in part (c).

(d) $\mathbb{E}(R_2) = \sum_{k=1}^{\infty} k \left(p^2 q^{k-1} + q^2 p^{k-1}\right) = \frac{p^2}{(1-q)^2} + \frac{q^2}{(1-p)^2} = 2.$

**Solution 4.18** A possible one-to-one mapping $f$ from $M_1$ to $M_2$ is provided by the *contraction mapping*

$$f((i_1, \ldots, i_r)) := (j_1, \ldots, j_r), \qquad (i_1, \ldots, i_r) \in M_1,$$

Solutions to the Exercises

where $j_\ell := i_\ell - (\ell-1)k$ for $\ell = 1, \ldots, r$. Due to the requirement $i_{s+1} - i_s \geq k+1$ for each $s = 1, \ldots, r-1$, we have $1 \leq j_1 < \ldots < j_r$, and $i_r \leq n-k+1$ results in $j_r \leq n - rk + 1$. Thus, $f$ is obviously an injective mapping from $M_1$ to $M_2$. $f$ is also surjective, because if $\mathbf{j} := (j_1, \ldots, j_r)$ is a tuple from $M_2$, then $\mathbf{i} := (i_1, \ldots, i_r)$ with $i_\ell := j_\ell + (\ell-1)k$ for $\ell = 1, \ldots, r$ is a tuple from $M_1$ with $f(\mathbf{i}) = \mathbf{j}$.

**Solution 4.19** Using (4.34) and $\mu := \mathbb{E}(X_1)$ as well as $\mathbb{E}(S_N) = \mu \mathbb{E}(N)$,

$$\mathbb{E}[S_N(S_N - 1)] = \mathbb{E}[N(N-1)]\mu^2 + \mathbb{E}(N)\mathbb{E}[X_1(X_1 - 1)],$$

and thus

$$\begin{aligned}\mathbb{V}(S_N) &= \mathbb{E}[S_N(S_N - 1)] + \mathbb{E}S_N - (\mathbb{E}S_N)^2 \\ &= \mathbb{E}[N(N-1)]\mu^2 + \mathbb{E}(N)\mathbb{E}[X_1(X_1 - 1)] + \mu\mathbb{E}(N) - \mu^2(\mathbb{E}N)^2 \\ &= \mathbb{V}(N)\mu^2 + (\mathbb{E}N)^2\mu^2 - (\mathbb{E}N)\mu^2 + \mathbb{E}(N)\mathbb{E}(X_1^2) \\ &\quad - \mathbb{E}(N)\mu + \mu\mathbb{E}(N) - \mu^2(\mathbb{E}N)^2 \\ &= \mathbb{V}(N)\mu^2 + \mathbb{E}(N)\mathbb{V}(X_1).\end{aligned}$$

**Solution 4.20** According to the definition of $\sigma^2$, $\mathbb{V}(M_1) = \sigma^2$, thus establishing the base case for a proof by mathematical induction. For the induction step $n \mapsto n+1$, we use the equation

$$\mathbb{V}(M_{n+1}) = \mathbb{V}(M_n)\mu^2 + \mathbb{E}(M_n)\sigma^2$$

(cf. (4.35)) and distinguish between the cases where $\mu = 1$ and $\mu \neq 1$. If $\mu = 1$, this equation and the induction hypothesis yield $\mathbb{V}(M_{n+1}) = n\sigma^2 + \sigma^2 = (n+1)\sigma^2$, which was to be shown. If $\mu \neq 1$, the induction hypothesis provides

$$\mathbb{V}(M_{n+1}) = \frac{\sigma^2\mu^{n-1}(\mu^n - 1)}{\mu - 1} \cdot \mu^2 + \mu^n\sigma^2 = \frac{\sigma^2\mu^n(\mu^{n+1} - 1)}{\mu - 1},$$

thus completing the proof.

**Solution 4.21** Since $p_j = \frac{1}{j+1} - \frac{1}{j+2}$,

$$g(t) = \sum_{j=0}^{\infty} p_j t^j = \sum_{j=0}^{\infty} \frac{t^j}{t+1} - \sum_{j=0}^{\infty} \frac{t^j}{j+2} = \frac{1}{t}\sum_{k=1}^{\infty} \frac{t^k}{k} - \frac{1}{t^2}\sum_{k=2}^{\infty} \frac{t^k}{k}, \quad |t| < 1,\ t \neq 0.$$

Using the fact that $-\log(1-t) = \sum_{k=1}^{\infty} t^k/k$ if $|t| < 1$ (see, e.g., [TA], p. 80), it follows that

$$g(t) = \frac{1}{t}\bigl(-\log(1-t)\bigr) - \frac{1}{t^2}\bigl(-\log(1-t) - t\bigr) = \frac{1}{t}\left(1 + \left(\frac{1}{t} - 1\right)\log(1-t)\right).$$

Furthermore, $g(0) = p_0 = 1 - 1/2 = 1/2$.

**Solution 4.22** Since $w < 1$, the hint gives

$$\mu - 1 = g'(1) - g(1) \le g'(1) - g'(w) = \int_w^1 g''(t)\,dt \le \int_w^1 g''(1)\,dt = (1-w)g''(1),$$

and thus the assertion.

**Solution 4.23** In the case of the geometric offspring distribution, $\sigma^2 = \mu^2 + \mu$. Since $w = \frac{1}{\mu}$, the inequality from Exercise 4.22 turns into

$$\frac{1}{\mu} \le 1 - \frac{\mu-1}{2\mu^2},$$

which is equivalent to $\frac{3}{2} \le \mu + \frac{1}{2\mu}$.

**Solution 4.24** For the extinction probability $w$, we have $w = g(w)$. Because $g(0) = \exp(-\mu) > 0$, the inequality $w < \frac{1}{\mu}$ is equivalent to $g(\frac{1}{\mu}) < \frac{1}{\mu}$. The latter inequality is equivalent to $e^{\mu-1} > \mu$. With the hint, the assertion follows.

**Solution 5.1** Due to symmetry, $\mathbb{E}(U_n^2 + V_n^2) = 2\mathbb{E}(U_n^2)$. Let $A_j$ denote the event that the $j$-th step of the random walk is in the direction of the first coordinate axis ($j = 1, \ldots, n$). Let $Y_1, \ldots, Y_n$ be independent random variables with $\mathbb{P}(Y_j = \pm 1) = \frac{1}{2}$ for $j = 1, \ldots, n$, which are also independent of $\mathbf{1}\{A_1\}, \ldots, \mathbf{1}\{A_n\}$. We then have the equality in distribution

$$U_n \sim \sum_{j=1}^n \mathbf{1}\{A_j\} Y_j.$$

Since $\mathbb{E}(U_n) = n\mathbb{E}(\mathbf{1}\{A_1\})\mathbb{E}(Y_1) = 0$, it follows that $\mathbb{E}(U_n^2) = \mathbb{V}(U_n)$. Because the summands of the above sum are i.i.d. random variables, $\mathbb{V}(U_n) = n\mathbb{V}(\mathbf{1}\{A_1\}Y_1)$. The latter variance is equal to $\mathbb{E}(\mathbf{1}\{A_1\}Y_1^2) = \mathbb{E}(\mathbf{1}\{A_1\}) = \mathbb{P}(A_1) = \frac{1}{2}$ due to $Y_1^2 = 1$, and thus the assertion follows.

**Solution 5.2** Since each step is independent of the others and occurs with equal probability $\frac{1}{2}$ in the first or in the second coordinate, the random number of steps in the first coordinate, denoted by $R_n$, has the binomial distribution $\text{Bin}(2n, \frac{1}{2})$. Since $R_n$ can only take on even values and a total of $2n$ steps is taken, the number of steps performed in the second coordinate is also an even number. Thus, the desired probability is equal to

$$\mathbb{P}(R_n \text{ even}) = \sum_{j=0}^{n} \binom{2n}{2j} \left(\frac{1}{2}\right)^{2n} = \left(\frac{1}{2}\right)^{2n} \sum_{j=0}^{n} \binom{2n}{2j} = \left(\frac{1}{2}\right)^{2n} 2^{2n-1} = \frac{1}{2}.$$

The last equality follows from $\binom{2n}{2j} = \binom{2n-1}{2j} + \binom{2n-1}{2j-1}$ for $j = 1, \ldots, n-1$.

**Solution 5.3**

(a) Letting $X_j =: (X_{j,1}, \ldots, X_{j,d})^\top$ for $j = 1, 2$,

$$\mathbb{E}(X_1^\top X_2) = \mathbb{E}\left(\sum_{\ell=1}^{d} X_{1,\ell} X_{2,\ell}\right) = \sum_{\ell=1}^{d} \mathbb{E}(X_{1,\ell} X_{2,\ell}).$$

The independence of $X_1$ and $X_2$ carries over to $X_{1,\ell}$ and $X_{2,\ell}$, and thus $\mathbb{E}(X_{1,\ell} X_{2,\ell}) = \mathbb{E}(X_{1,\ell})\mathbb{E}(X_{2,\ell}) = 0$.

(b) Note that $\|S_n\|^2 = \sum_{k=1}^{n} \sum_{\ell=1}^{n} X_k^\top X_\ell$. Taking expectations and observing part (a) as well as $X_j^\top X_j = \|X_j\|^2 = 1$ for $j \in \{1, \ldots, d\}$, the assertion follows.

**Solution 5.4** Because $X_1, X_2, \ldots$ are independent, it follows that

$$\mathbb{P}(S_{n+k} = y | S_n = x) = \mathbb{P}\left(S_n + \sum_{j=n+1}^{n+k} X_j = y \Big| S_n = x\right)$$

$$= \mathbb{P}\left(x + \sum_{j=n+1}^{n+k} X_j = y\right)$$

$$= \mathbb{P}\left(\sum_{j=n+1}^{n+k} X_j = y - x\right).$$

Similarly, $\mathbb{P}(S_{n+k} = x | S_n = y) = \mathbb{P}\left(\sum_{j=n+1}^{n+k} X_j = x - y\right)$. Since $\sum_{j=n+1}^{n+k} X_j$ has a uniform distribution over all linear combinations $\sum_{\ell=1}^{k} c_\ell e_\ell$ with $c_\ell \in \{1, -1\}$ for each $\ell = 1, \ldots, k$, the assertion follows.

**Solution 5.5** Since $X_1, X_2, \ldots$ are independent,

$$\mathbb{P}(S_{n+k} = y | S_n = x) = \mathbb{P}\left(S_n + \sum_{j=n+1}^{n+k} X_j = y \Big| S_n = x\right) = \mathbb{P}\left(x + \sum_{j=n+1}^{n+k} X_j = y\right)$$

$$= \mathbb{P}\left(\sum_{j=n+1}^{n+k} X_j = y - x\right).$$

Similarly, $\mathbb{P}(S_{m+k} = y | S_m = x) = \mathbb{P}\left(\sum_{j=m+1}^{m+k} X_j = y - x\right)$. Since $\sum_{j=n+1}^{n+k} X_j$ and $\sum_{j=m+1}^{m+k} X_j$ have the same distribution, the assertion follows.

**Solution 5.6**

(a) Note that $S_2 = \mathbf{0}$ if and only if $X_2 = -X_1$. For this to be true, we must have $X_1 = e_j$ and $X_2 = -e_j$ or $X_1 = -e_j$ and $X_2 = e_j$ for exactly one $j \in \{1, \ldots, d\}$.
(b) Let $A$ be the event described in Problem 5.6 (b), and define $A_n := \{S_{n+2} = S_n\}$ for $n \geq 0$. Since $A_n = \{X_{n+1} + X_{n+2} = \mathbf{0}\}$, part (a) yields $\mathbb{P}(A_n) = 2\sum_{j=1}^{d} p_j q_j > 0$ for $n \geq 0$. The events $A_1, A_3, A_5, \ldots$ are independent, and $\sum_{j=0}^{\infty} \mathbb{P}(A_{2j+1}) = \infty$. According to part (b) of the Borel–Cantelli lemma in Sect. 7.13, it follows that $\mathbb{P}(\limsup_{j \to \infty} A_{2j+1}) = 1$. Since the event $A$ is a superset of this limit superior, the assertion follows.

**Solution 5.7** If you rotate the $(s, t)$-coordinate system by 45 degrees counterclockwise, the old $(s, t)$-coordinates and new $(u, v)$-coordinates of the random walk after the first step are represented as follows:

| $(s, t)$-coordinates | $(u, v)$-coordinates |
| --- | --- |
| $(0, 1)$ | $(1/\sqrt{2}, 1/\sqrt{2})$ |
| $(0, -1)$ | $(-1/\sqrt{2}, -1/\sqrt{2})$ |
| $(1, 0)$ | $(1/\sqrt{2}, -1/\sqrt{2})$ |
| $(-1, 0)$ | $(-1/\sqrt{2}, 1/\sqrt{2})$ |

If two particles, each starting at 0, independently perform a symmetric one-dimensional random walk with step size $1/\sqrt{2}$, one on the $u$-axis and the other on the $v$-axis, their four equally probable positions after the first step in the $(u, v)$-coordinate system would coincide with those of the two-dimensional random walk after its first step. Thus, a symmetric random walk on $\mathbb{Z}^2$ cannot be distinguished from two independent one-dimensional symmetric random walks. Since the probability that both random walks visit 0 after $2n$ steps does not depend on the size of the equally long steps, the assertion follows.

**Solution 5.8**

(a) Suppose the probability stated in Problem 5.8 takes its maximum for $(k_1, \ldots, k_d) \in \mathbb{N}_0^d$ with $k_1 + \ldots + k_d = n$. If $k_j \geq 1$ and $k_i \leq n-1$, replacing $k_j$ with $k_j - 1$ and $k_i$ with $k_i + 1$ results in a probability that is at most as large. If we take the quotient of both probabilities, almost everything cancels out, and we obtain the inequality

$$p_i k_j \leq p_j (k_i + 1). \qquad (*)$$

This also holds for $k_j = 0$ and $k_i = n$ (since then $k_j = 0$) and thus for all pairs $(i, j)$ with $i, j \in \{1, \ldots, d\}$.

(b) Adding both sides of the inequalities $(*)$ obtained in (a) over all $j$, and noting that the less-than sign holds for $j = i$, the inequality $np_i - 1 < k_i$ follows due to $\sum_{\ell=1}^d p_\ell = 1$ and $\sum_{\ell=1}^d k_\ell = n$. Summing over all $i$ with $i \neq j$, the result is $(1 - p_j)k_j \leq p_j(n - k_j + d - 1)$ and thus $k_j \leq (n + d - 1)p_j$.

**Solution 5.9** Let $d \geq 4$. The event $\{S_{2n} = 0\}$ implies that, after $2n$ steps, the first three components of $S_{2n}$ must be equal to the zero vector in $\mathbb{R}^3$. If we denote this event by $A_{2n}$, we know from the proof of the transience of the symmetric random on $\mathbb{Z}^3$ (Theorem 5.1 (b)) that $\sum_{n=1}^\infty \mathbb{P}(A_{2n}) < \infty$. This also implies that $\sum_{n=1}^\infty \mathbb{P}(S_{2n} = 0) < \infty$, which was to be shown.

**Solution 5.10** After 3 steps, there are a total of 64 different paths. Four different positions are visited when the random walk ends at the points $(3, 0)$, $(-3, 0)$, $(0, 3)$, or $(0, -3)$ after three steps, with one path each leading to these points. The "knight's move" position $(2, 1)$, which also leads to four visited points, can be reached in three ways: $\rightarrow\rightarrow\uparrow$, $\rightarrow\uparrow\rightarrow$, or $\uparrow\rightarrow\rightarrow$. The same applies to the 7 other knight's move positions: $(2, -1)$, $(1, 2)$, $(1, -2)$, $(-1, 2)$, $(-2, 1)$, $(-2, -1)$, and $(-1, -2)$. Additionally, there are 8 more paths that lead to 4 different visited positions each, namely "three-quarter square runs" of the type $\rightarrow\uparrow\leftarrow$ or $\leftarrow\uparrow\rightarrow$. Since no other path leads to 4 different visited positions, $\mathbb{P}(R_3 = 4) = \frac{36}{64}$.

To systematically count the paths with $R_3 = 3$, we consider the four possibilities for the first step, which, without loss of generality, goes to the right. From there, there are the 6 possibilities $\rightarrow\leftarrow$, $\uparrow\downarrow$, $\downarrow\uparrow$, $\leftarrow\leftarrow$, $\leftarrow\uparrow$, and $\leftarrow\downarrow$, to visit exactly one more point. Thus, $\mathbb{P}(R_3 = 3) = \frac{24}{64}$. Exactly two different points are reached in the four cases where the random walk reverses the initial step and then performs it again. Thus, $\mathbb{P}(R_3 = 2) = \frac{4}{64}$.

**Solution 5.11** In total, there are $(2d)^2$ equally likely paths of length 2. The event $\{R_2 = 2\}$ occurs if and only if the second step is opposite to the first, leading to $S_1 \in M := \{e_1, -e_1, \ldots, e_d, -e_d\}$, and thus $S_2 = 0$. There are $2d$ possibilities for the first step, so $\mathbb{P}(R_2 = 2) = 2d/(2d)^2 = \frac{1}{2d}$. The event $\{R_2 = 3\}$ occurs if and only if exactly one more point is visited after the first step. Assuming the first step leads to the point $e_1$, there are $2(d-1)+1$ ways for a second step to result in $R_2 = 3$, namely $e_1 \pm e_j$ for $j = 2, \ldots, d$, and $2e_1$. This proves the assertion. The probability $\mathbb{P}(R_2 = 3)$ can also be obtained using $\mathbb{P}(R_2 = 3) = 1 - \mathbb{P}(R_2 = 2)$.

**Solution 5.12** $N = \{0, e_1, -e_1, e_2, -e_2, e_1 + e_2, e_1 - e_2, -e_1 + e_2, -e_1 - e_2\}$.

**Solution 5.13** Let $F, G \in V$ and $\alpha, \beta \in \mathbb{R}$. Since $F, G \in V$ is equivalent to $F$ and $G$ satisfying the mean value property, it follows for every $x \in M$:

$$(\alpha F + \beta G)(x) = \alpha \cdot \frac{1}{2d} \sum_{y \in \mathcal{N}(x)} F(y) + \beta \cdot \frac{1}{2d} \sum_{y \in \mathcal{N}(x)} H(y)$$

$$= \frac{1}{2d} \sum_{y \in \mathcal{N}(x)} (\alpha F + \beta G)(y).$$

Thus, $\alpha F + \beta G$ also satisfies the mean property.

**Solution 5.14** For $y_0 \in \partial M$, let $G_{y_0} : \partial M \to \mathbb{R}$ be defined by $G_{y_0}(x) := 1$ if $x = y_0$, and $G_{y_0}(x) := 0$ if $x \in \partial M \setminus \{y_0\}$. A function $H_{y_0} : \overline{M} \to \mathbb{R}$ is a solution to the Dirichlet problem for $G_{y_0}$ if $H_{y_0}(x) = G_{y_0}(x)$ for every $x \in \partial M$, and if $H_{y_0}$ satisfies the mean value property

$$H_{y_0}(x) = \frac{1}{2d} \sum_{z \in \mathcal{N}(x)} H_{y_0}(z), \qquad x \in M.$$

Any function $F_0 : \overline{M} \to \mathbb{R}$ has the representation

$$F_0 = \sum_{y_0 \in \partial M} F_0(y_0) \cdot G_{y_0}.$$

Then the function $F : \overline{M} \to \mathbb{R}$ defined by

$$F(x) := \sum_{y_0 \in \partial M} F_0(y_0) H_{y_0}(x), \qquad x \in \overline{M},$$

is a solution to the Dirichlet problem for $F_0$. According to Exercise 5.13, $F$ is harmonic as a linear combination of the (mean value property possessing and thus harmonic) functions $\{H_{y_0} : y_0 \in \partial M\}$ and thus has the mean value property. Furthermore, for each $x \in \partial M$,

$$F(x) = \sum_{y_0 \in \partial M} F_0(y_0) H_{y_0}(x) = \sum_{y_0 \in \partial M} F_0(y_0) G_{y_0}(x) = F_0(x),$$

which was to be shown.

# Solutions to the Exercises

**Solution 5.15** For every $a \in \mathbb{R}$, the solution to this discrete Dirichlet problem is given by $F(x) := ax$, $x \in \overline{M}$, because $F$ is an extension of $F_0$ on $\overline{M}$ that has the mean value property. The finiteness of $M$ required in Theorem 5.9 is essential for the uniqueness of the solution (cf. Theorem 5.4 and Corollary 5.5). Note that the function $F(x) = ax$ for $a > 0$ has no maximum and for $a < 0$ has no minimum on $\overline{A}$.

**Solution 5.16**

(a) Let $S$ be the random sum of the numbers when both dice are thrown simultaneously. A possible solution for four equally probable step directions is as follows: If $S \in \{5, 6\}$, the player moves to the right, if $S \in \{4, 7\}$ to the left, if $S \in \{8, 9\}$, the player moves upwards; and in all other cases, the player moves downwards.

(b) For the game to be fair, the expected win when playing this game must be 1 euro. If one proceeds as in the answer to Self-Assessment Question 8, the game ends with probability $\frac{4}{15}$ at one of the points $(0, 1)$, $(-1, 0)$ and $(0, -1)$, and with probability $\frac{1}{15}$ at one of the other boundary points. According to Theorem 5.9, the expectation of the win, denoted by E, is equal to

$$\mathrm{E} = -2 \cdot \frac{12}{15} + 3 \cdot \frac{2}{15} + g \cdot \frac{1}{15}.$$

We have $\mathrm{E} = 1$ if and only if $g = 33$.

**Solution 5.17**

(a) We distinguish cases according to the position of the random walk after $n$ steps. Since $X_1, X_2, \ldots$ are independent, the law of total probability yields

$$\mathbb{P}(S_{2n}^0 = x) = \sum_{y \in M(0,n)} \mathbb{P}(S_n^0 = y) \, \mathbb{P}(S_n^y = x).$$

Because $\mathbb{P}(S_n^y = x) = \mathbb{P}(S_n^x = y)$ (symmetry!) the assertion follows.

(b) Using part (a) and the Cauchy–Schwarz inequality, it follows that

$$\mathbb{P}(S_{2n}^0 = x) \le \sqrt{\sum_{y \in M(0,n)} \mathbb{P}(S_n^0 = y) \mathbb{P}(S_n^0 = y)}$$

$$\times \sqrt{\sum_{y \in M(0,n)} \mathbb{P}(S_n^x = y) \mathbb{P}(S_n^x = y)}$$

$$= \sqrt{\sum_{y \in M(0,n)} \mathbb{P}(S_n^0 = y) \mathbb{P}(S_n^y = 0)}$$

$$\times \sqrt{\sum_{y \in M(0,n)} \mathbb{P}(S_n^x = y)\mathbb{P}(S_n^y = x)} \qquad (*)$$

$$\leq \sqrt{\mathbb{P}(S_{2n}^0 = \mathbf{0})}\sqrt{\mathbb{P}(S_{2n}^x = x)}$$

$$= \mathbb{P}(S_{2n}^0 = \mathbf{0}).$$

The second less-than-or-equal-to sign holds because $\mathbb{P}(S_{2n}^x = x)$ equals the second sum in $(*)$ when instead of summing over $y \in M(\mathbf{0}, n)$ we sum over $y \in M(x, n)$.

# Bibliography

[AEB] *Aebly, J.* (1923): Démonstration du problème du scrutin par des considérations géométriques. L'Enseignement Mathématique 23, 13–14.
[AND] *André, D.* (1887): Solution directe du problème résolu par M. Bertrand. Compt. Rend. Acad. Sci. Paris 105, 436–437.
[AP] *Apostol, T.M.* (1981): Mathematical Analysis. 2nd Edition, 5th Printing. Addison-Wesley. Reading. Massachusetts.
[AN] *Arthreya, K.B., and Ney, P.E.* (2004): Branching Processes. Dover Publications, Mineola, N.Y.
[BEL] *Bell, W.W.* (2004): Special Functions for Scientists and Engineers. Dover Publications Inc., Mineola, New York.
[BE] *Bertrand, J.* (1887): Solution d'un problème. Comptes Rendus de l'Académie des Sciences, Paris 105, p. 369.
[BT] *Bertsekas, D.P., and Tsitsiklis, J.N.* (2008): Introduction to Probability. 2nd Edition. Athena Scientific. Nashua, U.S.A.
[BI] *Billingsley, P.* (1999): Convergence of Probability Measures. 2nd Edition. Wiley, New York.
[BI2] *Billingsley, P.* (1995): Probability and Measure. 3rd Edition. Wiley, New York.
[BHS] *Blom, G., Holst, L., and Sandell, D.* (1991): Problems and Snapshots from the World of Probability. Springer, New York.
[BO] *Borovkov, A.A.* (2020): Asymptotic Analysis of Random Walks. Light-tailed Distributions. Encyclopedia of Mathematics and its Applications 176. Cambridge University Press, Cambridge.
[BOB] *Borovkov, A.A., and Borovkov, K.A.* (2008): Asymptotic Analysis of Random Walks. Heavy-tailed Distributions. Encyclopedia of Mathematics and its Applications 118. Cambridge University Press, Cambridge.
[BU] *Butler, C.* (1969): A Test for Symmetry Using the Sample Distribution Function. Ann. Math. Statist. 40, 2209–2210.
[CH] *Chandra, T.K.* (2012): The Borel–Cantelli Lemma. Springer Briefs in Statistics. Springer India, Heidelbarg.
[CGS] *Chen, L.Y.H., Goldstein, L., and Shao, Q.-M.* (2011): Normal Approximation by Stein's Method. Springer, Berlin.
[CA] *Chung, K.L., and AitSahlia, F.* (2003): Elementary Probability Theory. With Stochastic Processes and an Introduction to Mathematical Finance. 4th Edition. Springer, New York.
[CF] *Chung, K.L., and Feller, W.* (1949): On Fluctuations in Coin-Tossing. Proc. Nat. Acad. Sci. U.S.A. 35, 605–608.
[CO] *Cohen, J.W.* (1992): Analysis of Random Walks. Studies in Probability, Optimization and Statistics. 2. IOS Press, Amsterdam.

[COL] *Coleman, A.J.* (1951): A Simple Proof of Stirling's Formula. Amer. Mathem. Monthly 58(5), 334–336.
[CV] *Csáki, E., and Vincze, I.* (1961): On Some Problems Connected with the Galton Test. Publ. Math. Inst. Hungar. Acad. Sci. 6, 97–109.
[DA] *Darwin, Ch.* (1876): The Effect of Cross and Self Fertilization in the Vegetable Kingdom. John Murray, London.
[DF] *Downham, D.Y., and Fotopoulos, S.B.* (1988): The Transient Behaviour of the Simple Random Walk in the Plane. Journal of Appl. Probab. 25, 58–69.
[DS] *Doyle, P., and Snell, J.L.* (1984). Random Walks and Electric Networks. The Carus Mathematical Monographs, 22. Wiley & Sons, New York.
[DUS] *Duminil-Copin, H., and Smirnov, S.* (2012): The Connective Constant of the Honeycomb Lattice is $\sqrt{2+\sqrt{2}}$. Annals of Mathematics 175(2), 1653–1665.
[DU] *Durrett, R.* (2010): Probability. Theory and Examples. 4th Edition. Cambridge University Press, New York.
[DE] *Dvoretzky, A., and Erdös, P.* (1951): Some Problems on the Random Walk in Space. Proc. Second Berkeley Symp. Math. Statist. Probab., 353–367.
[DW] *Dwass, M.* (1967). Simple Random Walk and Rank Order Statistics. Ann. Math. Statist. 38, 1042–1053.
[EKM] *Embrechts, P., Klüppelberg, C., and Mikosch, Th.* (2003): Modelling Extremal Events: for Insurance and Finance. Corrected 4th printing. Springer, Berlin, Heidelberg, New York.
[EK] *Erdös, P., and Kac, M.* (1946): On certain limit theorems in the theory of probability. Bull. Amer. Math. Soc. 52, 292–302-
[ER] *Erdös, P., and Révesz, P.* (1970): On a New Law of Large Numbers. J. Analyse Math. 22, 103–111.
[ET] *Ethier, St.N.* (2010): The Doctrine of Chances. Probabilistic Aspects of Gambling. Springer, Berlin, Heidelberg, New York.
[FG1] *Feige, U.* (1995): A Tight Lower Bound on the Cover Time for Random Walks on Graphs. Random Structures and Algorithms 6(4), 433–438.
[FG2] *Feige, U.* (1995): A Tight Upper Bound on the Cover Time for Random Walks on Graphs. Random Structures and Algorithms 6(1), 51–54.
[FE1] *Feller, W.* (1970): An Introduction to Probability Theory and Its Applications Vol.1, 3rd Edition. Wiley, New York.
[FE2] *Feller, W.* (1957). The Number of Zeroes and Changes of Sign in a Simple Random Walk. L'enseignement Mathématique, tome III, fasc. 3, 229–235.
[GK] *Gnedenko, B.V., and Korolyuk, V.S.* (1951): On the Maximum Discrepancy between Two Empirical Distributions. Dokl. Akad. Nauk SSSR. 80, 525–528. English translation in: Selected Translations in Mathematical Statistics and Probability Vol.1, Amer. Math. Soc., Providence, Rhode Island 1961, 13–16.
[GM] *Gnedenko, B.V., and Mihalevič, V.S.* (1952): On the Distribution of the Number of Excesses of One Empirical Distribution Function over Another. Dokl. Akad. Nauk SSSR. 82, 841–843. English translation in: Selected Translations in Mathematical Statistics and Probability Vol.1, Amer. Math. Soc., Providence, Rhode Island 1961, 83–86.
[GRV] *Gnedenko, B.V., and Rvačeva, E.L.* (1952): On a Problem of the Comparison of Two Empirical Distributions. Dokl. Akad. Nauk SSSR. 82, 513–516. English translation in: Selected Translations in Mathematical Statistics and Probability Vol.1, Amer. Math. Soc., Providence, Rhode Island 1961, 69–72.
[GR] *Griffin, P.* (1990): Accelerating Beyond the Third Dimension: Returning to the Origin in Simple Random Walk. Math. Scientist 15, 24–35.
[GRS] *Grimmet, G., and Stirzaker, D.* (2001): Probability and Random Processes. 3rd Edition. Oxford University Press. Oxford.
[GSN] *Grinstead, Ch.M., and Snell, J.L.* (1997): Introduction to Probability. 2nd Revised Edition. American Mathematical Society. Providence, Rhode Island.

[GS] *Guillotin-Plantard, N., and Schott, R.* (2006): Dynamic Random Walks. Theory and Applications. Elsevier, Amsterdam.
[GU] *Gut, A.* (2009): Stopped Random Walks. Limit Theorems and Applications, 2nd Edition. Springer Series in Operations Research and Financial Engineering. Springer, New York.
[HA] *Hand, D.J. et al. (eds.)* (1994): A Handbook of Small Data Sets. Chapman & Hall, London, New York.
[HAR] *Harris, T.E.* (1989): The Theory of Branching Processes. Dover Publications, Mineola, New York.
[HO] *Hodges, J.L. Jr.* (1955): Galton's Rank-Order Test. Biometrika 42, No. 1/2, 261–262.
[HU] *Hughes, B.D.* (1995): Random Walks and Random Environments. Vol. 1: Random walks. Clarendon Press xxi, Oxford.
[JO] *Jain, N.C., and S. Orey* (1968): On the Range of Random Walk. Israel J. Mathem. 6, 373–380.
[JP1] *Jain, N.C., and Pruitt, W.E.* (1971): The Range of Transient Random Walk. J. Analyse Math. 24, 369–393.
[JP2] *Jain, N.C., and Pruitt, W.E.* (1973): The Range of Random Walk. Proc. 6th Berkeley Symp. on Mathem. Statist. and Probab., Berkeley, University of California Press, Vo. 3, 31–50. J. Analyse Math. 24, 369–393.
[KP] *Katzenbeisser, W., and Panny, W.* (1984): Asymptotic Results on the Maximal Deviation of Simple Random Walks. Stoch. Proc. Appl. 18, 263–275.
[KP2] *Katzenbeisser, W., and Panny, W.* (2002): The Maximal Height of Simple Random Walks Revisited. Journ. Statist. Plann. Infer. 101, 149–161.
[KH] *Khintchin, A.Y.* (1924). Über einen Satz der Wahrscheinlichkeitsrechnung. Fund. Mathem. 6, 9–20.
[KS] *Klafter, J., and Sokolov, I.M.* (2011): First Steps in Random Walks. From Tools to Applications. Oxford University Press, Oxford.
[KSC] *Köckler, N., and Schwarz, H. R.* (2011): Numerische Mathematik. 8th Edition. Vieweg+Teubner, Wiesbaden.
[KN] *Knopp, K.* (1956): Infinite Sequences and Series. Dover Publications, New York.
[LA1] *Lawler, G.F.* (2006): Introduction to Stochastic Processes. 2nd Edition. Chapman & Hall/CRC. New York.
[LA] *Lawler, G.F.* (1996): Intersections of Random Walks. Birkhäuser, Boston, MA.
[LAW] *Lawler, G.F.* (2010): Random Walk and the Heat Equation. Student Mathematical Library 55. American Mathematical Society. Providence, RI.
[LL] *Lawler, G.F., and Limic, V.* (2010): Random Walk. A Modern Introduction. Cambridge Studies in Advanced Mathematics 123. Cambridge University press, Cambridge.
[LE] *Levasseur, K.M.* (1988): How to Beat your Kids at their own Game. Mathematics Magazine 61, 301–305.
[LG] *Le Gall, J-F.* (1985): Un théorème central limite pour le nombre de points visités par une marche aléatoire plane récurrente. Comptes Rend. Acad. Sci. Paris 300, 505–508.
[LEV] *Lévy, P.* (1940): Sur certains processus stochastiques homogènes. Compositio Mathematica 7, 283–330.
[MS] *Madras, N., and Slade, G.* (1993): The Self-Avoiding Walk. Birkäuser, Boston.
[NO] *Nolan, J.P.* (2020): Univariate Stable Distributions. Models for Heavy Tailed Data. Springer. Cham, Switzerland.
[PM] *Philippou, A.N. and Makri, F.S.* (1986): Successes, Runs and Longest Runs. Statist. & Probab. Lett. 4, 211–215.
[PE] *Pearson. K.* (1905): The Problem of the Random Walk. Nature 72, 294.
[PO] *Pólya, G.* (1921): Über eine Aufgabe der Wahrscheinlichkeitsrechnung betreffend die Irrfahrt im Straßennetz. Mathem. Annalen 84, 149–160.
[RE1] *Rényi, A.* (1957): On the Distribution Function $L(z)$. Magyar. Tud. Akad. Kutató Int. Közl. 2, 43–50.
[REN] *Rényi, A.* (1970): Probability Theory. North Holland, Amsterdam, London.

[RE] *Révész, P.* (2005): Random Walk in Random and Non-random Environments. 2nd Edition. Singapore etc.: World Scientific. xiv.
[RI] *Riordan, J.* (1968): Combinatorial Identities. John Wiley & Sons, New York.
[RO] *Ross, S.* (2019): A First Course in Probability, Global Edition. 10th Edition. Pearson.
[SM] *Smirnov, N.V.* (1939): On the Estimation of the Discrepancy Between Empirical Curves of Distribution for Two Independent Samples. Bull. Moskov. Gos. Univ. Sect. A 2.
[SP] *Spitzer, F.* (1976): Principles of Random Walk. 2nd Edition. Graduate Texts in Mathematics 34. Springer-Verlag, New York.
[ST] *Stirzaker, D.* (2010): Elementary Probability. 2nd Edition. Cambridge University Press, Cambridge.
[TA] *Tao, T.* (2022): Analysis II. Texts and Readings in Mathematics 38. 4th Edition. Hindustan Book Agency, New Delhi.
[TE] *Telcs, A.* (2006). The Art of Random Walks. Lecture Notes in Mathematics 185. Springer, Berlin.
[VH] *Van der Hofstadt, R.* (2016): Random Graphs and Complex Networks. Volume 1. Cambridge Series in Statistical and Probabilistic Mathematics, Vol. 43. Cambridge University Press. Cambridge, UK.
[WI] *Whitworth, W.A.* (1878): Arrangements of $m$ Things of one Sort and $n$ Things of another Sort, Under Certain Conditions of Priority. Messenger of Math. 8(2), 105–114.
[WO] *Woess, W.* (2000): Random Walks on Infinite Graphs and Groups. Cambridge Tracts in Mathematics 138. Cambridge University Press, Cambridge.
[ZA] *Zagier, D.* (1990): How Often should you Beat your Kids? Mathematics Magazine 63, 89–92.

# Index

**A**
Absorption probability, 191
Arcsine distribution, 27, 49, 67, 110
  discrete, 25
Arcsine law
  first maximizer of a SSRW, 67
  last visit to zero, 19, 25, 109
  sojourn time, 49, 110
Asymmetric random walk, 177
Asymptotic equality, 3

**B**
Ballot problem, 14, 117
Banach, S., 38
Banach's matchbox problem, 38
Bernoulli random walk, 5, 177
Biased random walk, 177
  distribution of number of zeros, 189
  expectation of number of zeros, 188
Binomial distribution, 7, 91
Binomial series, 267
Bonferroni, C.E., 95
Borel–Cantelli lemma, 270
Bridge, 13, 121
Bridge - first return time
  definition, 142
  distribution, 142
  expectation, 142
  variance, 142
Bridge - last visit to zero
  definition, 141
  distribution, 145
  expectation, 145
  variance, 145
Bridge - maximum
  definition, 145

  distribution, 146
  expectation, 147
  limit distribution, 148
  variance, 147
Bridge - maximum modulus
  definition, 158
  distribution, 160
  limit distribution, 161
Bridge - number of changes of sign
  definition, 151
  distribution, 154
  expectation, 154
  limit distribution, 156
  variance, 155
Bridge - number of interior zeros
  definition, 124
  distribution, 126
  expectation, 126
  limit distribution, 129
  variance, 126
Bridge - random maximizer
  definition, 150
  distribution, 150
Bridge - sojourn time
  definition, 138
  distribution, 138
  limit distribution, 141
Brownian bridge, 171
  distribution of maximum, 172
  distribution of maximum modulus, 172
  distribution of sojourn time, 172
Brown, R., 108
Brown–Wiener process, 108, 169
  distribution of last zero, 109
  distribution of maximum, 110
  distribution of maximum modulus, 110
  distribution of sojourn time, 110

## C

Cantell, F.P.i, 100, 165
Ceiling function, 2
Central limit theorem
  de Moivre–Laplace, 7, 54, 136, 256
  Lindeberg–Lévy, 8, 110, 256
Change of sign, 86, 151
Collisions of random walks, 77
Conditional probability, 253
Connected set, 230
Connective constant, 244
Continuity
  from above, 252
  from below, 252
Convergence in distribution, 254
Cover time, 247

## D

Darwin, C., 123
Decreasing sequence of events, 252
De Moivre, 7
De Moivre–Laplace central limit theorem, 7, 54, 136, 256
Dirichlet, 231
Dirichlet problem (discrete), 231
Discrete arcsine law for last visit, 19
Distribution
  arcsine, 27, 49, 67, 110
  arcsine, discrete, 25
  binomial, 7, 91
  changes of sign in a SSRW, 88
  changes of sign of a bridge, 154
  first maximizer, 66
  first-passage time $V_1$ of a biased random walk, 178
  first-passage time $V_k$ of a biased random walk, 184
  geometric, 65, 189
  first return time of a biased random walk, 186
  first return time of a bridge, 142
  half-normal, 37, 54
  hypergeometric, 176
  Kolmogorov, 163, 168
  $k$-th return time of a biased random walk, 187
  Lévy, 74, 84
  last interior zero of a bridge, 145
  length of the longest upward run, 198
  maximum modulus of a bridge, 160
  maximum modulus of the Brown–Wiener process, 110
  maximum of a bridge, 146
  maximum of a SRRW, 52
  maximum of the Brown–Wiener process, 110
  number of interior zeros, 126
  number of zeros of a SSRW, 29
  number of zeros of a transient walk, 189
  Poisson, 215
  Rényi, A., 98
  sojourn time of a bridge, 138
  sojourn time of a SSRW, 47
  uniform, 141
  Weibull, 128, 129, 148, 157
Donsker, M.D., 110
Donsker's invariance principle, 110
Downward run, 197
Duality, 103
Dual random walk, 104
  maxima at time $n$ and first-passage times, 105
  record at time $n$, 104
  zeros and final height visits, 106

## E

Empirical distribution function, 100, 165
Erdös, P., 200
Event, 251
Expectation
  of an $\mathbb{N}_0$-valued random variable, 257
Extinction-explosion dichotomy, 208
Extinction probability, 206

## F

First entry time, 232
First maximizer, 65
First-passage time, 68, 178
First return time, 40, 142, 186, 218
Floor function, 2

## G

Galton, F., 123
Galton, s rank order test, 123
Galton–Watson process, 202
  critical, 203
  expectation, 205
  extinction-explosion dichotomy, 208
  extinction probability, 206
  offspring distribution, 202
  subcritical, 203
  supercritical, 203
  variance, 205
Gambler's ruin problem, 190

# Index

Generating function, 80
  of a randomized sum, 204
  of a random variable, 263
  of a sequence of numbers, 262
Geometric distribution, 65, 189
Glivenko, V.I., 100, 165
Glivenko–Cantelli theorem, 100, 165
Graph, 246

## H
Half-normal distribution, 37, 54
Harmonic function, 229
Homogeneity
  spatial, 10
  temporal, 10, 240
Hypergeometric distribution, 176

## I
Identities for binomial coefficients, 264
Increasing sequence of events, 252
Independence
  of events, 253
  of random vectors, 253
Independent and identically distributed (i.i.d), 7
Inequalities for the logarithm, 256
Interior zero of a bridge, 124
Invariance principle, 110

## K
Khintchin, A., 112
Kolmogorov, A.N., 163
Kolmogorov distribution, 163, 168
Kolmogorov's axiomatic system, 251
Kolmogorov–Smirnov test, 164

## L
Laplace, P.S., 7
Laplace filter, 229
Laplace operator, 229
Last visit to zero, 19
  arcsine law, 25
Law of the iterated logarithm, 112
Law of total probability, 253
Legendre, A.M., 268
Legendre polynomial, 268
Lévy, P., 8
Lévy distribution, 74, 84
Limit distribution
  changes of sign of a bridge, 156
  changes of sign of a SSRW, 92
  first maximizer of a SSRW, 67
  first-passage time $V_k$ of a biased random walk, 184
  last zero of a SSRW, 25
  $2k$-meeting time, 84
  $k$-th passage time, 74
  $k$-th return time, 76
  maximum modulus of a bridge, 161
  maximum modulus of a SSRW, 96
  maximum of a bridge, 148
  maximum of a SSRW, 53
  number of interior zeros, 129
  number of maximizers of a SRRW, 65
  number of zeros of a SSRW, 34
  sojourn time of a bridge, 141
  sojourn time of a SRRW, 49
Limit superior of events, 270
Lindeberg, J.W., 8, 110
Lindeberg–Lévy central limit theorem, 8, 110, 256
Longest upward run, 198

## M
Markov, A.A., 10
Markov property, 10
Matchbox problem, 38
Maximum modulus, 93
Mean value property, 228
Meeting time
  $2k$-, 78

## N
Neighboring point, 217, 228
Nelson, E., 16

## O
Offspring distribution, 202
  geometric, 207, 292
  Poisson, 215

## P
Parity, 13
Partial sum process, 107
Path, 6, 107
  negative, 13
  non-negative, 13
  non-positive, 13
  positive, 13
  zero-avoiding, 13

Poisson distribution, 215
Pólya, G., 218
Probability integral transform, 166
Probability space, 251

**R**
Random walk
   with absorbing boundaries, 190
   asymmetric, 177
   biased, 177
   on $\mathbb{Z}^d$, 217
   simple on a graph, 246
   simple symmetric, 5
   symmetric Bernoulli, 5
Randomized sum, 89, 204
   expectation, 205
   generating function, 204
   variance, 205
Recurrence, 39, 42
Recurrence of the symmetric random walk for $d \leq 2$, 219
Reflection principle, 14, 50, 64, 76, 138, 145, 182, 274, 278
Rényi, A., 98
Rényi distribution, 98
Révész, P., 200
Reproduction equation, 202
Return probability of symmetric random walk, 223
Return time
   biased random walk, 187
Roulette, 199
Ruin probabilities, 191
Run, 214

**S**
Second factorial moment, 264
Self-avoiding walk, 243
$\sigma$-algebra, 251
Simple symmetric random walk (SSRW), 5
Smirnov, N.W., 147
Sojourn time, 45, 138
Spatial homogeneity, 10
Spectral norm, 236
SSRW - duality
   definition, 104
   maxima at time $n$ and first-passage times, 105
   record at time $n$, 104
   zero and final height visits, 106
SSRW - first maximizer
   arcsine law, 67

   definition, 65
   distribution, 66
   limit distribution, 67
SSRW - first return time
   definition, 40
   distribution, 40
SSRW - first-passage increments
   definition, 68
   independence, 72
SSRW - first-passage time
   and maxima, 69
   definition, 68
   distribution, 68
SSRW - $2k$-meeting time
   definition, 77
   distribution, 83
   limit distribution, 84
SSRW - $k$-th passage time
   definition, 68
   distribution, 71
   limit distribution, 74
SSRW - $k$-th return time
   definition, 75
   distribution, 76
   limit distribution, 76
SSRW - last visit to zero
   definition, 19
   discrete arcsine law, 19
   distribution, 19
   expectation, 20
   limit distribution, 25
   variance, 20
SSRW - maximum
   connection with visits to zero, 55
   definition, 50
   distribution, 52
   expectation, 52
   limit distribution, 53
   variance, 52
SSRW - maximum modulus
   definition, 93
   distribution, 95
   limit distribution, 96
SSRW - 2-meeting time
   distribution, 82
SSRW - number of changes of sign
   definition, 86
   distribution, 88
   expectation, 88
   limit distribution, 92
   variance, 88
SSRW - number of maximizers
   definition, 58
   distribution, 59

expectation, 60
  limit distribution, 65
  variance, 60
SSRW - number of visits to zero
  definition, 29
  expectation, 29
  variance, 29
SSRW - number of zeros
  distribution, 29
  limit distribution, 34
SSRW - sojourn time
  arcsine law, 49
  definition, 45
  distribution, 47
  limit distribution, 49
Standard normal distribution, 34
Stirling, J., 260
Stirling's formula, 25, 35, 36, 260
Stopping time, 232
Symmetric Bernoulli random walk, 5
Symmetry test, 99

**T**
Temporal homogeneity, 10, 240
Test
  Kolmogorov–Smirnov, 164
  of symmetry, 99
  two-sample, 165
Tied-down random walk, see bridge, 121

Time until absorption
  expectation, 194
Total probability law, 253
Transience, 42
  asymmetric random walk, 185
  symmetric random walk for $d \geq 3$, 219
Two-sample problem, 121, 164
Two-sample test, 165

**U**
Uniform distribution, 141
Upward run, 197
Urn model, 144, 171

**V**
Vandermonde identity, 117
Variance
  of an $\mathbb{N}_0$-valued random variable, 258

**W**
Wallis, J., 258
Wallis product for $\pi$, 258
Watson, H.W., 202
Weibull, E.H.W., 128
Weibull distribution, 128, 129, 148, 157
Wiener, N., 108

The manufacturer's authorised representative in the EU is Springer Nature Customer Service Centre GmbH, Europaplatz 3, 69115 Heidelberg, Germany. If you have any concerns regarding our products, please contact ProductSafety@springernature.com

Printed and bound by CPI Group (UK) Ltd, Croydon, CR0 4YY

26/03/2026

02078953-0009